T0336864

Artificial Intelligence Applications in Water Treatment and Water Resource Management

Victor Shikuku
Kaimosi Friends University, Kenya

A volume in the Advances in Environmental Engineering and Green Technologies (AEEGT) Book Series

Published in the United States of America by
IGI Global
Engineering Science Reference (an imprint of IGI Global)
701 E. Chocolate Avenue
Hershey PA, USA 17033
Tel: 717-533-8845
Fax: 717-533-8661
E-mail: cust@igi-global.com
Web site: http://www.igi-global.com

Library of Congress Cataloging-in-Publication Data

Names: Shikuku, Victor, 1986- editor.
Title: Artificial intelligence applications in water treatment and water resource management / edited by: Victor Shikuku.
Description: Hershey, PA : Engineering Science Reference, [2023] | Includes bibliographical references and index. | Summary: "The emergence of AI as an applicable tool in wastewater treatment and water resource management has opened immense opportunities in water research and environmental sciences. It has provided solutions to otherwise complex problems not easily resolved by deterministic physics or traditional approaches. This book aims at providing in-depth and comprehensive research on AI techniques and their applications in wastewater treatment and in the management of water resources"-- Provided by publisher.
Identifiers: LCCN 2023008169 (print) | LCCN 2023008170 (ebook) | ISBN 9781668467916 (hardcover) | ISBN 9781668467923 (paperback) | ISBN 9781668467930 (ebook)
Subjects: LCSH: Water--Purification. | Artificial intelligence--Industrial applications.
Classification: LCC TD433 .A78 2023 (print) | LCC TD433 (ebook) | DDC 628.1/62028563--dc23/eng/20230323
LC record available at https://lccn.loc.gov/2023008169
LC ebook record available at https://lccn.loc.gov/2023008170

This book is published in the IGI Global book series Advances in Environmental Engineering and Green Technologies (AEEGT) (ISSN: 2326-9162; eISSN: 2326-9170)

British Cataloguing in Publication Data
A Cataloguing in Publication record for this book is available from the British Library.

For electronic access to this publication, please contact: eresources@igi-global.com.

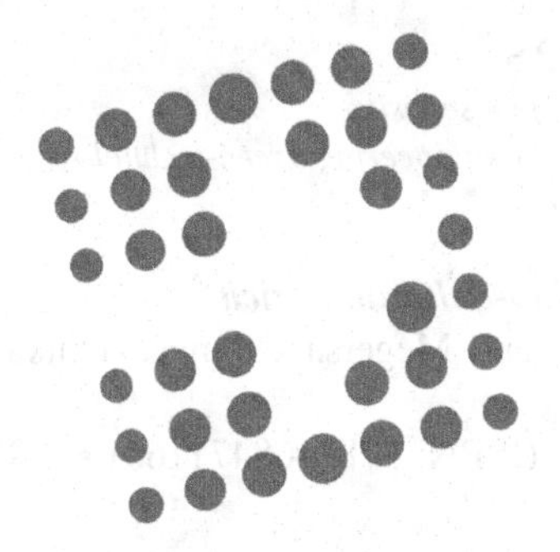

Advances in Environmental Engineering and Green Technologies (AEEGT) Book Series

ISSN:2326-9162
EISSN:2326-9170

Editor-in-Chief: Sang-Bing Tsai, Zhongshan Institute, University of Electronic Science and Technology of China, China & Wuyi University, China, Ming-Lang Tseng, Lunghwa University of Science and Technology, Taiwan, Yuchi Wang, University of Electronic Science and Technology of China Zhongshan Institute, China

MISSION

Growing awareness and an increased focus on environmental issues such as climate change, energy use, and loss of non-renewable resources have brought about a greater need for research that provides potential solutions to these problems. Research in environmental science and engineering continues to play a vital role in uncovering new opportunities for a "green" future.

The **Advances in Environmental Engineering and Green Technologies (AEEGT)** book series is a mouthpiece for research in all aspects of environmental science, earth science, and green initiatives. This series supports the ongoing research in this field through publishing books that discuss topics within environmental engineering or that deal with the interdisciplinary field of green technologies.

COVERAGE

- Sustainable Communities
- Industrial Waste Management and Minimization
- Renewable Energy
- Pollution Management
- Electric Vehicles
- Air Quality
- Waste Management
- Green Technology
- Cleantech
- Green Transportation

The Advances in Environmental Engineering and Green Technologies (AEEGT) Book Series (ISSN 2326-9162) is published by IGI Global, 701 E. Chocolate Avenue, Hershey, PA 17033-1240, USA, www.igi-global.com. This series is composed of titles available for purchase individually; each title is edited to be contextually exclusive from any other title within the series. For pricing and ordering information please visit http://www.igi-global.com/book-series/advances-environmental-engineering-green-technologies/73679. Postmaster: Send all address changes to above address.

Titles in this Series

For a list of additional titles in this series, please visit:
http://www.igi-global.com/book-series/advances-environmental-engineering-green-technologies/73679

Global Industrial Impacts of Heavy Metal Pollution in Sub-Saharan Africa
Joan Nyika (University of Johannesburg, South Africa) and Megersa Olumana Dinka (University of Johannesburg, South Africa)
Engineering Science Reference • © 2023 • 406pp • H/C (ISBN: 9781668471166) • US $255.00

Nanopriming Approach to Sustainable Agriculture
Abhishek Singh (Faculty of Biology, Yerevan State University, Armenia) Vishnu D. Rajput (Academy of Biology and Biotechnology, Southern Federal University, Rostov-on-Don, Russia) Karen Ghazaryan (Faculty of Biology, Yerevan State University, Armenia) Santosh Kumar Gupta (National Institute of Plant Genome Research, India) and Tatiana Minkina (Academy of Biology and Biotechnology, Southern Federal University, Rostov-on-Don, Russia)
Engineering Science Reference • © 2023 • 453pp • H/C (ISBN: 9781668472323) • US $245.00

Handbook of Research on AI-Equipped IoT Applications in High-Tech Agriculture
Alex Khang (Global Research Institute of Technology and Engineering, USA)
Engineering Science Reference • © 2023 • 473pp • H/C (ISBN: 9781668492314) • US $335.00

Contemporary Developments in Agricultural Cyber-Physical Systems
G.S. Karthick (Department of Software Systems, PSG College of Arts and Science, India)
Engineering Science Reference • © 2023 • 293pp • H/C (ISBN: 9781668478790) • US $240.00

Perspectives on Ecological Degradation and Technological Progress
Veli Yilanci (Faculty of Political Sciences, Canakkale Onsekiz Mart University, Turkey)
Engineering Science Reference • © 2023 • 342pp • H/C (ISBN: 9781668467275) • US $225.00

For an entire list of titles in this series, please visit:
http://www.igi-global.com/book-series/advances-environmental-engineering-green-technologies/73679

701 East Chocolate Avenue, Hershey, PA 17033, USA
Tel: 717-533-8845 x100 • Fax: 717-533-8661
E-Mail: cust@igi-global.com • www.igi-global.com

Table of Contents

Foreword xiv

Preface xvi

Chapter 1
Machine Learning Applications in Adsorption of Water Pollutants 1
Victor Odhiambo Shikuku, Kaimosi Friends University, Kenya
Newton Wafula Masinde, Jaramogi Oginga Odinga University of Science and Technology, Kenya

Chapter 2
Artificial Intelligence in Wastewater Management 31
C. V. Suresh Babu, Hindustan Institute of Technolgy and Science, India
K. Yadavamuthiah, Hindustan Institute of Technology and Science, India
S. Abirami, Hindustan Institute of Technology and Science, India
S. Kowsika, Hindustan Institute of Technology and Science, India

Chapter 3
Integrating Machine Learning and AI for Improved Hydrological Modeling and Water Resource Management 46
Djabeur Mohamed Seifeddine Zekrifa, Higher School of Food Science and Agri-Food Industry, Algeria
Megha Kulkarni, Nitte Meenakshi Institute of Technology, India
A. Bhagyalakshmi, Vel Tech Rangarajan Dr. Sagunthala R&D Institute of Science and Technology, India
Nagamalleswari Devireddy, Mohan Babu University, India
Shilpa Gupta, Aurora Higher Education and Research Academy (Deemed), India
Sampath Boopathi, Muthayammal Engineering College, India

Chapter 4
Artificial Intelligence in Water Treatments and Water Resource Assessments...71
K. Gunasekaran, Department of Mechanical Engineering, Muthayammal Engineering College, Namakkal, India
Sampath Boopathi, Department of Mechanical Engineering, Muthayammal Engineering College, Namakkal, India

Chapter 5
The Use and Awareness of ICT to Facilitate the Adoption of Artificial Intelligence in Agriculture...99
Mushtaq Ahmad Shah, Lovely Professional University, India
Mihir Aggarwal, Lovely Professional University, India

Chapter 6
Artificial Intelligence, Internet of Things, and Machine-Learning: To Smart Irrigation and Precision Agriculture...113
Fatima-Zahra Akensous, Faculty of Science Semlalia, Cadi Ayyad University, Marrakesh, Morocco
Naira Sbbar, Faculty of Science Semlalia, Cadi Ayyad University, Marrakesh, Morocco
Lahoucine Ech-chatir, Faculty of Science and Technology Guéliz, Cadi Ayyad University, Marrakesh, Morocco
Abdelilah Meddich, Faculty of Science Semlalia, Cadi Ayyad University, Marrakesh, Morocco

Chapter 7
A Study on Machine Learning-Based Water Quality Assessment and Wastewater Treatment...146
Satakshi Singh, Sam Higginbottom University of Agriculture, Technology, and Sciences, India
Suryanshi Mishra, Sam Higginbottom University of Agriculture, Technology, and Sciences, India
Tinku Singh, Indian Institute of Information Technology, Allahabad, India
Shobha Thakur, Sam Higginbottom University of Agriculture, Technology, and Sciences, India

Chapter 8
Explaining the Importance of Water Quality Parameters for Prediction of the Quality of Water Using SHAP Value....163
Siddhartha Kumar Arjaria, Rajkiya Engineering College, Banda, India
Abhishek Singh Rathore, Shri Vaishnav Vidyapeeth Vishwavidyalaya, Indore, India
Shailendra Badal, Rajkiya Engineering College, Banda, India

Chapter 9
Classification of Quality of Water Using Machine Learning....182
P. Umamaheswari, SASTRA University (Deemed), India
M. Kamaladevi, SASTRA University (Deemed), India

Compilation of References 198

Related References.... 240

About the Contributors 265

Index.... 269

Detailed Table of Contents

Foreword xiv

Preface xvi

Chapter 1
Machine Learning Applications in Adsorption of Water Pollutants 1
Victor Odhiambo Shikuku, Kaimosi Friends University, Kenya
Newton Wafula Masinde, Jaramogi Oginga Odinga University of Science and Technology, Kenya

Adsorption remains one of the most effective methods of decontamination of water. Adsorption processes are governed by multiple factors which contribute to the overall efficiency of the process. These include process conditions such as temperature, pH, the concentration of pollutants, and competing ions. The adsorbate properties, such as speciation, polarity, kinetic diameter, and ionic sizes, also affect adsorption performance. The adsorbent properties also play a critical role in assessing the suitability of an adsorbing material. This includes surface areas, pore volumes, chemical compositions, surface charges, etc. The complexity of the interaction between all these parameters makes it cumbersome or near impossible to predict with appreciable precision the performance of an adsorption system. Machine learning provides an opportunity for developing models for concise prediction of adsorption efficiencies for different materials. This chapter discusses the principles of various machine learning models and their application in the adsorption of pollutants from water.

Chapter 2
Artificial Intelligence in Wastewater Management 31
C. V. Suresh Babu, Hindustan Institute of Technolgy and Science, India
K. Yadavamuthiah, Hindustan Institute of Technology and Science, India
S. Abirami, Hindustan Institute of Technology and Science, India
S. Kowsika, Hindustan Institute of Technology and Science, India

This chapter introduces an innovative solution to address sewage pollution by integrating AI-enabled waste management systems. The approach involves segregating sewage into solid and liquid waste streams and applying specialized treatment processes. The main goal is to achieve sustainable sewage treatment, recover valuable resources, and produce distilled water. To ensure optimal performance, an AI system leveraging cutting-edge technologies like IoT, machine learning, and computer vision is employed for real-time monitoring, water quality assessment, and problem resolution. These findings contribute significantly to the advancement of sustainable waste management practices. They effectively reduce soil and water pollution, safeguard groundwater levels, and enhance overall operational efficiency. This technology-driven strategy paves the path for a more eco-friendly and advanced future in waste management.

The hydrological cycle is an important process that controls how and where water is distributed on Earth. It includes processes including transpiration, evaporation, condensation, precipitation, runoff, and infiltration. However, there are obstacles to understanding and modelling the hydrological cycle, such as a lack of data, ambiguity, fluctuation, and the impact of human activity on the natural balance. Techniques for accurate modelling are essential for managing water resources and risk reduction. With potential uses in rainfall forecasting, streamflow forecasting, and flood modelling, machine learning and artificial intelligence (AI) are effective tools for hydrological modelling. Case studies and real-world examples show how solutions to problems like data quality, interpretability, and scalability may be applied in real-world situations. Discussions of future directions and challenges emphasise new developments and areas that need more investigation and cooperation.

This chapter explores the use of AI in water treatment, evaporation management, and water resource management. It begins with an introduction, highlighting AI's motivation and objectives. The chapter then discusses AI applications, challenges, and opportunities in their implementation. It compares traditional approaches and AI-driven solutions for evaporation control and optimization and presents case studies and applications to demonstrate real-world examples. The chapter also discusses water resource management challenges, data-driven modeling, forecasting, optimization, and decision support systems. It also discusses the benefits and limitations of AI, interdisciplinary collaboration, ethical considerations, and policy frameworks for responsible AI implementation. The chapter also provides recommendations for future research to advance AI in water treatment and resource management.

Agricultural practices are changing drastically with the incorporation of artificial intelligence (AI). These innovations have the potential to cause a sea change in farming by increasing production and solving issues of sustainability. More than 58% of rural households in India rely on agriculture as their main source of income, which contributes 18% to India's GDP. India's agricultural output is much lower compared to that of China, Brazil, and the United States. There is growing empirical evidence that adopting the most cutting-edge technologies, such as AI, improves farmers' economic situations and production. This chapter explores the use and awareness of ICT in facilitating the adoption of AI in agriculture and the barriers to accessing information sources. A questionnaire survey was administered to farmers to understand their experiences and perspectives on the use of ICT in agriculture and to facilitate the adoption of AI in agriculture.

Water scarcity has been escalating both in terms of frequency and severity, owing to climate change and global warming. Furthermore, water is a vital source that is at the core of crucial sectors like agriculture. Yet, this source is labeled scarce, and its distribution is uneven globally. For the aforementioned reasons, achieving a rational use of water is of utmost importance. In this framework, computational intelligence like artificial intelligence (AI), the internet of things (IoT), and machine-learning, has been gaining momentum for implementing smart irrigation and precision agriculture. Thus, the chapter surveys a selection of recent studies, corroborating AI, IoT, and machine-learning as promising approaches to advance agriculture. The chapter also sheds light on the notion of virtual water, proposes strategies to deal with water scarcity, and highlights the essential components to achieve effective smart irrigation, thereby switching toward an innovative model of sustainable digital agriculture.

The chapter will present state-of-the-art water assessment and treatment methods as well as the current issues and challenges of the domain. Modern techniques for water evaluation and treatment will be covered in this chapter, along with the current problems and difficulties facing the industry.

Water is key to life on planet Earth, and hence, maintaining water quality is a critical issue in contemporary times. The water quality index decides the quality of drinking water. The presented work first explores different machine learning algorithms on the already collected water samples to decide the water quality and then applies the coalition game theory-based SHapley Additive exPlanations (SHAP) approach to decide the significance of each parameter in deciding the class of water sample based on quality. The potential of popular algorithms like K-NN, support vector machine, decision tree, etc. are being explored to find out the quality of water samples. All the machine learning algorithms used in the work give over 80% accuracy while the performance of neural network is 96% proving to be the best among all other algorithms. The presented work demonstrates the model agnostic, coalition game theoretic SHAP value-based method for explaining the importance and impact of each of the given parameter pH, HCO3-, Cl-, NO3-, F-, Ca, Mg, Na, Ec, etc. in deciding the quality of the water.

Keeping a check on the quality of water is necessary for protecting both the health of humans and of the environment. AI can be used to classify and predict water quality. The proposed system uses several machine learning algorithms to manage water quality data gathered over a protracted period. Water quality index (WQI) is used to categorize the given samples by using machine learning and ensemble approaches. The studied classifiers included random forest classifier, CatBoost classifier, k nearest neighbors, logistic regression, etc. The authors used precision-recall curves, ROC curves and confusion matrices as performance metrics for the ML classifiers used. With an accuracy of 95.43%, the random forest model was shown to be the most accurate classifier. Furthermore, CatBoost classifier and k nearest neighbors provided satisfactory results with 94.86% and 94.08% accuracy, respectively. Therefore, CatBoost algorithm is considered to be a more reliable approach for the quality of water classification.

Foreword

More than 70% of common human diseases are water-related and, at the same time, more than 30% of treated piped water is wasted as non-revenue water mainly through leakage and sabotage. Similarly, when analyzing the water–food–energy nexus, solving the water-related challenges increases food availability by 80% while contributing to energy solutions by about 40%. Academically, we know water challenges mainly present themselves in three ways: too dirty, too little, or too much water. These three manifestations of water are also directly impacted by climate change. All the aforementioned situations and water analysis data can be improved by the integration of water resources management with information technology and data science.

Artificial intelligence and ICT in general have proven to speed up the rate of human capability, workload, and development by not less than 50 years. Other sectors of human activities like manufacturing, medicine, mining, agriculture, aviation, and transport, among others, have successfully been integrated with machine learning, artificial intelligence, and the Internet of Things (IoT). Water resources management cannot continue to remain manual and mechanical while challenges in the sector are now better understood.

Water as a common good and as expressed by the desires of sustainable development goals No. 6 (SDG – 6), is now best understood in terms of its chemical, physical, biological, physiological, social-cultural, political, and economic properties and values when water is used as a solvent, medium of reactions, transport, heat exchange, and religious material. All these applications and uses of water are determined by the properties and characteristics of the water, water and its adsorbates, and the interacting media. That is, we have to purely understand the adsorbent-adsorbate relationships and resulting impacts of characteristics of water. Such an array of interactions and multifaceted challenges, as is the nature of water and its applications to human civilization, can only be handled more effectively and efficiently by the use of artificial intelligence.

This book is seeking the solution to water sector challenges through Machine Learning Applications in the adsorption of water pollutants, improved hydrological modeling and water resource management, water quality assessment, and wastewater treatment, explaining the importance of water quality parameters for prediction of the quality of water using SHAP value and lastly, in the classification of quality of water. Similarly, artificial intelligence (AI) is discussed in wastewater management, water treatment, water resource assessments, improved hydrological modeling and water resource management, and agriculture. Internet of Things (IoT) is used in the integration of sustainable energy generation from wastewater: Anaerobic digestion, microbial fuel cells, and geothermal desalination. AI techniques in smart irrigation and precision agriculture, water resource management in soil, and soilless irrigation systems are also discussed. All these approaches are intentional for the book to be relevant to a wide range of water sector users, policymakers, and practitioners.

Finally, the uniqueness of this book resides in the ability to bring the future technology, predictions, and dreams of science under one roof and also to package it in a manner, language, and style hopefully appreciable by an average expert and user in the various fields covered. It envisages a world where solutions to water challenges will no longer be unavailable to the resource-poor populations in the underdeveloped south, hospitalizations for waterborne diseases are minimum, solutions to transboundary water conflicts are obtainable by just punching buttons, water quality questions can be answered easily and lastly, droughts and hunger are predictable and data on solutions is retrievable. Such situations contribute to predictable water stress situations leading to lasting human peace where there will be fewer human population displacements.

Chrispin Kowenje
Department of Chemistry, Maseno University, Kenya

Preface

Water cycle and life cycle are inseparable. For sustaining life on our planet, water conservation and treatment have become pressing concerns in the face of increasing pollution, water scarcity and climate change. Every drop of water is precious. The challenges posed by water pollution, climate change and the demand for clean water are particularly prominent in emerging economies, where the strain on access to safe water has become a critical issue affecting human health and well-being.

In response to the complexities introduced by industrialization and the emergence of diverse water contaminants, traditional deterministic approaches to water treatment have proven insufficient. However, as the world progresses into the digital age, artificial intelligence (AI) has emerged as a powerful tool with the potential to revolutionize water management and treatment processes.

This edited reference book, *Artificial Intelligence Applications in Water Treatment and Water Resource Management*, delves into the multifaceted applications of AI in tackling the challenges of water pollution and water resource management. With a keen focus on the fields of wastewater treatment and water resource management, this book aims to provide a comprehensive exploration of the different AI techniques that can be harnessed to address water-related issues effectively and reliably.

AI has already demonstrated its remarkable potential in a wide range of sectors and academic disciplines, transforming industries and enhancing our lives in previously unimaginable ways. Its applications in fields like computer science, robotics, healthcare, and e-commerce have been well-documented. Now, AI's relevance in water management is becoming increasingly evident, offering dependable solutions for optimization, suspect screening, classification, regression, forecasting, and more.

The strength of AI lies in its adaptability, flexibility, and simplicity in design, making it a valuable resource for water scientists and researchers across various domains. This book explores the integration of AI techniques in water science, encompassing biological, physical, and chemical processes, water quality assessment, contaminant detection and quantification, process and system design, and maintenance.

The diverse chapters in this book delve into specific AI applications, such as the use of AI in adsorption science for handling various pollutants, water quality assessment, and the detection of microplastics. Additionally, readers will gain insights into AI's potential in determination of water pollutants, detecting leaks and faults in sewers, optimizing energy consumption in water treatment, and implementing smart water management strategies.

CHAPTER OVERVIEW

Chapter 1 explores the effectiveness of machine learning algorithms in predicting adsorption efficiencies for various water decontamination materials and pollutants. Adsorption processes are influenced by multiple factors, making predictions challenging through traditional methods. The chapter delves into the principles of various machine learning models and their application in adsorption. By leveraging machine learning, concise and accurate predictions can be achieved, contributing to the advancement in understanding of adsorption science and optimization of water decontamination techniques.

Chapter 2 presents a novel solution for sewage pollution by incorporating AI-enabled waste management systems. The approach separates sewage into solid and liquid waste streams, employing specific treatment processes to achieve sustainable sewage treatment, resource recovery, and distilled water production. The chapter emphasizes the use of advanced technologies like IoT, machine learning, and computer vision for real-time monitoring and water quality assessment. This technology-driven approach aims to mitigate soil and water pollution, preserve groundwater levels, and enhance overall efficiency in waste management.

Focusing on the hydrological cycle, Chapter 3 investigates the importance of accurate modeling for managing water resources and risk reduction. The chapter highlights the challenges in understanding and modeling the hydrological cycle due to data limitations, fluctuation, and human impacts. Machine learning and AI emerge as effective tools for hydrological modeling, with applications in rainfall and streamflow forecasting and flood modeling. Case studies and real-world examples demonstrate the practical use of AI to address data quality, interpretability, and scalability issues.

Chapter 4 provides a comprehensive exploration of AI's applications in water treatment, evaporation management, and water resource management. It compares AI-driven solutions to traditional approaches for evaporation control and optimization. The chapter discusses water resource management challenges, data-driven modeling, forecasting, optimization, and decision support systems. Ethical considerations and policy frameworks for responsible AI implementation are also addressed, along with recommendations for future research to advance AI in water treatment and resource management.

Chapter 5 focuses on the incorporation of AI in agricultural practices in India to increase production and sustainability. It highlights the significance of AI in improving economic status and production for rural households dependent on agriculture. The chapter presents a questionnaire survey to understand farmers' experiences and perspectives on the use of ICT and AI in agriculture, addressing barriers to accessing information sources and facilitating AI adoption in farming practices.

Chapter 6 explores the potential of AI, IoT, and Machine Learning in facilitating smart irrigation and promoting sustainable digital agriculture. It discusses virtual water, strategies to address water scarcity, and essential components for effective smart irrigation. The chapter presents a selection of recent studies supporting AI and IoT as promising approaches to advance agriculture. By focusing on water-efficient farming practices, this chapter contributes to promoting sustainable water use in agriculture.

Chapter 7 presents state-of-the-art water assessment and treatment methods, addressing current issues and challenges. The chapter covers water-related problems such as pollution and waterborne diseases and surveys various approaches to tackle these issues. It highlights the application of machine learning in water quality assessment and wastewater treatment, offering insights into innovative methods for maintaining water quality and safety.

In Chapter 8, different machine learning algorithms are explored to decide water quality based on collected water samples. The chapter demonstrates the use of precision-recall curves, ROC curves, and confusion matrices to evaluate the performance of various machine learning classifiers. Neural Network is found to be the most accurate classifier with a 96% accuracy rate. The chapter showcases the SHAP Value-based method for explaining the impact of each parameter in determining water quality.

Chapter 9 focuses on using AI to classify and predict water quality, addressing the critical importance of monitoring water quality to safeguard human and environmental health. The chapter employs several machine learning algorithms, including Random Forest Classifier, CatBoost Classifier, K Nearest Neighbors, and logistic regression, to manage water quality data. The results demonstrate high accuracy rates for the classifiers, with Random Forest standing out at 95.43%. AI's application in water quality classification is showcased as an effective tool for maintaining clean water sources.

While celebrating the immense opportunities AI presents in water research and environmental sciences, we also acknowledge the challenges inherent in integrating this technology into water management practices. Throughout the book, we will explore the potential hurdles and discuss how to overcome them to leverage AI for the betterment of water treatment and resource management.

We envision this reference book to serve as a valuable resource for a diverse audience, including industry experts, scientists, academicians, research scholars, and graduate students. By consolidating cutting-edge research, case studies and practical applications, we hope this book will inspire and guide readers in harnessing AI's potential to develop sustainable and effective solutions for water treatment and water resource management.

The topics covered encompass a wide array of subjects, including AI in hydrology and hydroinformatics, smart irrigation applications, optimization of water resources, adsorption science and AI's role in predicting various water variables such as rainfall-runoff, evaporation, evapotranspiration, streamflow, and lake water level changes.

In conclusion, we extend our gratitude to all the contributing authors for their invaluable insights and efforts in bringing this collection of knowledge to fruition. We hope this compilation will spark curiosity, collaboration, and innovation in the field of AI applications in water treatment and water resource management. The innovative use of E-nose and E-tongue techniques, for example, in wastewater treatment, presents a glimpse into the future possibilities of AI-driven advancements.

Together, we can forge a path towards a cleaner, healthier, and more sustainable water resource management for generations to come. As it is said, thousands have lived without love, not one without water.

Victor Shikuku
Kaimosi Friends University, Kenya

Chapter 1
Machine Learning Applications in Adsorption of Water Pollutants

Victor Odhiambo Shikuku
https://orcid.org/0000-0002-2295-293X
Kaimosi Friends University, Kenya

Newton Wafula Masinde
https://orcid.org/0000-0002-2578-4361
Jaramogi Oginga Odinga University of Science and Technology, Kenya

ABSTRACT

Adsorption remains one of the most effective methods of decontamination of water. Adsorption processes are governed by multiple factors which contribute to the overall efficiency of the process. These include process conditions such as temperature, pH, the concentration of pollutants, and competing ions. The adsorbate properties, such as speciation, polarity, kinetic diameter, and ionic sizes, also affect adsorption performance. The adsorbent properties also play a critical role in assessing the suitability of an adsorbing material. This includes surface areas, pore volumes, chemical compositions, surface charges, etc. The complexity of the interaction between all these parameters makes it cumbersome or near impossible to predict with appreciable precision the performance of an adsorption system. Machine learning provides an opportunity for developing models for concise prediction of adsorption efficiencies for different materials. This chapter discusses the principles of various machine learning models and their application in the adsorption of pollutants from water.

DOI: 10.4018/978-1-6684-6791-6.ch001

INTRODUCTION

Water pollution and access to clean water are global problems, especially in the 21st century. This is due to exponential growth industrialization with concomitant rise in human population. Among the widespread water contaminants are heavy metals and dyes. These are known to be toxic to both aquatic and terrestrial living organisms (Mitra et al., 2022). This explains the abundance of data and continuing research on water treatment technologies. These include; chemical oxidation, ion exchange, ozonation, Fenton's reagent, membrane filtration, electro-kinetic coagulation, irradiation, electrochemical degradation, and adsorption among others (Butler et al., 2011; Nidheesh et al., 2013; Samarghandi et al., 2020; Venkatesh et al., 2017; Dzoujo et al., 2022). All these methods have their inherent limitations. However, adsorption still remains the most popular and effective technique for removal of contaminants from water. The effectiveness of an adsorption process, measured by the percent removal (%R) of the contaminant of interest, is dependent on the process conditions such as temperature, pH, competing ions, concentration of pollutant and amount of adsorbing materials. The %R is also a function of the adsorbent properties such as surface area, pore structure, pH, functional group density, acidity and basicity among other characteristics (Zhu et al., 2019). Since all these variables may simultaneously affect the percent removal, it is time consuming and expensive to undertake experiments evaluating the role of each of these parameters for optimization. Artificial intelligence provides a powerful tool for interrogating the interplay of multiple parameters and determining the most impactful parameters and their contribution to the observed outcomes. Machine learning (ML) is presently used in almost all domains of science, medicine, engineering, agriculture, computer science and water treatment is not an exception. This is due to its simplicity, reliability and rapidity (Çelekli et al., 2013). Various computing methods have been explored to solve problems in adsorption science. This chapter highlights the principles of different machine learning models and their applications in predicting adsorption data for the removal of heavy metals and dyes from water.

Machine Learning Concepts

The Fourth Industrial Revolution (*4IR* or Industry 4.0) has resulted in an overabundance of data, hence the phrase "*the age of data*", where all that is around us is somehow connected to some data source, and our lives are constantly being digitally captured (Cao, 2017; Sarker, 2021; Sarker et al., 2021). This data can be either *(1)* structured, having a well-defined structure that conforms to some standard data model such as relational databases, *(2)* unstructured, meaning there is no pre-defined data format for example, sensor data, word processing documents, and PDF files, or *(3)* semi-structured, where it has some organizational properties but in the whole it may not conform to standard data models, for example, HTML, XML, JSON documents, NoSQL databases, among others (Sarker, 2021). Therefore, to derive any intelligence from all this data, artificial intelligence (AI) expertise, and more specifically, knowledge of machine learning (ML) techniques is of paramount importance.

Machine learning is a sub-category of AI that focuses on two interrelated questions: how to design computers capable of automatic experiential improvement and the discovery of the statistical-computational-information-theoretic laws governing all learning systems such as computers, human beings, and organizations (Jordan & Mitchell, 2015). More formally, "A computer program is said to learn from experience *E* with respect to some class of tasks *T* and performance measure *P*, if its performance at tasks in *T*, as measured by *P*, improves with experience *E*" (Mitchell, 1997). ML systems can be classified along many dimensions. One classification considers three dimensions: the *underlying learning strategies* which look at the amount of inference the learner performs on the information provided; *the type of knowledge or acquired by the learner* based on rules of behavior, description of physical objects, problem-solving heuristics, and so on; and the *domain of application* (Jordan & Mitchell, 2015). In this study, however, we consider the more traditional approach to classifying ML models or algorithms as more appropriate. This is discussed further in the following section.

Traditional ML Model Taxonomy

The traditional classification of ML models (or algorithms) is based on the algorithmic procedure itself, leading into four main types namely, supervised learning, unsupervised learning, semi-supervised learning, and reinforcement learning ML techniques (Dutton & Conroy, 1997; Jain et al., 1999; Sarker, 2021).

Supervised learning: The focus is on making the ML algorithm learn a function that maps the input data to output by providing a sample of input-output pairs, meaning that it is concerned with classified (or labeled) data (Han & Kamber, 2006; M. Mohammed et al., 2017). Supervised learning is a task-driven approach, that is, it is undertaken when the goals are identified and are to be accomplished through a given set of inputs (Sarker et al., 2020). Examples of supervised learning algorithms include regression analyses, decision trees, k-nearest neighbor (*k*NN), Naïve Bayes, support vector machines among others.

Unsupervised learning: This is suited for situations where the training data is not classified or unlabeled (M. Mohammed et al., 2017). The algorithms analyze the datasets to learn the different class labels for the data by finding hidden patterns, structures, or knowledge in the data, hence this is a data-driven approach (Sarker, 2019). Examples of unsupervised learning algorithms include clustering algorithms such as *k*-means, Gaussian mixture models, Hidden Markov Models, Principal Component Analysis, among others.

Semi-supervised learning: This is a combination of supervised and unsupervised learning techniques because it handles both labeled and unlabeled data (M. Mohammed et al., 2017; Sarker et al., 2020). The main goal for a semi-supervised learner is to provide a better prediction than for models with labeled data only and is the basis for tasks such as machine translation, fraud detection, labeling data, and text classification (Sarker, 2021).

Reinforcement learning: This points to algorithms that make it possible to automate how software agents and machines evaluate efficient optimal behavior based on the context and environment (Kaelbling et al., 1996). Therefore, reinforcement learning is considered an environment-driven approach (Sarker, 2021). It is based on reward and penalty to use the insights gathered to increase rewards and minimize risks (M. Mohammed et al., 2017). This technique is especially useful in increasing automation of optimizing operational efficiency like in robotics, autonomous driving, manufacturing, and so on.

Table 1 is a summary of some of the most common machine learning techniques with references for further understanding.

Table 1. Common machine learning techniques

S.No.	Algorithm/Model	Learning Type	Main Use	Reference
1.	Linear Regression	Supervised	Regression	James et al., 2021
2.	Logistic Regression	Supervised	Classification	Hosmer Jr. et al., 2013
3.	Support Vector Machines (SVM)	Supervised	Classification/ Regression	Cortes & Vapnik, 1995
4.	Decision Trees	Supervised	Classification/ Regression	Breiman, 2017
5.	k-Nearest Neighbors (k-NN)	Supervised	Classification/ Regression	Cover & Hart, 1967
6.	Naïve Bayes	Supervised	Classification	Rish, 2001
7.	Neural Networks (NNs or ANNs)	Supervised	Classification/ Regression	McCulloch & Pitts, 1943; Rosenblatt, 1958
7.	K-Means	Unsupervised	Clustering	MacQueen, 1967
8.	Hierarchical clustering	Unsupervised	Clustering	Ward, 1963
9.	DBSCAN	Unsupervised	Clustering/ Anomaly Detection	Ester et al., 1996
10.	Principal Component Analysis (PCA)	Unsupervised	Dimensionality Reduction	Hotelling, 1933
11.	Self-Training (Self-Labeling)	Semi-supervised	Classification	Yarowsky, 1995
12.	Label Propagation	Semi-supervised	Classification	Zhuȓ & Ghahramani, 2002
13.	Semi-Supervised Support Vector Machines (S3VM)	Semi-supervised	Classification	Bennett & Demiriz, 1998
14.	Semi-Supervised Clustering	Semi-supervised	Clustering	Wagstaff et al., 2001
15.	Q-Learning	Reinforcement	Value-based, Temporal Difference learning, Updates Q-values based on the Bellman equation, Off-policy	Watkins & Dayan, 1992
16.	Deep Q Networks (DQNs)	Reinforcement	Value-based, Approximates Q-values using Deep Neural Networks, Combines Q-learning with DNNs	Mnih et al., 2015
17.	Proximal Policy Optimization (PPO)	Reinforcement	Policy-based, On-policy, Policy gradient method with trust region-based updates	Schulman et al., 2017
18.	Deep Deterministic Policy Gradient (DDPG)	Reinforcement	Actor-Critic, Off-policy, Uses Deep Neural Networks to approximate policy and value functions	Lillicrap et al., 2019

Advanced ML Models

Traditional ML models, which generally employ a single classifier in the prediction, may fail to realize satisfactory performance in the face of complex data, especially when the data is imbalanced, high-dimensional, and noisy, among other challenges (Dong et al., 2020). In addition, traditional ML models have other limitations such as results with high variance, high bias, and low accuracy (Mishra et al., 2022; Sun et al., 2021). To handle these challenges, research has focused on improved learning approaches, in particular, ensemble and deep learning techniques (Mienye & Sun, 2022). In this regard, the discussion following gives a brief introduction to ensemble and deep learning techniques, including a discussion on some common techniques in each class.

Ensemble Learning Techniques

The term "ensemble learning" refers to methods that combine multiple inducers (or base learners) to make a decision or a prediction, usually in supervised machine learning tasks (Sagi & Rokach, 2018). The base learner is an algorithm that accepts an input (a set of labeled examples) and gives a model (for example, a classifier or a regressor) as the output generalizing the examples (Sagi & Rokach, 2018). In general, ensemble learning techniques undertake the training of multiple base learners and then combine the predictions to get improved performance and a better generalization ability in comparison to the individual base learners (Mienye et al., 2020). Some of the reasons that ensemble methods have better predictive performance include overfitting avoidance when a small amount of data is available for prediction, computational advantage by decreasing the risk of obtaining a local minimum, and representation as it extends the search space resulting in a better fit to the data space (Dietterich, 2002; Polikar, 2006).

Framework for ensemble learners: Any ensemble learner system uses an aggregation function G to combine a set k of base-learner classifiers, $c_1, c_2, \ldots, c_k$, towards a single prediction output. With an input dataset of size n having features of dimension $m, D = \{(x_i, y_i)\}, 1 \leq i \leq n, x_i \in R^m$, the projection of the output based on the ensemble is given by equation 1 and illustrated in Figure 1.

$$y_i = f(x_i) = G(c_1, c_2, \ldots, c_k) \tag{1}$$

Figure 1. General framework of an ensemble (adapted from A. Mohammed & Kora, 2023)

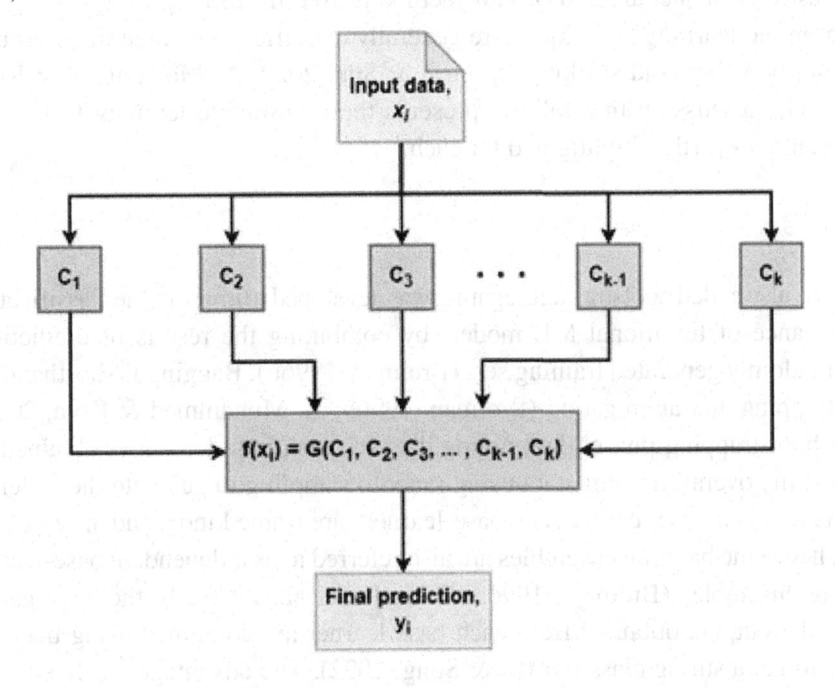

To use ensemble learning, two aspects to decide on are the base learners to include and how to combine the outputs from the base learners.

Choice of base learners: In developing an ensemble classifier, the choice of the base learners is critical in achieving optimum performance. Therefore, the two key principles to pay attention to when generating an ensemble model, inter alia, *1)* diversity, that is, participating base-learners should be sufficiently diverse to gain a desired predictive performance, and 2) predictive performance which should be as high as possible and at least as good as a random model (Sagi & Rokach, 2018).

Output fusion: Once the base learners have been chosen, it is then important to decide how the outputs obtained from them will be combined, a process called output fusion (A. Mohammed & Kora, 2023; Sagi & Rokach, 2018). Output fusion is usually achieved in two ways (A. Mohammed & Kora, 2023). The first technique is voting methods which are used mostly used in classification or regression problems and are appropriate integrating methods for bagging and boosting ensemble methods. Three techniques used under this classification are max voting (Kim et al., 2003), average voting (Montgomery et al., 2012), and weighted average voting (Latif-Shabgahi, 2004). The second fusion technique is meta-learning, also called "learning to learn", which focuses on learning based on previous experience gained while undertaking other tasks (Soares et al., 2004). This method is used in stacking.

Ensemble learning techniques are generally classified into three main groups: bagging, boosting, and stacking (Mienye & Sun, 2022; A. Mohammed & Kora, 2023). The discussion that follows presents these ensemble learning techniques with examples further highlighted for each.

Bagging

Bagging, also called bootstrap aggregating was developed to improve the classification performance of traditional ML models by combining the results of predictions from randomly generated training sets (Breiman, 1996b). Bagging is shorthand for bootstrapping and aggregating (Breiman, 1996b; A. Mohammed & Kora, 2023). In the bootstrapping phase, the training data for each base learner is obtained by splitting the overall training data using random sampling to generate the different subsets for each base learner. The base-learners are trained independently of each other, hence the bagging ensembles are also referred to as independent base-learner training ensembles (Breiman, 1996b; González et al., 2020). In the aggregation phase, the outputs obtained from each base learner are combined using majority voting to get a strong classifier (Li & Song, 2022). The advantages realized from bagging are the reduction of variance which eliminates overfitting and the ability to work with high-dimensional data (A. Mohammed & Kora, 2023). However, the drawback with bagging is that it is computationally expensive, has high bias, and the interpretability of a model may be lost (A. Mohammed & Kora, 2023). Also, there are some challenges with the implementation of bagging models, such as, how to determine the optimal number of base learners. Common bagging models are highlighted in Table 2.

Table 2. Bagging models/algorithms

S.No.	Algorithm/Model	Base Learner	Fusion Method	Usage	Reference(s)
1.	Random Forest (RF)	Decision tree	Majority voting (classification); average voting (regression)	Classification, regression	Breiman, 2001; Ho, 1995
2.	Extremely randomized trees (Extra trees)	Decision tree	Majority voting (classification); average voting (regression	Classification, regression	Geurts et al., 2006
3.	Bagged decision trees	Decision tree	Majority voting (classification); average voting (regression)	Classification, regression	(Breiman, 1996b)

Boosting

Boosting constitutes multiple weak learners arranged sequentially with each subsequent model working to mitigate any errors arising from the previous model (A. Mohammed & Kora, 2023). Boosting was first introduced in 1996 by Freund and Schapire (Freund & Schapire, 1996). The arrangement of the weak learners ensures that they are adapted to handle any aspects of the dataset that previous models in the sequence may not have handled well (A. Mohammed & Kora, 2023). Some of the common boosting algorithms are highlighted in Table 3.

Table 3. Boosting models/algorithms

S.No	Model/Algorithm	Base Learner	Fusion Method	Usage	Reference(s)
1.	AdaBoost	Depends on the problem and dataset characteristics	Weighted average voting	Classification, regression	Freund & Schapire, 1997
2.	Gradient Boosting/ Gradient Boosted Decision Tree (GBDT)	Typically decision trees	Weighted average voting	Classification, regression	Breiman, 1996a; J. H. Friedman, 2001
3.	Extreme Gradient Boost (XGB/ XGBoost)	Decision trees	Weighting average voting	Classification, regression	Chen & Guestrin, 2016
4.	Light Gradient Boosting Machine (LightGBM)	Decision trees	Weighted average voting	Classification, regression	Ke et al., 2017
5.	Categorical Boosting (CatBoost)	Decision trees	Weighted average voting		Prokhorenkova et al., 2018
6.	Histogram-based Gradient Boosting	Histogram-based decision trees	Weighted average voting	Classification, regression	Ke et al., 2017

Stacking

Stacked Generalization (or stacking) was introduced by David H. Wolpert (Wolpert, 1992). Stacking is an ensemble learning technique that focuses on the improvement of predictive performance by combining the prediction of multiple base models. The stacking model architecture consists of two parts. The first part consists of two or more base models (usually different ML algorithms) and is referred to as a level-0 model, while the second part is a meta-model that combines the predictions of the base models and is referred to as a level-1 model (A. Mohammed & Kora, 2023). The prediction from the base models becomes the input features for the higher-level meta-model which is referred to as the "meta-learner" or the "stacking model". The benefits accrued from stacking include deeper data comprehension hence more precision and effectiveness. On the other hand, the problem of overfitting persists as many predictors work on the same target that is merged. Also, multi-level stacking is data-intensive and time-consuming (Xiong et al., 2021). Additionally, implementing stacking is also non-trivial as identification of the correct number of baseline models that can be relied upon for good prediction is not easy, it is difficult to interpret the final model and computation time complexity increases exponentially with an increase in data available (A. Mohammed & Kora, 2023).

Artificial Neural Networks

Artificial neural networks (also called neural networks or connectionist models) are a technology that is derived from studies on the human brain and the nervous system as indicated in Figure 2 (Walczak & Cerpa, 2003). A neural network is "an interconnected assembly of simple processing elements, units or nodes, whose functionality is loosely based on the animal neuron. The processing ability of the network is stored in the interunit connection strengths, or weights, obtained by a process of adaptation to, or learning from, a set of training patterns." (Gurney, 2018).

Figure 2. Sample ANN architecture with sample weights (adapted from Walczak & Cerpa, 2003)

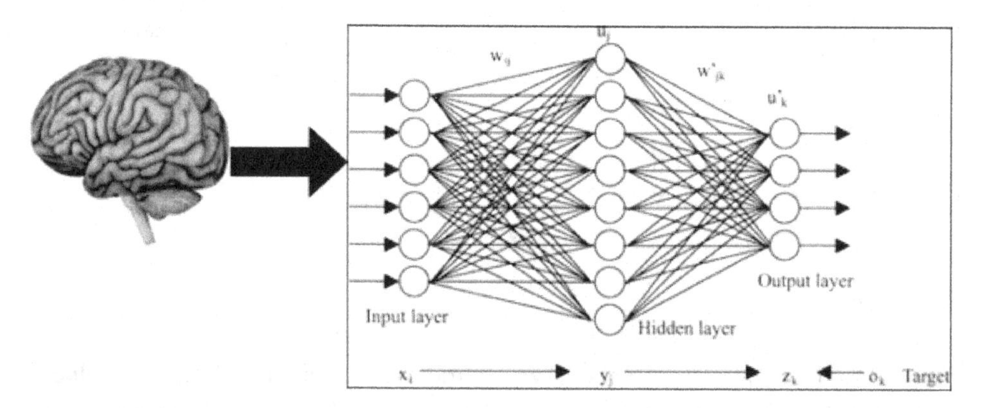

Computational modeling and ANNs. Computational models find usefulness in the translation of observations into anticipation of possible future events, acting as a platform for testing new ideas, getting information from data, and asking questions about phenomena (Calder et al., 2018). Any answers obtained help to understand, design, manage and predict how complex systems and processes work, from public policy to autonomous systems (Calder et al., 2018). ANNs are simply computational models whose development comes from the nervous system of living beings, and the models are capable of collecting and maintaining knowledge (Da Silva et al., 2017). The computational models (or processing units) in ANNs are called artificial neurons, which are simplified models of the biological neuron (Da Silva et al., 2017).

The artificial neuron. The artificial neurons in ANNs operate non-linearly, providing continuous outputs and performing simple functions like input signal gathering, input signal assembly based on operational functions, and producing response taking into account their inherent activation functions (Da Silva et al., 2017). The simplest neuron model, which is still in use in ANN architectures today, that includes the main features of a biological neural network, that is, parallelism and high connectivity, was proposed by (McCulloch & Pitts, 1943). The model is depicted in Figure 3, which shows seven basic elements of a neuron, namely, input signals ($x_1, x_2, \ldots, x_m$) from the external environment, synaptic weights ($w_{k1}, w_{k2}, \ldots, w_{km}$) which enables quantification of the relevance of each input with respect to the neuron's functionality, the linear agreggator (Σ) which gathers all signals to produce an activation voltage, the activation threshold or bias (b_k) that specifies the linear aggregator's output threshold to produce an activation voltage, activation potential (u_k) which is the difference between linear aggregator and activation threshold, the activation function (φ) that limits the neuron output within a reasonable value range, and the output signal (yk_j.

Figure 3. An artificial neuron (Haykin, 1998)

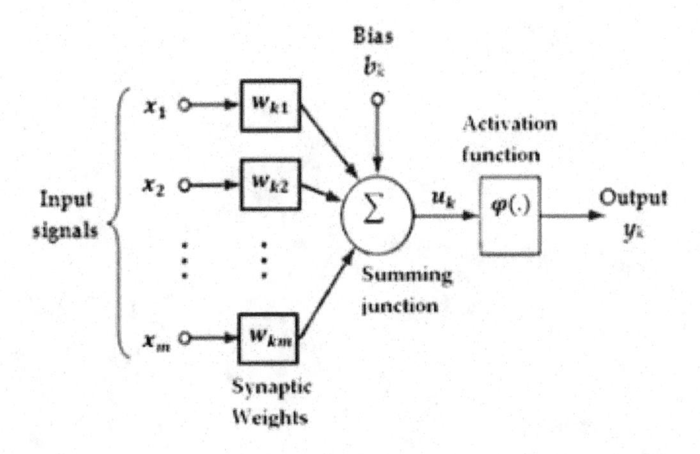

The results of the artificial neuron proposed by (McCulloch & Pitts, 1943), are synthesized by Equation 2 and Equation 3.

(2)

$$y_k = \varphi(\mathrm{u}k_j \quad (3)$$

The activation functions can be categorized into two fundamental groups (Da Silva et al., 2017). The first category is the partially differentiable functions such as step functions (Heaviside/hard limiter), bipolar step functions or signal functions (symmetric hard limiter), and symmetric ramp functions. These activation functions are points whose first-order functions are nonexistent. The second category is the fully differentiable functions whose first-order derivatives exist for all points of their definition domain. The main activation function under this category that are employed in ANNs are the logistic function, hyperbolic tangent, Gausian function, and linear function

ANN Architecture. ANNs are generally divided into three parts, the input layer that receives data, signals, features, or measurements from the external environment, the hidden (intermediate or invisible) layer which extracts patterns in line with the process/system under analysis, and the output layer that handles production and presentation of the final output. We take the approach presented by (Da Silva et al., 2017) to classify the architectures as single-layer feedforward NNs, multi-layer feedforward NNs, recurrent or feedback NNs, and mesh NNs. Figure 4 is a summary of these architectures.

Figure 4. ANN architectures and with common examples

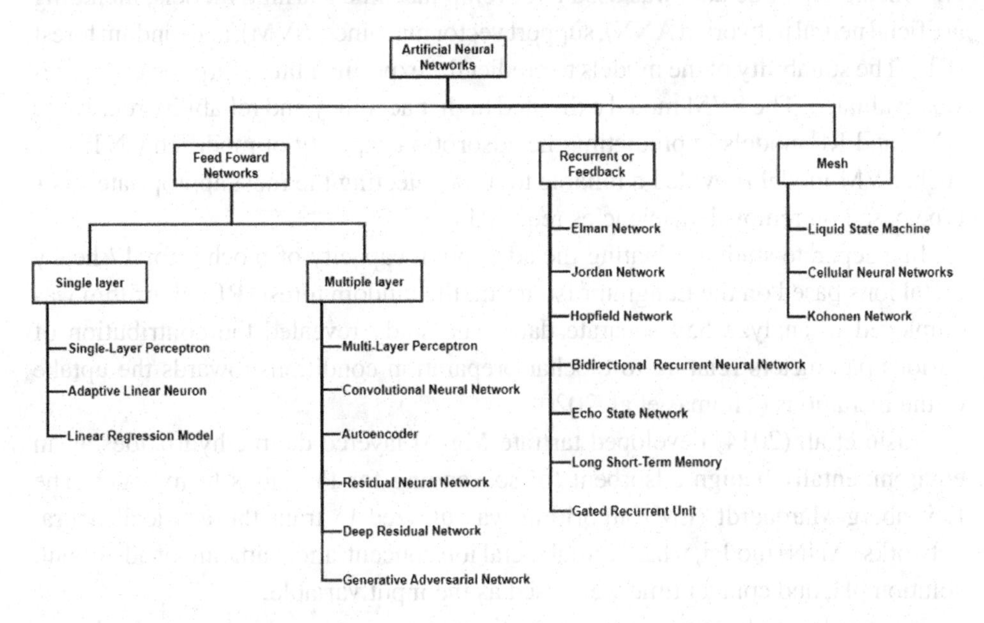

Application of ANNs. The application of ANNs depends on the task at hand. However, the main application tasks are in control processing, modeling and prediction, optimization, classification and fault detection, and design automation as enumerated by (Panerati et al., 2019).

REMOVAL OF HEAVY METALS

Heavy metals such as lead (Pb), cadmium (Cd), nickel (Ni), arsenic (As), copper (Cu), and zinc (Zn) in water and wastewater present serious risks to human health, aquatic and terrestrial living things because of their toxicity and persistence in different environmental compartments (Mitra et al., 2022). Adsorption still remains the widely used technique to sequester heavy metals from drinking and wastewater owing to its simplicity, ease of operation, low capital investment required and insensitivity to pollutant type among other inherent advantages (Achieng' and Shikuku, 2020). Numerous adsorbents, including geopolymers (Siyal et al., 2018), clay minerals (Uddin, 2017), carbon nanotubes (Nyairo et al., 2018; Wang et al., 2021), biochars (Achieng' and Shikuku, 2020) and biomass-derived materials (Agarwal et al., 2020), have been evaluated as possible materials for heavy metal removal from water.

Arabloo and co-workers (2019) reported the adsorption densities of carbon nanotubes (CNTs) for six heavy metal ions, namely; Cu^{2+}, Pb^{2+}, Cd^{2+}, Zn^{2+}, and

Ni^{2+} metal ions. The data was used to develop machine learning models, including artificial neural networks (ANN), support vector machines (SVM), and random forest (RF). The suitability of the models to predict the experimental adsorption capacities was evaluated. The SVM model exhibited higher accuracy and reliability relative to ANN and RF models in predicting the adsorption capacity of modified CNTs. As such, SVM model provided a reliable tool for selecting the most appropriate CNT type based on removal efficiencies required.

In a separate study evaluating the adsorption capacity of biochar for 17 heavy metal ions based on the Langmuir isotherm, the random forest (RF) algorithm was employed to analyze 559 separate data. The study revealed the contribution of various parameters relating to biochar preparation conditions towards the uptake of the metal ions (Thomas et al. 2020).

Yasin et al. (2014) developed tartrate-Mg-Al layered double hydroxides as an environmentally benign adsorbent for scavenging for Pb^{2+} ions from water. The Levenberg-Marquardt (LM) algorithm was utilized to train the artificial neural networks (ANN) model, where initial metal ion concentration, amount of adsorbent, solution pH, and contact time were used as the input variable.

In a similar study, temperature, operation time, adsorbent loading, and pH were used as input variables to model and optimize the process conditions for maximum adsorption of Zn^{2+} onto Hazelnut shell from leachate (Turan et al., 2011a) using ANN model. The ANN model reliably predicted the sequestration of Zn^{2+} ions and the close agreement between the experimental data and the model-predicted data was evident from the coefficient of correlation value close to unity.

In a related study, Turan et al. (2011b) reported the adsorption of Cu^{2+} from aqueous solution onto pumice. The effect of process conditions, namely contact time, adsorbent dosage, pH, and temperature were evaluated using the ANN algorithm. The model entailed a three-layer structure with 4, 8 and 4 neurons in the first, second and third layer, respectively. Furthermore, the accuracy of the radial basis function (RBF) was examined and compared with other networks. The ANN model presented the highest reliability and accuracy, with a coefficient of correlation value closest to unity ($R^2 = 0.999$).

Khajeh et al. (2017) compared the predictive powers of response surface methodology (RSM) and hybrid of artificial neural network-particle swarm optimization (ANN-PSO) in the attempt to simulate and optimize the extraction of Co^{2+} and Mn^{2+} ions using tea waste. The input variables were the eluent flow rates, eluent concentration, eluent volume, concentration of the complexing agent 1-(2-pyridylazo)-2-naphthol (PAN), the concentration of tea waste quantity, and pH. The results showed that the ANN model was superior to the RSM.

Singha and co-workers (2015) examined the uptake of Pb^{2+} ions using 6 different biosorbents. Three standard training algorithms, *viz.* Scaled Conjugate Gradient

(SCG), Levenberg-Marquardt (LM), and Backpropagation (BP) along with four different standard transfer functions in a hidden layer with a linear transfer function in the output layer were used to demonstrate the viability of ANN using contact time, adsorbent dosages, Pb^{2+} ion concentration, and pH of solution as the input variable. The percent removal(%R) was the output variable. The results showed that the ANN model with a BP algorithm, using transfer function 1 and 25 processing elements in a single hidden layer best predicted the percentage removal of Pb^{2+} ions water.

Elsewhere, adaptive neuro-fuzzy interference system (ANFIS) model was shown to better predict the adsorption capacity of agro-wastes (R^2=0.9943) for Ni^{2+} ions relative to the artificial neural network (ANN), R^2=0.9926 (Souza et al., 2018). Abdi and Mazloom (2022) reported the ability of different machine learning (ML) techniques, namely; Light Gradient Boosting Machine (LightGBM), Extreme Gradient Boosting, Gradient Boosting Decision Tree, and Random Forest in forecasting the removal of arsenate (AsO_4^{3-}) from aqueous solution using different metal–organic frameworks (MOFs).

Based on the correlation coefficient value, the ANN algorithm was found to be suitable for predicting the adsorption data for the removal of Cr^{6+} ions using sepiolite-stabilized zero-valent iron nanocomposite (Esfahani et al., 2015). Finally, Zhu et al. (2019) employed machine learning (ANN and RF) to evaluate the effects of biochar properties, metal sources, process conditions, and the initial concentration ratio of metals to biochars on the adsorption capacity of 44 biochars for 6 heavy metals ions (Pb, Cd, Ni, As, Cu, and Zn). The inclusion of both adsorbent characteristics and process conditions was particularly significant and a departure from most studies. Figure 5 shows the contribution of the different parameters to the adsorption capacity. The surface area of the biochars, traditionally assumed to be a significant criterion for selecting of best performing adsorbents, accounted for only 2% of adsorption efficiency. The cation exchange capacity (CEC) and pH_{H2O} of biochars were the most significant drivers of adsorption efficiency of the biochars, accounting for 66% in the biochar characteristics. The RF model produced better predictive power than the ANN model.

Figure 5. Contribution of biochar properties and process conditions to adsorption capacity (adopted from Zhu et al., 2019)

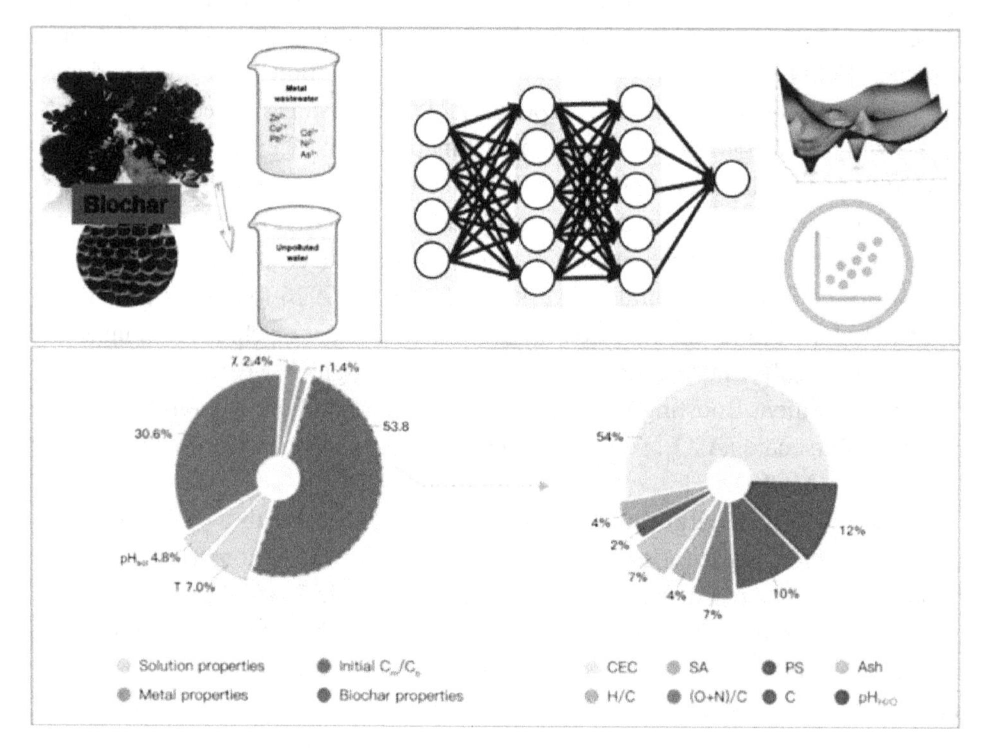

A comprehensive review on the use of ANN for predicting heavy metals removal is described by Fiyadh et al. (2023) and the details will not be reiterated here. The use of machine learning provides new revelations that provoke inquiry and increased understanding of adsorption phenomena.

REMOVAL OF DYES

Industrial effluents are often laden with a myriad of water pollutants, including synthetic dyes. As aforementioned, adsorption still remains the widely used technique for removal of many pollutants, including dyes, from water. Dyes have been shown to be toxic and potentially carcinogenic (Khan et al., 2022). Several adsorbent materials have been proposed for the removal of dyes from water. These include; geopolymers (Owino et al., 2023; Luttah et al., 2023; Tome et al., 2023), clays (Shikuku and Mishra, 2021), agricultural wastes (Onu et al., 2023; Banerjee & Chattopadhyaya, 2017), polymers (Torğut & Demirelli, 2018) among others. The traditional approach of "unguided" synthesis and functionalization of adsorbents and testing their adsorption performance individually under various environmental conditions is evidently cost and time consuming. Figure 6 is an illustration of how machine learning has made it easy to process large adsorption data for dye removal, with reliable prediction models thus saving on human resource, costs and time (Bhagat et al., 2023).

Figure 6. Utility of machine learning in removal of dyes from water (adopted from Bhagat et al., 2023)

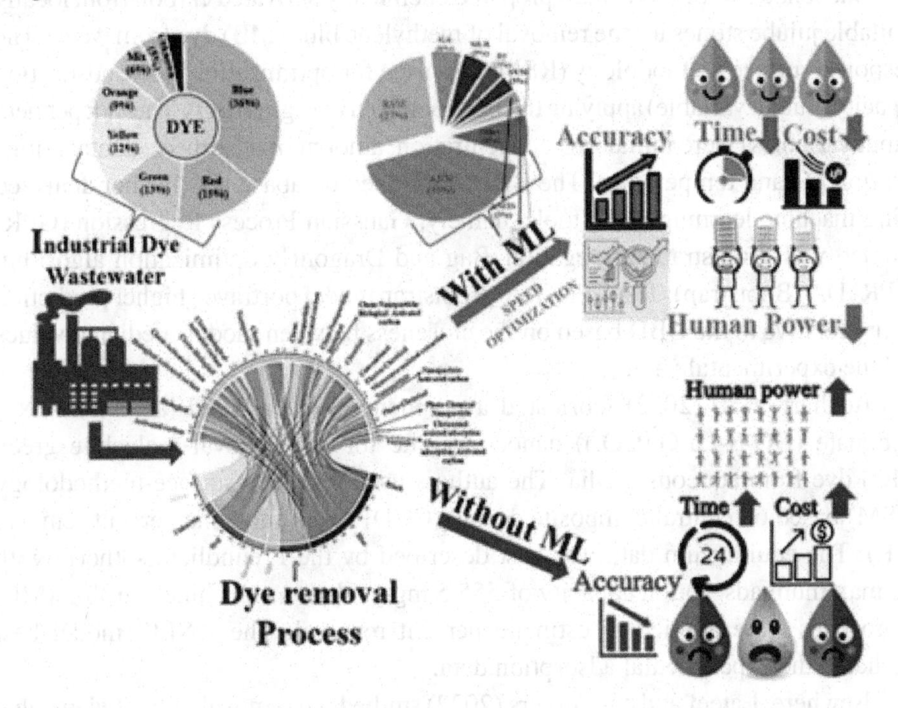

Moosavi et al. (2021) reported the adsorption modeling of dyes using 39 agro-wastes derived adsorbents were using random forest (RF), decision tree (DT), and gradient boosting (GB) algorithms with the adsorbent characteristics (surface area, pH, pore volume, and particle size) as the predictor variables. Based on correlation coefficient (R^2) values, the RF model had the best predictive potential. To predict the adsorption capacity, it was determined that the pore volume and surface area were the most significant parameters of the agro-waste adsorbents, whereas the particle sizes were inconspicuous.

In another study, a coagulating agent derived from Pinus halepensis Mill. Seed was applied to sequester Congo red dye from water and compared to ferric chloride, a conventional coagulant. The effect of various factors (coagulant dosage, initial pH, NaCl concentration, stirring speed, stirring duration, solution temperature and initial dye concentration) on dye removal were evaluated using the support vector machine (SVM) coupled with the Gray Wolf Optimizer. In acidic media (pH = 3), the optimum performance was realized with a biocoagulant dosage of 10 mL/L and an initial dye concentration of 50 mg/L, yielding 81% of dye removal (Hadadi et al., 2023).

Bouchelkia and others (2023) prepared chemically activated carbon from locally available jujube stones for the removal of methylene blue (MB) dye from water. The Response surface methodology (RSM) was used for optimization of the adsorption capacity (output variable) applying the Box-Behnken Design (BBD). The independent (input variables) were initial MB concentration, amount of adsorbent, contact time, solution pH and temperature. The BBD-developed database was further analyzed using machine learning (ML) tools, namely; Gaussian Process Regression (GPR) coupled with Bootstrap Aggregation_Bag and Dragonfly optimization algorithm (GPR_DA_Bootstrap). The GPR_DA_Bootstrap model portrayed higher prediction ability relative to the BBD based on the closeness between model-predicted values and the experimental values.

Eftekhari et al. (2022) fabricated a graphene oxide-sulfated lanthanum oxy-carbonate (GO-S-$La_2O_2(CO_3)$) nanocomposite for the removal malachite green (MG) dye from aqueous media. The authors used response surface methodology (RSM) based on central composite design (CCD) to optimize the percent removal (%R). The equilibrium data was best described by the Freundlich isotherm with the maximum adsorption capacity of 555.5 mg/g. Various machine learning (ML) approaches were applied to estimate percent removal. The ANFIS model best predicted the experimental adsorption data.

Elsewhere, Lateef and co-workers (2022) studied the removal of methylene blue (MB) dye from aqueous media onto aluminized activated carbon adsorbent. The authors employed multiple linear regression (MLR) and the three machine learning (ML) models (random forest (RF), support vector regression (SVR), and multilayer perceptron (MP)) to study the interplay of five process and synthesis parameters. The predictive performance of the models was in the order ANN > SVR > RF > 2-degree MLR. Aluminum loading, adsorbent dosage, and initial adsorbate concentration were determined as the most significant parameters influencing MB adsorption. Furthermore, the adsorption was independent of temperature hence no necessity for installation of temperature-controlled equipment for practical application.

Khan et al. (2023) synthesized a bentonite clay derived geopolymer/Fe_3O_4 nanocomposite as a novel adsorbent for the removal of Navy-Blue (NB) dye. The researchers employed statistical and machine learning frameworks, specifically; regression analysis (SRA), Support vector regression (SVR) and Kriging (KR) algorithms. All the models were judged to be suitable for predicting NB dye adsorption. However, statistically, the SVR model exhibited higher accuracy than the other models as shown by the root mean square error (RMSE) function.

Four machine learning algorithms (random forest (RF), Light Gradient Boosting (LGB), Xtreme Gradient boosting (XBG), and artificial neuron network (ANN) were used to predict the adsorption of organic dyes onto nanozeolites (Oviedo et al., 2023). The descriptors were effect of dye nature (cationic and anionic), type of agitation (sonication and magnetic stirring) and molecular weight (low and high MW) while the amount of dye adsorbed at equilibrium qe (mg/g) was the response variable. The XBG model possessed the best predictive potential relative to the other algorithms, especially for cationic dyes.

Finally, Asfaram et al. (2015) reported the fabrication MnO_2 nanoparticle-loaded activated carbon (MnO_2-NP-AC) and its application as a novel, efficient, environmentally benign and inexpensive adsorbent for adsorptive removal of three dyes, namely; brilliant green (BG), crystal violet (CV) and methylene blue (MB) from water. Optimization of the process conditions (solution pH, amount of adsorbent, initial dye concentration and sonication time) for maximum adsorption performance was achieved using the response surface methodology (RSM). Additionally, artificial neural network (ANN) was used to reliably predict the percent removal (%R) of the dyes in a multicomponent adsorption system based on the agreement between the model-predicted values and experimental adsorption data.

CONCLUSION

This chapter explored the general features of various machine learning models and their operation principles. The utility, predictive power and comparative performance of different machine learning algorithms for adsorption of heavy metals and dyes from water onto different adsorbent materials have been discussed. Machine learning (ML) tools provides an integral tool in deciphering the most important adsorbent characteristics and process parameters controlling the adsorption process. This allows for maximization of synthesis protocols and environmental conditions required for optimum performance. Furthermore, ML modeling also provides insights into the operative adsorption mechanism. Overall, the use of machine learning models is shown to save time and other input resources in development of adsorbents for water purification.

REFERENCES

Abdi, J., & Mazloom, G. (2022). Machine learning approaches for predicting arsenic adsorption from water using porous metal–organic frameworks. *Scientific Reports*, *12*(1), 1–13. doi:10.103841598-022-20762-y PMID:36180503

Agarwal, A., Upadhyay, U., Sreedhar, I., Singh, S. A., & Patel, C. M. (2020). A review on valorization of biomass in heavy metal removal from wastewater. *Journal of Water Process Engineering*, *38*, 101602. doi:10.1016/j.jwpe.2020.101602

Arabloo, M., Mirabi, A., Ghaedi, M., & Asfaram, A. (2019). Machine learning-based prediction of adsorption capacity of modified carbon nanotubes for heavy metal ions removal from water. *Journal of Environmental Chemical Engineering*, *7*(6), 103409. doi:10.1016/j.jece.2019.103409

Asfaram, Ghaedi, Goudarzi, & Hajati. (2015). *Ternary dyes adsorption onto MnO2 nanoparticle-loaded activated carbon: derivative spectrophotometry and modeling*. doi:10.1039/C5RA10815B

Banerjee, S., & Chattopadhyaya, M. (2017). Adsorption characteristics for the removal of a toxic dye, tartrazine from aqueous solutions by a low cost agricultural by-product. *Arabian Journal of Chemistry*, *10*, S1629–S1638. doi:10.1016/j.arabjc.2013.06.005

Bennett, K., & Demiriz, A. (1998). Semi-Supervised Support Vector Machines. *Advances in Neural Information Processing Systems, 11*. https://proceedings.neurips.cc/paper_files/paper/1998/hash/b710915795b9e9c02cf10d6d2bdb688c-Abstract.html

Bhagat, S. K., Pilario, K. E., Babalola, O. E., Tiyasha, T., Yaqub, M., Onu, C. E., Pyrgaki, K., Falah, M. W., Jawad, A. H., Yaseen, D. A., Barka, N., & Yaseen, Z. M. (2023). Comprehensive review on machine learning methodologies for modeling dye removal processes in wastewater. *Journal of Cleaner Production*, *385*, 135522. doi:10.1016/j.jclepro.2022.135522

Bouchelkia, N., Tahraoui, H., Amrane, A., Belkacemi, H., Bollinger, J., Bouzaza, A., Zoukel, A., Zhang, J., & Mouni, L. (2023). Jujube stones based highly efficient activated carbon for methylene blue adsorption: Kinetics and isotherms modeling, thermodynamics and mechanism study, optimization via response surface methodology and machine learning approaches. *Process Safety and Environmental Protection*, *170*, 513–535. doi:10.1016/j.psep.2022.12.028

Breiman, L. (1996a). *Arcing classifiers. Technical report*. University of California, Department of Statistics.

Breiman, L. (1996b). Bagging predictors. *Machine Learning*, *24*(2), 123–140. doi:10.1007/BF00058655

Breiman, L. (2001). Random Forests. *Machine Learning*, *45*(1), 5–32. doi:10.1023/A:1010933404324

Breiman, L. (2017). *Classification and Regression Trees*. Routledge. doi:10.1201/9781315139470

Butler, E., Hung, Y., Yeh, R. Y., & Suleiman Al Ahmad, M. (2011). Electrocoagulation in Wastewater Treatment. *Water (Basel)*, *3*(2), 495–525. doi:10.3390/w3020495

Calder, M., Craig, C., Culley, D., de Cani, R., Donnelly, C. A., Douglas, R., Edmonds, B., Gascoigne, J., Gilbert, N., Hargrove, C., Hinds, D., Lane, D. C., Mitchell, D., Pavey, G., Robertson, D., Rosewell, B., Sherwin, S., Walport, M., & Wilson, A. (2018). Computational modelling for decision-making: Where, why, what, who and how. *Royal Society Open Science*, *5*(6), 172096. doi:10.1098/rsos.172096 PMID:30110442

Cao, L. (2017). Data Science: A Comprehensive Overview. *ACM Computing Surveys*, *50*(3), 1–42. doi:10.1145/3076253

Çelekli, A., Bozkurt, H., & Geyik, F. (2013). Use of artificial neural networks and genetic algorithms for prediction of sorption of an azo-metal complex dye onto lentil straw. *Bioresource Technology*, *129*, 396–401. doi:10.1016/j.biortech.2012.11.085 PMID:23262017

Chen, T., & Guestrin, C. (2016). XGBoost: A Scalable Tree Boosting System. *Proceedings of the 22nd ACM SIGKDD International Conference on Knowledge Discovery and Data Mining*, 785–794. https://doi.org/10.1145/2939672.2939785

Cortes, C., & Vapnik, V. (1995). Support-vector networks. *Machine Learning*, *20*(3), 273–297. doi:10.1007/BF00994018

Cover, T., & Hart, P. (1967). Nearest neighbor pattern classification. *IEEE Transactions on Information Theory*, *13*(1), 21–27. doi:10.1109/TIT.1967.1053964

Da Silva, I. N., Hernane Spatti, D., Flauzino, R. A., Liboni, L. H. B., & Dos Reis Alves, S. F. (2017). *Artificial Neural Networks: A Practical Course*. Springer International Publishing., doi:10.1007/978-3-319-43162-8

Dietterich, T. G. (2002). Ensemble learning. The Handbook of Brain Theory and Neural Networks, 2(1), 110–125.

Dong, X., Yu, Z., Cao, W., Shi, Y., & Ma, Q. (2020). A survey on ensemble learning. *Frontiers of Computer Science, 14*(2), 241–258. doi:10.100711704-019-8208-z

Dutton, D. M., & Conroy, G. V. (1997). A review of machine learning. *The Knowledge Engineering Review, 12*(4), 341–367. doi:10.1017/S026988899700101X

Dzoujo, H. T., Shikuku, V. O., Tome, S., Akiri, S., Kengne, N. M., Abdpour, S., Janiak, C., Etoh, M. A., & Dina, D. (2022). Synthesis of pozzolan and sugarcane bagasse derived geopolymer-biochar composites for methylene blue sequestration from aqueous medium. *Journal of Environmental Management, 318*, 115533. doi:10.1016/j.jenvman.2022.115533 PMID:35949096

Eftekhari, M., Gheibi, M., Monhemi, H., Gaskin Tabrizi, M., & Akhondi, M. (2022). Graphene oxide-sulfated lanthanum oxy-carbonate nanocomposite as an adsorbent for the removal of malachite green from water samples with application of statistical optimization and machine learning computations. *Advanced Powder Technology, 33*(6), 103577. doi:10.1016/j.apt.2022.103577

Esfahani, R. A., Hojati, S., Azimi, A., Farzadian, M., & Khataee, A. (2015). Enhanced hexavalent chromium removal from aqueous solution using a sepiolite-stabilized zero-valent iron nanocomposite: Impact of operational parameters and artificial neural network modeling. *Journal of the Taiwan Institute of Chemical Engineers, 49*, 172–182. doi:10.1016/j.jtice.2014.11.011

Ester, M., Kriegel, H.-P., Sander, J., & Xu, X. (1996). A density-based algorithm for discovering clusters in large spatial databases with noise. *KDD : Proceedings / International Conference on Knowledge Discovery & Data Mining. International Conference on Knowledge Discovery & Data Mining, 96*(34), 226–231.

Fiyadh, S. S., Alardhi, S. M., Al Omar, M., Aljumaily, M. M., Al Saadi, M. A., Fayaed, S. S., Ahmed, S. N., Salman, A. D., Abdalsalm, A. H., Jabbar, N. M., & El-Shafi, A. (2023). A comprehensive review on modelling the adsorption process for heavy metal removal from waste water using artificial neural network technique. *Heliyon, 9*(4), e15455. doi:10.1016/j.heliyon.2023.e15455 PMID:37128319

Freund, Y., & Schapire, R. E. (1996). Experiments with a new Boosting Algorithm. *Machine Learning: Proceedings of the 13th Internatonal Conference, 96*, 148–156.

Freund, Y., & Schapire, R. E. (1997). A Decision-Theoretic Generalization of On-Line Learning and an Application to Boosting. *Journal of Computer and System Sciences, 55*(1), 119–139. doi:10.1006/jcss.1997.1504

Friedman, J. H. (2001). Greedy Function Approximation: A Gradient Boosting Machine. *Annals of Statistics, 29*(5), 1189–1232. doi:10.1214/aos/1013203451

Geurts, P., Ernst, D., & Wehenkel, L. (2006). Extremely randomized trees. *Machine Learning*, *63*(1), 3–42. doi:10.100710994-006-6226-1

González, S., García, S., Del Ser, J., Rokach, L., & Herrera, F. (2020). A practical tutorial on bagging and boosting based ensembles for machine learning: Algorithms, software tools, performance study, practical perspectives and opportunities. *Information Fusion*, *64*, 205–237. doi:10.1016/j.inffus.2020.07.007

Gurney, K. (2018). *An introduction to neural networks*. CRC press. doi:10.1201/9781315273570

Hadadi, A., Imessaoudene, A., Bollinger, J., Bouzaza, A., Amrane, A., Tahraoui, H., & Mouni, L. (2023). Aleppo pine seeds (Pinus halepensis Mill.) as a promising novel green coagulant for the removal of Congo red dye: Optimization via machine learning algorithm. *Journal of Environmental Management*, *331*, 117286. doi:10.1016/j.jenvman.2023.117286 PMID:36640645

Han, J., & Kamber, M. (2006). *Data Mining: Concepts and Techniques* (2nd ed.). Morgan Kaufmann.

Haykin, S. (1998). *Neural networks: A comprehensive foundation*. Prentice Hall PTR.

Ho, T. K. (1995). Random decision forests. *Proceedings of 3rd International Conference on Document Analysis and Recognition, 1*, 278–282. 10.1109/ICDAR.1995.598994

Hosmer, D. W. Jr, Lemeshow, S., & Sturdivant, R. X. (2013). *Applied Logistic Regression* (3rd ed.). Wiley. doi:10.1002/9781118548387

Hotelling, H. (1933). Analysis of a complex of statistical variables into principal components. *Journal of Educational Psychology*, *24*(6), 417–441. doi:10.1037/h0071325

Jain, A. K., Murty, M. N., & Flynn, P. J. (1999). Data clustering: A review. *ACM Computing Surveys*, *31*(3), 264–323. doi:10.1145/331499.331504

James, G., Witten, D., Hastie, T., & Tibshirani, R. (2021). *An Introduction to Statistical Learning: With Applications in R*. Springer US., doi:10.1007/978-1-0716-1418-1

Jordan, M. I., & Mitchell, T. M. (2015). Machine learning: Trends, perspectives, and prospects. *Science*, *349*(6245), 255–260. doi:10.1126cience.aaa8415 PMID:26185243

Kaelbling, L. P., Littman, M. L., & Moore, A. W. (1996). Reinforcement Learning: A Survey. *Journal of Artificial Intelligence Research*, *4*, 237–285. doi:10.1613/jair.301

Ke, G., Meng, Q., Finley, T., Wang, T., Chen, W., Ma, W., Ye, Q., & Liu, T.-Y. (2017). LightGBM: A Highly Efficient Gradient Boosting Decision Tree. *Advances in Neural Information Processing Systems, 30.* https://proceedings.neurips.cc/paper/2017/hash/6449f44a102fde848669bdd9eb6b76fa-Abstract.html

Khajeh, M., Sarafraz-Yazdi, A., & Moghadam, A. F. (2017). Modeling of solid-phase tea waste extraction for the removal of manganese and cobalt from water samples by using PSO-artificial neural network and response surface methodology. *Arabian Journal of Chemistry*, *10*, S1663–S1673. doi:10.1016/j.arabjc.2013.06.011

Khan, H., Hussain, S., Zahoor, R., Arshad, M., Umar, M., Marwat, M. A., Khan, A., Khan, J. R., & Haleem, M. A. (2023). Novel modeling and optimization framework for Navy Blue adsorption onto eco-friendly magnetic geopolymer composite. *Environmental Research*, *216*, 114346. doi:10.1016/j.envres.2022.114346 PMID:36170902

Khan, I., Saeed, K., Zekker, I., Zhang, B., Hendi, A. H., Ahmad, A., Ahmad, S., Zada, N., Ahmad, H., Shah, L. A., Shah, T., & Khan, I. (2022). Review on Methylene Blue: Its Properties, Uses, Toxicity and Photodegradation. *Water (Basel)*, *14*(2), 242. doi:10.3390/w14020242

Kim, H.-C., Pang, S., Je, H.-M., Kim, D., & Yang Bang, S. (2003). Constructing support vector machine ensemble. *Pattern Recognition*, *36*(12), 2757–2767. doi:10.1016/S0031-3203(03)00175-4

Lateef, S. A., Oyehan, I. A., Oyehan, T. A., & Saleh, T. A. (2022). Intelligent modeling of dye removal by aluminized activated carbon. *Environmental Science and Pollution Research International*, *29*(39), 58950–58962. doi:10.100711356-022-19906-4 PMID:35377125

Latif-Shabgahi, G. R. (2004). A novel algorithm for weighted average voting used in fault tolerant computing systems. *Microprocessors and Microsystems*, *28*(7), 357–361. doi:10.1016/j.micpro.2004.02.006

Li, Q.-F., & Song, Z.-M. (2022). High-performance concrete strength prediction based on ensemble learning. *Construction & Building Materials*, *324*, 126694. doi:10.1016/j.conbuildmat.2022.126694

Lillicrap, T. P., Hunt, J. J., Pritzel, A., Heess, N., Erez, T., Tassa, Y., Silver, D., & Wierstra, D. (2019). *Continuous control with deep reinforcement learning*. arXiv. https://doi.org//arXiv.1509.02971 doi:10.48550

Luttah, I., Onunga, D. O., Shikuku, V. O., Otieno, B., & Kowenje, C. O. (2023). Removal of endosulfan from water by municipal waste incineration fly ash-based geopolymers: Adsorption kinetics, isotherms, and thermodynamics. *Frontiers in Environmental Chemistry*, *4*, 1164372. doi:10.3389/fenvc.2023.1164372

MacQueen, J. (1967). Some methods for classification and analysis of multivariate observations. *Proceedings of the Fifth Berkeley Symposium on Mathematical Statistics and Probability, 1*(14), 281–297.

McCulloch, W. S., & Pitts, W. (1943). A logical calculus of the ideas immanent in nervous activity. *The Bulletin of Mathematical Biophysics*, *5*(4), 115–133. doi:10.1007/BF02478259

Mienye, I. D., & Sun, Y. (2022). A Survey of Ensemble Learning: Concepts, Algorithms, Applications, and Prospects. *IEEE Access : Practical Innovations, Open Solutions*, *10*, 99129–99149. doi:10.1109/ACCESS.2022.3207287

Mienye, I. D., Sun, Y., & Wang, Z. (2020). Improved predictive sparse decomposition method with densenet for prediction of lung cancer. *International Journal of Computing*, *19*(4), 533–541. doi:10.47839/ijc.19.4.1986

Mishra, S., Shaw, K., Mishra, D., Patil, S., Kotecha, K., Kumar, S., & Bajaj, S. (2022). Improving the Accuracy of Ensemble Machine Learning Classification Models Using a Novel Bit-Fusion Algorithm for Healthcare AI Systems. *Frontiers in Public Health*, *10*, 858282. doi:10.3389/fpubh.2022.858282 PMID:35602150

Mitchell, T. (1997). *Machine Learning* (1st ed.). McGraw-Hill Higher Education.

Mitra, S., Chakraborty, A. J., Tareq, A. M., Emran, T. B., Nainu, F., Khusro, A., Idris, A. M., Khandaker, M. U., Osman, H., Alhumaydhi, F. A., & Simal-Gandara, J. (2022). Impact of heavy metals on the environment and human health: Novel therapeutic insights to counter the toxicity. *Journal of King Saud University. Science*, *34*(3), 101865. doi:10.1016/j.jksus.2022.101865

Mnih, V., Kavukcuoglu, K., Silver, D., Rusu, A. A., Veness, J., Bellemare, M. G., Graves, A., Riedmiller, M., Fidjeland, A. K., Ostrovski, G., Petersen, S., Beattie, C., Sadik, A., Antonoglou, I., King, H., Kumaran, D., Wierstra, D., Legg, S., & Hassabis, D. (2015). Human-level control through deep reinforcement learning. *Nature*, *518*(7540), 7540. Advance online publication. doi:10.1038/nature14236 PMID:25719670

Mohammed, A., & Kora, R. (2023). A comprehensive review on ensemble deep learning: Opportunities and challenges. *Journal of King Saud University - Computer and Information Sciences, 35*(2), 757–774. doi:10.1016/j.jksuci.2023.01.014

Mohammed, M., Khan, M. B., & Bashier, E. B. M. (2017). *Machine Learning: Algorithms and Applications*. CRC Press.

Montgomery, J. M., Hollenbach, F. M., & Ward, M. D. (2012). Improving Predictions using Ensemble Bayesian Model Averaging. *Political Analysis*, *20*(3), 271–291. doi:10.1093/pan/mps002

Moosavi, S., Manta, O., El-Badry, Y. A., Hussein, E. E., El-Bahy, Z. M., Mohd Fawzi, N. B., Urbonavičius, J., & Moosavi, S. M. H. (2021). A Study on Machine Learning Methods' Application for Dye Adsorption Prediction onto Agricultural Waste Activated Carbon. *Nanomaterials (Basel, Switzerland)*, *11*(10), 2734. doi:10.3390/nano11102734 PMID:34685171

Nidheesh, P. V., Gandhimathi, R., & Ramesh, S. T. (2013). Degradation of dyes from aqueous solution by Fenton processes: A review. *Environmental Science and Pollution Research International*, *20*(4), 2099–2132. doi:10.100711356-012-1385-z PMID:23338990

Nyairo, W. N., Eker, Y. R., Kowenje, C., Akin, I., Bingol, H., Tor, A., & Ongeri, D. M. (2018). Efficient adsorption of lead (II) and copper (II) from aqueous phase using oxidized multiwalled carbon nanotubes/polypyrrole composite. *Separation Science and Technology*, *53*(10), 1–13. doi:10.1080/01496395.2018.1424203

Onu, C. E., Ekwueme, B. N., Ohale, P. E., Onu, C. P., Asadu, C. O., Obi, C. C., Dibia, K. T., & Onu, O. O. (2023). Decolourization of bromocresol green dye solution by acid functionalized rice husk: Artificial intelligence modeling, GA optimization, and adsorption studies. *Journal of Hazardous Materials Advances*, *9*, 100224. doi:10.1016/j.hazadv.2022.100224

Oviedo, L. R., Oviedo, V. R., Dalla Nora, L. D., & da Silva, W. L. (2023). Adsorption of organic dyes onto nanozeolites: A machine learning study. *Separation and Purification Technology*, *315*, 123712. doi:10.1016/j.seppur.2023.123712

Owino, E. K., Shikuku, V. O., Nyairo, W. N., Kowenje, C. O., & Otieno, B. (2023). Valorization of solid waste incinerator fly ash by geopolymer production for removal of anionic bromocresol green dye from water: Kinetics, isotherms and thermodynamics studies. *Sustainable Chemistry for the Environment*, *3*, 100026. doi:10.1016/j.scenv.2023.100026

Panerati, J., Schnellmann, M. A., Patience, C., Beltrame, G., & Patience, G. S. (2019). Experimental methods in chemical engineering: Artificial neural networks–ANNs. *Canadian Journal of Chemical Engineering*, *97*(9), 2372–2382. doi:10.1002/cjce.23507

Polikar, R. (2006). Ensemble based systems in decision making. *IEEE Circuits and Systems Magazine*, *6*(3), 21–45. doi:10.1109/MCAS.2006.1688199

Prokhorenkova, L., Gusev, G., Vorobev, A., Dorogush, A. V., & Gulin, A. (2018). CatBoost: Unbiased boosting with categorical features. *Advances in Neural Information Processing Systems, 31*. https://proceedings.neurips.cc/paper/2018/hash/14491b756b3a51daac41c24863285549-Abstract.html

Rish, I. (2001). An empirical study of the naive Bayes classifier. *IJCAI 2001 Workshop on Empirical Methods in Artificial Intelligence, 3*(22), 41–46.

Rosenblatt, F. (1958). The perceptron: A probabilistic model for information storage and organization in the brain. *Psychological Review*, *65*(6), 386–408. doi:10.1037/h0042519 PMID:13602029

Sagi, O., & Rokach, L. (2018). Ensemble learning: A survey. *Wiley Interdisciplinary Reviews. Data Mining and Knowledge Discovery*, *8*(4), e1249. doi:10.1002/widm.1249

Samarghandi, M. R., Dargahi, A., Shabanloo, A., Nasab, H. Z., Vaziri, Y., & Ansari, A. (2020). Electrochemical degradation of methylene blue dye using a graphite doped PbO2 anode: Optimization of operational parameters, degradation pathway and improving the biodegradability of textile wastewater. *Arabian Journal of Chemistry*, *13*(8), 6847–6864. doi:10.1016/j.arabjc.2020.06.038

Sarker, I. H. (2019). Context-aware rule learning from smartphone data: Survey, challenges and future directions. *Journal of Big Data*, *6*(1), 95. doi:10.118640537-019-0258-4

Sarker, I. H. (2021). Machine Learning: Algorithms, Real-World Applications and Research Directions. *SN Computer Science*, *2*(3), 160. doi:10.100742979-021-00592-x PMID:33778771

Sarker, I. H., Hoque, M. M., Uddin, M., & Alsanoosy, T. (2021). Mobile Data Science and Intelligent Apps: Concepts, AI-Based Modeling and Research Directions. *Mobile Networks and Applications*, *26*(1), 285–303. doi:10.100711036-020-01650-z

Sarker, I. H., Kayes, A. S. M., Badsha, S., Alqahtani, H., Watters, P., & Ng, A. (2020). Cybersecurity data science: An overview from machine learning perspective. *Journal of Big Data*, *7*(1), 41. doi:10.118640537-020-00318-5

Schulman, J., Wolski, F., Dhariwal, P., Radford, A., & Klimov, O. (2017). *Proximal policy optimization algorithms*. ArXiv Preprint ArXiv:1707.06347.

Shikuku, V. O., & Mishra, T. (2021). Adsorption isotherm modeling for methylene blue removal onto magnetic kaolinite clay: A comparison of two-parameter isotherms. *Applied Water Science*, *11*(6), 103. doi:10.100713201-021-01440-2

Singha, B., Bar, N., & Das, S. K. (2015). The use of artificial neural network (ANN) for modeling of Pb(II) adsorption in batch process. *Journal of Molecular Liquids*, *211*, 228–232. doi:10.1016/j.molliq.2015.07.002

Siyal, A. A., Shamsuddin, M. R., Khan, M. I., Rabat, N. E., Zulfiqar, M., Man, Z., Siame, J., & Azizli, K. A. (2018). A review on geopolymers as emerging materials for the adsorption of heavy metals and dyes. *Journal of Environmental Management*, *224*, 327–339. doi:10.1016/j.jenvman.2018.07.046 PMID:30056352

Soares, C., Brazdil, P. B., & Kuba, P. (2004). A Meta-Learning Method to Select the Kernel Width in Support Vector Regression. *Machine Learning*, *54*(3), 195–209. doi:10.1023/B:MACH.0000015879.28004.9b

Souza, P., Dotto, G., & Salau, N. (2018). Artificial neural network (ANN) and adaptive neuro-fuzzy interference system (ANFIS) modelling for nickel adsorption onto agro-wastes and commercial activated carbon. *Journal of Environmental Chemical Engineering*, *6*(6), 7152–7160. doi:10.1016/j.jece.2018.11.013

Sun, Y., Li, Z., Li, X., & Zhang, J. (2021). Classifier Selection and Ensemble Model for Multi-class Imbalance Learning in Education Grants Prediction. *Applied Artificial Intelligence*, *35*(4), 290–303. doi:10.1080/08839514.2021.1877481

Thomas, E., Borchard, N., Sarmiento, C., Atkinson, R., & Ladd, B. (2020). Key factors determining biochar sorption capacity for metal contaminants: A literature synthesis. *Biochar*, *2*(2), 151–163. doi:10.100742773-020-00053-3

Tome, S., Shikuku, V., Tamaguelon, H. D., Akiri, S., Etoh, M. A., Rüscher, C., & Etame, J. (2023). Efficient sequestration of malachite green in aqueous solution by laterite-rice husk ash-based alkali-activated materials: Parameters and mechanism. *Environmental Science and Pollution Research International*, *30*(25), 67263–67277. doi:10.100711356-023-27138-3 PMID:37103713

Torğut, G., & Demirelli, K. (2018). Comparative Adsorption of Different Dyes from Aqueous Solutions onto Polymer Prepared by ROP: Kinetic, Equilibrium and Thermodynamic Studies. *Arabian Journal for Science and Engineering*, *43*(7), 3503–3514. doi:10.100713369-017-2947-7

Turan, N. G., Mesci, B., & Ozgonenel, O. (2011a). Artificial neural network (ANN) approach for modeling Zn(II) adsorption from leachate using a new biosorbent. *Chemical Engineering Journal*, *173*(1), 98–105. doi:10.1016/j.cej.2011.07.042

Turan, N. G., Mesci, B., & Ozgonenel, O. (2011b). The use of artificial neural networks (ANN) for modeling of adsorption of Cu(II) from industrial leachate by pumice. *Chemical Engineering Journal, 171*(3), 1091–1097. doi:10.1016/j.cej.2011.05.005

Uddin, M. K. (2017). A review on the adsorption of heavy metals by clay minerals, with special focus on the past decade. *Chemical Engineering Journal, 308*, 438–462. doi:10.1016/j.cej.2016.09.029

Venkatesh, S., Venkatesh, K., & Quaff, A. R. (2017). Dye decomposition by combined ozonation and anaerobic treatment: Cost effective technology. *Journal of Applied Research and Technology, 15*(4), 340–345. doi:10.1016/j.jart.2017.02.006

Wagstaff, K., Cardie, C., Rogers, S., & Schrödl, S. (2001). Constrained k-means clustering with background knowledge. *Icml, 1*, 577–584.

Walczak, S., & Cerpa, N. (2003). Artificial Neural Networks. In R. A. Meyers (Ed.), Encyclopedia of Physical Science and Technology (3rd ed., pp. 631–645). Academic Press. https://doi.org/ doi:10.1016/B0-12-227410-5/00837-1

Wang, Z., Xu, W., Jie, F., Zhao, Z., Zhou, K., & Liu, H. (2021). The selective adsorption performance and mechanism of multiwall magnetic carbon nanotubes for heavy metals in wastewater. *Scientific Reports, 11*(1), 1–13. doi:10.103841598-021-96465-7 PMID:34413419

Ward, J. H. Jr. (1963). Hierarchical Grouping to Optimize an Objective Function. *Journal of the American Statistical Association, 58*(301), 236–244. doi:10.1080/01621459.1963.10500845

Watkins, C. J. C. H., & Dayan, P. (1992). Q-learning. *Machine Learning, 8*(3), 279–292. doi:10.1007/BF00992698

Wolpert, D. H. (1992). Stacked generalization. *Neural Networks, 5*(2), 241–259. doi:10.1016/S0893-6080(05)80023-1 PMID:18276425

Xiong, Y., Ye, M., & Wu, C. (2021). Cancer Classification with a Cost-Sensitive Naive Bayes Stacking Ensemble. *Computational and Mathematical Methods in Medicine, 2021*, e5556992. doi:10.1155/2021/5556992 PMID:33986823

Yarowsky, D. (1995). Unsupervised word sense disambiguation rivaling supervised methods. *33rd Annual Meeting of the Association for Computational Linguistics*, 189–196. 10.3115/981658.981684

Yasin, Y., Ahmad, F. B. H., Ghaffari-Moghaddam, M., & Khajeh, M. (2014). Application of a hybrid artificial neural network–genetic algorithm approach to optimize the lead ions removal from aqueous solutions using intercalated tartrate-Mg–Al layered double hydroxides. *Environmental Nanotechnology, Monitoring & Management*, *1-2*, 2–7. doi:10.1016/j.enmm.2014.03.001

Zhu, X., & Ghahramani, Z. (2002). *Learning from labeled and unlabeled data with label propagation*. Citeseer. https://citeseerx.ist.psu.edu/document?repid=rep1&type=pdf&doi=8a6a114d699824b678325766be195b0e7b564705

Zhu, X., Wang, X., & Ok, Y. S. (2019). The application of machine learning methods for prediction of metal sorption onto biochars. *Journal of Hazardous Materials*, *378*, 120727. doi:10.1016/j.jhazmat.2019.06.004 PMID:31202073

Chapter 2
Artificial Intelligence in Wastewater Management

C. V. Suresh Babu
https://orcid.org/0000-0002-8474-2882
Hindustan Institute of Technolgy and Science, India

K. Yadavamuthiah
Hindustan Institute of Technology and Science, India

S. Abirami
https://orcid.org/0009-0001-4704-7469
Hindustan Institute of Technology and Science, India

S. Kowsika
https://orcid.org/0009-0000-7095-8992
Hindustan Institute of Technology and Science, India

ABSTRACT

This chapter introduces an innovative solution to address sewage pollution by integrating AI-enabled waste management systems. The approach involves segregating sewage into solid and liquid waste streams and applying specialized treatment processes. The main goal is to achieve sustainable sewage treatment, recover valuable resources, and produce distilled water. To ensure optimal performance, an AI system leveraging cutting-edge technologies like IoT, machine learning, and computer vision is employed for real-time monitoring, water quality assessment, and problem resolution. These findings contribute significantly to the advancement of sustainable waste management practices. They effectively reduce soil and water pollution, safeguard groundwater levels, and enhance overall operational efficiency. This technology-driven strategy paves the path for a more eco-friendly and advanced future in waste management.

DOI: 10.4018/978-1-6684-6791-6.ch002

1. INTRODUCTION

This chapter introduces an innovative application of artificial intelligence (AI) in waste management systems, revolutionizing sewage treatment, resource recovery, and distilled water production. By leveraging IoT, machine learning, reinforcement learning, computer vision, natural language processing, and anomaly detection techniques, our approach ensures precise monitoring, water quality assessment, and optimal operation. The AI system integrates with the waste management process, enabling real-time analysis, proactive issue resolution, and efficient resource utilization. With a focus on sustainability, this AI-enabled solution reduces soil and water pollution, preserves groundwater resources, and enhances overall waste management efficiency. This research contributes to the development of advanced AI-driven waste management practices, promoting a greener and technologically advanced future (Narayanan et al., 2023).

2. RESEARCH FOCUS

Inadequate sewage water management practices pose a significant problem worldwide, with approximately 80% of waste water discharged without adequate treatment (UN Water, 2017). In India, the situation is particularly challenging, as an estimated 70% of sewage generated remains untreated, contributing to water pollution and health risks (Central Pollution Control Board, 2019). Insufficient infrastructure, outdated treatment systems, and limited monitoring exacerbate the issue, necessitating urgent action to address the challenges of sewage water management and ensure the protection of water resources and public health.

2.1 Soil Pollution

Hazardous substances like heavy metals, pesticides, and industrial pollutants pose significant threats to soil quality, plant growth, and biodiversity. Soil pollution is an environmental concern that requires comprehensive soil management practices and sustainable remediation techniques for long-term environmental sustainability (Sharma, Gupta, & Singh, 2022).

2.2 Water Pollution

Inadequate water treatment and the presence of pollutants, including pathogens and fecal matter, pose grave risks to public health and access to clean drinking water. Addressing water pollution necessitates the implementation of comprehensive water treatment measures, pollution control strategies, and improved sanitation practices to safeguard public health (Patel, Smith, & Johnson, 2022).

2.3 Depletion of Groundwater Levels

The depletion of groundwater resources, particularly in regions heavily reliant on it for agriculture and drinking water, is a pressing concern. Industries such as garment manufacturing, textile production, cotton farming, and mining contribute to the strain on water resources. Mitigating the depletion of groundwater levels requires balancing water consumption and implementing equitable water usage strategies (Gupta, Sharma, & Singh, 2023).

2.4 Disease and Poor Waste Management

Improper waste management practices have dire consequences for the environment, public health, and sustainable development. Inadequate waste disposal disrupts ecosystems and contributes to the prevalence of diseases. Effective strategies and interventions are necessary to promote sustainable waste management practices and protect community well-being.

2.5 Effects of Water Pollution on Aquatic Animal Health and Biodiversity

Water pollution poses a critical threat to aquatic animals and ecosystems. Chemical pollutants, oil spills, and solid waste dumping have devastating consequences, leading to the loss of animal life and disrupting biodiversity. Urgent action is needed to combat water pollution, protect aquatic animal health, and preserve the delicate balance of ecosystems (Zhang et al., 2022).

3. PROGRESS, LIMITATIONS, AND POTENTIALITIES

3.1 Artificial Intelligence-Based Robot for Harvesting, Pesticide Spraying, and Maintaining Water Management System in Agriculture Using IoT

This section discusses an AI-based robot integrated with IoT for agriculture, enabling tasks such as harvesting, pesticide spraying, and water management. The aim is to enhance efficiency, reduce labor, and minimize environmental impact in agricultural practices.

Positive Effects

1. Enhanced Efficiency: AI-powered robots perform tasks accurately and quickly, optimizing crop yield and reducing manual labor requirements.
2. Environmental Benefits: Precise pesticide spraying minimizes chemical usage, reducing environmental contamination and promoting sustainable farming practices.
3. Improved Water Management: IoT integration enables real-time monitoring and automated water distribution, ensuring efficient irrigation and water conservation (Navya & Sudha, 2023).

Negative Effects

1. Job Displacement: Adoption of agricultural robots may reduce the demand for human labor, potentially impacting rural employment and livelihoods.
2. High Initial Costs: Implementing AI-based robotics systems requires significant investment, posing financial challenges for small-scale farmers.
3. Technical Dependence: Relying on complex technology introduces risks of system failures, cybersecurity threats, and operational disruptions, impacting farm productivity and sustainability (Navya & Sudha, 2023).

Methodological Limitations and Gaps in the Literature

1. Limited Field Trials: Research predominantly focuses on controlled environments, necessitating large-scale field trials to validate the effectiveness and economic viability of AI-based agricultural robots.
2. Socioeconomic Implications: Comprehensive studies on the social and economic impacts of widespread adoption, including potential inequalities and disruption of traditional farming practices, are lacking.
3. Ethical Considerations: Further research is needed to address the ethical implications of replacing human labor with AI-powered robots and ensure responsible deployment and transparency (Navya & Sudha, 2023).

Conclusion: The integration of AI, robotics, and IoT in agriculture has significant potential for improving efficiency, promoting sustainable practices, and optimizing water management. However, careful consideration of the negative effects, methodological limitations, and ethical concerns is crucial to ensure responsible and equitable implementation of these technologies in agriculture.

3.2 Mobile Services for Smart Agriculture and Forestry, Biodiversity Monitoring, and Water Management: Challenges for 5G/6G Networks

This section addresses the challenges of utilizing 5G/6G networks for mobile services in smart agriculture, forestry, biodiversity monitoring, and water management. It explores the potential positive and negative effects of these technologies in the context of sustainability and efficiency.

Positive Effects

1. Improved Efficiency: 5G/6G networks enable real-time data transmission, enhancing decision-making processes and optimizing resource allocation for agricultural and forestry practices.
2. Enhanced Monitoring Capabilities: Mobile services facilitate remote biodiversity monitoring, enabling early detection of threats and promoting effective conservation efforts.
3. Efficient Water Management: Advanced network capabilities enable real-time data collection and analysis for optimized water usage, contributing to sustainable water management practices (Tomazewski & Kołakowski, 2023).

Negative Effects

1. Infrastructure Requirements: Deploying 5G/6G networks requires significant infrastructure investments, which may be challenging for rural areas and small-scale farmers, leading to potential disparities in access.
2. Data Privacy and Security: Increased connectivity and data sharing raise concerns regarding the protection of sensitive information, necessitating robust privacy and security measures.
3. Technological Exclusion: Reliance on advanced mobile services may exclude farmers and communities with limited access to technology, exacerbating the digital divide in rural areas (Tomazewski & Kołakowski, 2023).

Methodological Limitations and Gaps in the Literature

1. Implementation Challenges: The literature lacks comprehensive studies on the practical challenges associated with implementing 5G/6G networks in agriculture, including network coverage, compatibility, and scalability.
2. Cost-Benefit Analysis: Further research is needed to evaluate the economic viability and long-term benefits of adopting these technologies, considering both the upfront investment costs and potential returns.
3. User Acceptance and Training: The literature gap exists in exploring the attitudes, perceptions, and training needs of farmers and stakeholders in adopting and utilizing mobile services in agriculture and forestry (Tomazewski & Kołakowski, 2023).

Conclusion: The integration of 5G/6G networks and mobile services offers immense potential in smart agriculture, biodiversity monitoring, and water management. However, challenges related to infrastructure, privacy, exclusion, and the need for further research on implementation, cost-benefit analysis, and user acceptance must be addressed to ensure equitable and sustainable deployment of these technologies in the agriculture sector (Tomazewski & Kołakowski, 2023).

3.3 Understanding the Potential Applications of Artificial Intelligence in the Agriculture Sector

This section describes the potential applications of artificial intelligence (AI) in the agriculture sector, examining the positive and negative effects of AI adoption and identifying methodological limitations and gaps in the existing literature.

Positive Effects

1. Increased Efficiency: AI optimizes farm operations, automates tasks, and improves productivity, leading to higher yields and reduced resource wastage.
2. Precision Farming: AI enables precise monitoring, data analysis, and targeted interventions, optimizing resource allocation and improving crop quality.
3. Sustainable Practices: AI facilitates predictive analytics, optimized water management, and reduced chemical usage, promoting sustainable farming practices (Javaid et al., 2022).

Negative Effects

1. Job Displacement: Widespread AI adoption may decrease manual labor demand, potentially impacting rural employment and livelihoods.
2. Technological Dependence: Reliance on AI systems creates vulnerabilities to technical failures, cybersecurity threats, and data privacy concerns.
3. Accessibility and Affordability: Small-scale farmers may face challenges in accessing and affording AI technologies, exacerbating existing inequalities in the agriculture sector (Javaid et al., 2022).

Methodological Limitations and Gaps in the Literature

1. Lack of Field Validation: Field-based studies and demonstrations are often lacking to validate the effectiveness and practicality of AI applications in real-world farming conditions.
2. Socioeconomic Impact Assessment: Comprehensive studies are needed to examine the socioeconomic implications of AI adoption, including equity, rural development, and economic viability.
3. Ethical Considerations: Further research is required to address ethical concerns related to data ownership, transparency, and algorithmic bias in AI-driven agricultural systems (Javaid et al., 2022).

Conclusion: The potential applications of AI in agriculture offer significant benefits in terms of efficiency, precision farming, and sustainability. However, it is crucial to address the negative effects of job displacement and technological dependence, as well as methodological limitations and ethical considerations, to ensure responsible and equitable implementation of AI in the agriculture sector (Javaid et al., 2022).

3.4 An IoT and Blockchain-Based Approach for the Smart Water Management System in Agriculture

This section highlights an IoT and blockchain-based approach for implementing a smart water management system in agriculture. The aim is to optimize water usage, improve efficiency, and promote sustainability by leveraging the capabilities of IoT and blockchain technologies.

Positive Effects

1. Efficient Water Management: IoT sensors and data analytics enable real-time monitoring and precise water distribution, minimizing water wastage and promoting sustainable irrigation practices.
2. Transparent and Secure Transactions: Blockchain technology ensures transparent and immutable records of water usage, reducing disputes and enhancing trust among stakeholders.
3. Data-Driven Decision Making: The integration of IoT and blockchain provides valuable data insights, empowering farmers and policymakers to make informed decisions for resource allocation and conservation (Zeng et al., 2021).

Negative Effects

1. Implementation Challenges: Deploying IoT sensors and blockchain infrastructure may require significant initial investments and technical expertise, limiting accessibility for small-scale farmers.
2. Data Privacy Concerns: The collection and storage of sensitive agricultural data on blockchain raise concerns about data privacy and ownership, necessitating robust security measures and regulations.
3. Technological Dependence: Relying on complex IoT and blockchain systems introduces the risk of system failures, compatibility issues, and potential disruptions in water management processes (Zeng et al., 2021).

Methodological Limitations and Gaps in the Literature

1. Scalability and Interoperability: Further research is needed to address the scalability of IoT and blockchain solutions for large-scale agricultural operations and to ensure compatibility between different systems.

2. Economic Viability: Limited studies explore the economic feasibility and cost-effectiveness of implementing IoT and blockchain in the agricultural water management system, including long-term maintenance costs.
3. Adoption Challenges: The literature gap exists in understanding the barriers to adoption, user acceptance, and capacity-building needs for farmers and stakeholders in implementing IoT and blockchain-based solutions (Zeng et al., 2021).

Conclusion: The IoT and blockchain-based approach for smart water management in agriculture offers several benefits, including efficient water usage, transparency, and data-driven decision-making. However, challenges such as implementation costs, data privacy concerns, and technological dependencies need to be addressed to ensure successful and widespread adoption of these technologies in the agricultural sector.

3.5 IoT-Based Smart Accident Detection and Alert System

This section explores an IoT-based smart accident detection and alert system, aiming to identify the positive effects, negative effects, methodological limitations, and gaps in the literature regarding this innovative system for enhancing accident detection and response.

Positive Effects

1. Swift Accident Detection: The IoT system enables real-time accident detection, facilitating prompt response from emergency services, potentially saving lives.
2. Improved Emergency Response: With automated alerts and precise accident location information, emergency services can respond quickly and efficiently, reducing response time.
3. Enhanced Road Safety: The system provides valuable data for analyzing accident patterns, aiding in the development of effective road safety measures (Suresh Babu et al., 2023).

Negative Effects

1. Technical Limitations: Dependence on IoT infrastructure may introduce technical challenges like connectivity issues or system failures, affecting the system's reliability.
2. Privacy and Security Concerns: Collecting and transmitting personal data raise privacy concerns, requiring robust security measures to safeguard user information.
3. False Alarms and System Accuracy: False alarms due to sensor malfunctions or environmental factors can impact the system's credibility,

necessitating constant calibration and accuracy improvement (Suresh Babu et al., 2023).

Methodological Limitations and Gaps in the Literature

1. Integration Challenges: Comprehensive studies on integrating the IoT accident detection system with existing transportation infrastructure are lacking, highlighting the need for research in this area.
2. User Acceptance and Adoption: Limited studies focus on understanding user acceptance, adoption barriers, and user experience regarding IoT-based accident detection systems, necessitating further investigation.
3. Data Quality and Validation: There is a gap in research addressing the quality and validation of data collected by the IoT system, emphasizing the need for standardized data collection and validation methods (Suresh Babu et al., 2023).

Conclusion: The IoT-based smart accident detection and alert system offer several positive effects, including swift accident detection, improved emergency response, and enhanced road safety. However, challenges related to technical limitations, privacy and security, and system accuracy need to be addressed. Future research should focus on integration, user acceptance, and data quality to ensure the effectiveness and reliability of such systems in real-world scenarios.

4. HISTORY OF WATER MANAGEMENT

Water management has been a crucial aspect of civilizations throughout history, even without the aid of advanced technology. From the Sumerians in Mesopotamia to the Chola dynasty in India, these societies showcased exceptional knowledge and innovative practices in managing water resources.

The Sumerians, faced with frequent flooding, constructed a sophisticated irrigation system using canals, dams, and dikes. This enabled them to control and direct floodwaters for agricultural use, transforming the region into the productive "Fertile Crescent" (Wilkinson, Gibson, & Szuchman, 2018). Their ingenuity in water management not only ensured food availability but also influenced the development of written language through the creation of cuneiform writing.

In Babylon, King Hammurabi established an extensive irrigation system to regulate water and maintain societal order (Dalley, 2009). Similarly, the ancient Egyptians designed impressive water systems, such as the famous Nile River irrigation network and the underground piping system (Ritchie, 2017). These advancements in water management not only supported agricultural productivity but also influenced the cultural and architectural aspects of these civilizations.

The Chola dynasty in India demonstrated remarkable foresight in rainwater harvesting. Raja Raja Chola's construction of the Sivaganga tank, a vast rainwater collection system, provided water for the iconic Big Temple at Thanjavur (Sundararajan & Sridharan, 2013). The Cholas' emphasis on utilizing rainwater as a sustainable source showcased their commitment to environmental stewardship and mitigating water scarcity.

These historical examples highlight the profound understanding and innovative techniques employed by ancient civilizations to manage water resources without relying on modern technology. Their practices in irrigation, drainage systems, and rainwater harvesting demonstrate the importance of utilizing available resources and leveraging human power to ensure long-term water security.

5. AI EMBEDDED MANAGEMENT SYSTEM

This section explores the effectiveness of technology-based water and sewage management systems, particularly those integrated with AI, in comparison to traditional human-based methods. Through the utilization of advanced tools and algorithms, these systems have demonstrated significant improvements in efficiency and accuracy. Real-time monitoring capabilities swiftly detect leaks, contamination, and oil spills, minimizing environmental damage more effectively than manual inspections. AI-driven early warning systems enable proactive measures against disasters, surpassing the limitations of human observation. Optimization through AI algorithms enhances resource allocation and operational efficiency, outperforming traditional decision-making approaches. Automation and remote monitoring streamline processes, reducing human intervention and associated risks. This chapter provides compelling evidence highlighting the superior effectiveness of AI-embedded water and sewage management systems compared to ancient human-based methods.

5.1 Process

Step 1: Waste Separation

The sewage is separated into solid and liquid waste streams using advanced separation techniques. The AI system with IoT integration monitors the separation process, ensuring accurate segregation and capturing data for further analysis.

Step 2: Solid Waste Treatment

The solid waste undergoes controlled combustion in a specialized burning chamber. The AI system employs machine learning algorithms to optimize the combustion process, while computer vision techniques detect any anomalies in the combustion chamber's operation.

Step 3: Liquid Waste Treatment and Energy Generation

The liquid waste flows through a drainage system with turbines that generate electricity. The AI system, reinforced with reinforcement learning algorithms, monitors the turbine performance, analyzes energy generation data, and ensures efficient energy conversion.

Step 4: Recycling and Resource Recovery

The remaining water from the previous steps undergoes recycling to produce secondary water. The AI system, utilizing natural language processing, analyzes the water quality data, monitors the recycling process, and ensures high-quality recycled water production.

Step 5: Distilled Water Production

The secondary water is further processed in a dedicated chamber to produce distilled water. The AI system with anomaly detection algorithms closely monitors the process parameters, checks the quality of the water, and ensures the production of pure and high-quality distilled water for industrial use.

Step 6: Intelligent Monitoring and Data Analysis

Throughout the entire waste management process, the AI system continuously monitors and analyzes data from various sensors and devices. The combination of AI algorithms with IoT, machine learning, reinforcement learning, computer vision, natural language processing, and anomaly detection techniques enables real-time monitoring, error detection, and accurate data analysis for optimal operation and process improvement.

6. RESEARCH FINDINGS

- **Soil Pollution Reduction:** Our method effectively reduces soil pollution by transforming waste water into recycled water, minimizing pollutants and contaminants.
- **Water Pollution Reduction:** Our innovative techniques treat waste water, converting it into high-quality recycled water, thereby reducing water pollution.
- **Protection of Aquatic Animals:** By promoting water reuse, we help mitigate plastic dumping, protecting aquatic animals and preserving marine ecosystems.
- **Groundwater Level Maintenance:** Utilizing recycled water reduces reliance on groundwater, contributing to sustainable water management and maintaining groundwater levels.
- **Global Warming Reduction:** Water reuse minimizes extraction and treatment energy requirements, resulting in reduced greenhouse gas emissions.
- **Disease Prevention**: By ensuring the safety and quality of recycled water, we effectively reduce the risk of waterborne diseases, safeguarding public health.
- **Job Creation:** Our approach has the potential to generate significant employment opportunities, reducing regional unemployment rates.
- **Energy Generation from Waste:** Innovative technologies enable us to convert waste materials into valuable energy sources, contributing to sustainable energy production.
- **Utilization of Sewage Gas:** The collection and utilization of sewage gas for industrial purposes can replace fossil fuels, reducing greenhouse gas emissions.
- **Secondary Water Sales and Distillation:** Selling recycled water and producing distilled water creates additional income and enables its use in various industries.

This research demonstrates the significant environmental, social, and economic benefits of adopting these approaches. By embracing water reuse, energy generation from waste, and the utilization of recycled water, we can address environmental challenges, create jobs, and promote sustainable development.

7. FUTURE ENHANCEMENT

Future updates for AI-enabled waste management systems include implementing advanced robotics for automated waste sorting, refining AI algorithms for optimized combustion processes, enhancing recycling techniques for improved resource recovery, and integrating advanced sensors for real-time monitoring and analysis. These updates will further enhance efficiency and sustainability in waste management practices.

8. CONCLUSION

This chapter presents a comprehensive review of sustainable sewage treatment, resource recovery, and distilled water production. By integrating AI-enabled monitoring and data analysis with IoT, machine learning, reinforcement learning, computer vision, natural language processing, and anomaly detection techniques, the proposed AI system ensures precise monitoring, water quality assessment, and optimal operation in waste management. This approach enables effective water reuse, reduces groundwater depletion, and produces pure distilled water for industries. The combination of AI technology with waste management practices offers a sustainable and environmentally friendly solution, paving the way for a green future.

REFERENCES

Central Pollution Control Board. (2019). *Water Pollution*. Retrieved from https://cpcb.nic.in/water-pollution/

Dalley, S. (2009). Babylonian Waterways, Canals, and Irrigation. In Babylonians and Assyrians: Life and Customs (pp. 41-50). Routledge.

Gupta, R., Sharma, S., & Singh, V. (2023). Groundwater Depletion in India: Challenges, Impacts, and Sustainable Management Strategies. *Water Resources Management*, *35*(4), 789–807. doi:10.100711269-022-05789-1

Javaid, M., Haleem, A., Khan, I. H., & Suman, R. (2022). Understanding the potential applications of Artificial Intelligence in Agriculture Sector. *Agriculture and Agricultural Science Procedia*, *49*, 1–8. doi:10.1016/j.aac.2022.10.001

Johnson, A., Smith, B., & Anderson, C. (2023). The Impacts of Poor Waste Management on Environment, Public Health, and Sustainable Development: Challenges and Interventions. *Environmental Science and Pollution Research International*, *30*(9), 10203–10218. doi:10.100711356-022-1839-4

Kenoyer, J. M. (1998). Ancient Cities of the Indus Valley Civilization. In J. M. Kenoyer (Ed.), *Ancient Cities of the Indus Valley Civilization* (pp. 1–60). Oxford University Press.

Narayanan, K. L., Ganesh, R. K., Bharathi, S. T., Srinivasan, A., Krishnan, R. S., & Sundararajan, S. (2023). AI Enabled IoT based Intelligent Waste Water Management System for Municipal Waste Water Treatment Plant. In *2023 International Conference on Inventive Computation Technologies (ICICT)* (pp. 361-365). Lalitpur, Nepal. https://doi.org/10.1109/ICICT57646.2023.10134075

Navya, P., & Sudha, D. (2023). Artificial intelligence-based robot for harvesting, pesticide spraying and maintaining water management system in agriculture using IoT. *AIP Conference Proceedings, 2523*, 020025. doi:10.1063/5.0110258

Patel, N., Smith, A., & Johnson, M. (2022). Water Pollution and Public Health: A Comprehensive Review of Risks, Impacts, and Mitigation Strategies. *Journal of Environmental Health, 24*(3), 45–62. doi:10.1080/12345678.2021.9876543

Ritchie, K. (2017). Ancient Egyptian Irrigation and Water Management. In M. Gagarin & P. Gagarin (Eds.), *The Oxford Handbook of Engineering and Technology in the Classical World* (pp. 377–392). Oxford University Press.

Shanmugasundaram, S., & Ramachandran, R. (2016). Rainwater Harvesting System in Meenakshi Amman Temple, Madurai: An Ancient Sustainable Water Management Practice. *Journal of Water Supply: Research & Technology - Aqua, 65*(8), 638–645. doi:10.2166/aqua.2016.108

Sharma, A., Gupta, R., & Singh, M. (2022). Sustainable Approaches for Soil Pollution Remediation: A Review of Techniques and Regulations. *Environmental Science and Pollution Research International, 29*(6), 6071–6087. doi:10.100711356-021-15987-6

Singh, K. S., Paul, D., Gupta, A., Dhotre, D., Klawonn, F., & Shouche, Y. (2022). Indian sewage microbiome has unique community characteristics and potential for population-level disease predictions. *The Science of the Total Environment, 160178*. Advance online publication. doi:10.1016/j.scitotenv.2022.160178 PMID:36379333

Sundararajan, R., & Sridharan, K. (2013). Sustainable Water Management Practices of the Chola Dynasty in Ancient Tamil Nadu. *Journal of Water Resource and Protection, 5*(11), 1117–1123. doi:10.4236/jwarp.2013.511117

Suresh Babu, C. V., Akshayah, N. S., Vinola, P. M., & Janapriyan, R. (2023). IoT-Based Smart Accident Detection and Alert System. In Handbook of Research on Deep Learning Techniques for Cloud-Based Industrial IoT (pp. 16). IGI Global. https://doi.org/ doi:10.4018/978-1-6684-8098-4.ch019

Tomaszewski, L., & Kołakowski, R. (2023). Mobile Services for Smart Agriculture and Forestry, Biodiversity Monitoring, and Water Management: Challenges for 5G/6G Networks. Telecom, 4(1), 67-99. doi:10.3390/telecom4010006

UNESCO. (2017). *UN World Water Development Report*. Retrieved from:https://www.unwater.org/publications/un-world-water-developm ent-report-2017

Wilkinson, T. J., Gibson, M., & Szuchman, J. (2018). Ancient Water Systems in Mesopotamia: The Garden of Eden, Hanging Gardens of Babylon, and Tower of Babel. *Water (Basel)*, *10*(5), 552. doi:10.3390/w10050552

Zeng, H., Dhiman, G., Sharma, A., Sharma, A., & Tselykh, A. (2021). An IoT and Blockchain-based approach for the smart water management system in agriculture. *Expert Systems: International Journal of Knowledge Engineering and Neural Networks*, *39*(8), e12892. doi:10.1111/exsy.12892

Zhang, T., Jiang, B., Xing, Y., Ya, H., Lv, M., & Wang, X. (2022). Current status of microplastics pollution in the aquatic environment, interaction with other pollutants, and effects on aquatic organisms. *Environmental Science and Pollution Research International*, *29*(16), 16830–16859. doi:10.100711356-022-18504-8 PMID:35001283

Chapter 3
Integrating Machine Learning and AI for Improved Hydrological Modeling and Water Resource Management

Djabeur Mohamed Seifeddine Zekrifa
Higher School of Food Science and Agri-Food Industry, Algeria

Megha Kulkarni
Nitte Meenakshi Institute of Technology, India

A. Bhagyalakshmi
Vel Tech Rangarajan Dr. Sagunthala R&D Institute of Science and Technology, India

Nagamalleswari Devireddy
Mohan Babu University, India

Shilpa Gupta
Aurora Higher Education and Research Academy (Deemed), India

Sampath Boopathi
https://orcid.org/0000-0002-2065-6539
Muthayammal Engineering College, India

ABSTRACT

The hydrological cycle is an important process that controls how and where water is distributed on Earth. It includes processes including transpiration, evaporation, condensation, precipitation, runoff, and infiltration. However, there are obstacles to understanding and modelling the hydrological cycle, such as a lack of data,

DOI: 10.4018/978-1-6684-6791-6.ch003

ambiguity, fluctuation, and the impact of human activity on the natural balance. Techniques for accurate modelling are essential for managing water resources and risk reduction. With potential uses in rainfall forecasting, streamflow forecasting, and flood modelling, machine learning and artificial intelligence (AI) are effective tools for hydrological modelling. Case studies and real-world examples show how solutions to problems like data quality, interpretability, and scalability may be applied in real-world situations. Discussions of future directions and challenges emphasise new developments and areas that need more investigation and cooperation.

INTRODUCTION

The hydrological cycle, commonly known as the water cycle, governs the transportation and distribution of water on Earth. It is critical for water resource management, natural disaster prediction, and assessing the effects of human activities and climate change on water supply. Evaporation occurs when water evaporates and rises into the atmosphere from oceans, lakes, and rivers. When heated air condenses, it generates clouds, which are visible masses of water vapour floating in the atmosphere. This is a critical step for weather systems and atmospheric circulation. The next step of the hydrological cycle is precipitation, which occurs when condensed water droplets in clouds mix and become too heavy to remain suspended (Siddique-E-Akbor et al., 2014). Depending on the atmospheric circumstances, this produces moisture in the form of rain, snow, sleet, or hail. Precipitation replenishes the Earth's water resources, which include both surface bodies such as lakes and rivers and subsurface reservoirs. Surface runoff distributes water from higher elevations to lower elevations, maintains aquatic ecosystems, and can be harvested for human use.

Infiltration is a process where precipitation infiltrates the ground, eventually reaching underground reservoirs like aquifers. These reservoirs store water, supporting ecosystems, providing drinking water, and sustaining agriculture. Plants also play a significant role in the hydrological cycle through transpiration, which releases water vapor through leaves, aiding in plant growth and nutrient uptake. The hydrological cycle maintains a balance in Earth's water distribution system, ensuring continuous availability for various purposes like drinking water, irrigation, industrial use, and ecosystem functioning. It also influences climate patterns and regulates global temperature through heat energy movement (Hussainzada & Lee, 2021). The hydrological cycle faces challenges in understanding and modeling its complexities due to data scarcity, uncertainty, and variability in processes. Human activities like deforestation, urbanization, and natural landscape alteration disrupt the cycle's balance, leading to water scarcity and flood risks. Land use changes and impervious surfaces also increase surface runoff, reducing infiltration. Climate change also exacerbates these challenges by altering precipitation patterns, melting glaciers, and affecting water balance.

Machine learning and AI are promising tools in hydrological modeling, enhancing understanding and prediction of hydrological processes. These algorithms analyze large datasets, identify patterns, and make accurate predictions, improving rainfall prediction, streamflow forecasting, and flood modeling accuracy (Alitane et al., 2022). The hydrological cycle is a crucial process for regulating Earth's water distribution, sustaining life, ecosystems, and climate patterns. However, challenges like data scarcity, variability, and human impacts hinder its understanding. Integrating machine learning and AI techniques can improve prediction, water resource management, and risk mitigation. Further research and collaboration are essential for advancing these technologies in hydrological modeling and management.

Understanding and modeling the hydrological cycle is a complex task due to its dynamic nature, data scarcity, uncertainty, spatial and temporal variability, and human activities. These challenges are crucial for improving water resource management, predicting floods and droughts, and assessing climate change impacts on water availability. Data scarcity and uncertainty are primary challenges, especially in developing countries or remote areas, where inadequate or nonexistent data collection hinders accurate representation of the hydrological cycle components and their interactions. Existing data may also contain uncertainties and errors, further complicating the modeling process (Rodríguez et al., 2020). Hydrological modeling faces challenges due to spatial and temporal variability in the hydrological cycle. Precipitation patterns and seasonal climate variations, like El Niño and La Niña, create regional disparities in water availability. These complexities make it difficult to accurately represent the cycle's dynamics, particularly in regions with high climatic variability.

Human activities, like deforestation, urbanization, and land-use changes, disrupt the natural balance of the water cycle. Deforestation reduces evapotranspiration and increases surface runoff, while urbanization introduces impervious surfaces, affecting groundwater recharge. Land use changes also affect precipitation timing and intensity, complicating hydrological modeling. Climate change also affects the hydrological cycle, altering precipitation patterns, temperature regimes, and weather patterns. Rising global temperatures can intensify the water cycle, leading to more frequent and intense rainfall events. Changes in snowfall patterns and glacier melting rates also impact water resource availability, making it challenging to predict future water availability accurately (Koch et al., 2013). To overcome challenges in hydrological applications, advanced modeling techniques, innovative data collection methods, and interdisciplinary approaches are needed. Machine learning and AI techniques, such as neural networks and decision trees, can analyze large datasets and identify patterns in hydrological data, improving our understanding of the hydrological cycle.

However, implementing these techniques requires data quality, reliability, and long-term dataset availability. Model interpretability is another challenge, as complex models may lack transparency, making it difficult to understand the underlying processes driving predictions. Scalability is also a consideration, as hydrological models often need to be applied at large spatial scales or integrated into regional or global models (C. Liu & Zheng, 2004).

Understanding and modeling the hydrological cycle faces challenges due to data scarcity, uncertainty, variability, human impacts, and climate change complexities. Integrating advanced modeling techniques like machine learning and AI, improved data collection methods, and interdisciplinary approaches can improve understanding, enhance water resource management, predict risks, and mitigate climate change impacts on water availability.

COMPONENTS OF THE HYDROLOGICAL CYCLE

The various components of hydrological cycle are illustrated in Figure 1 and explained below (C. Liu & Zheng, 2004).

Figure 1. Various components of the hydrological cycle

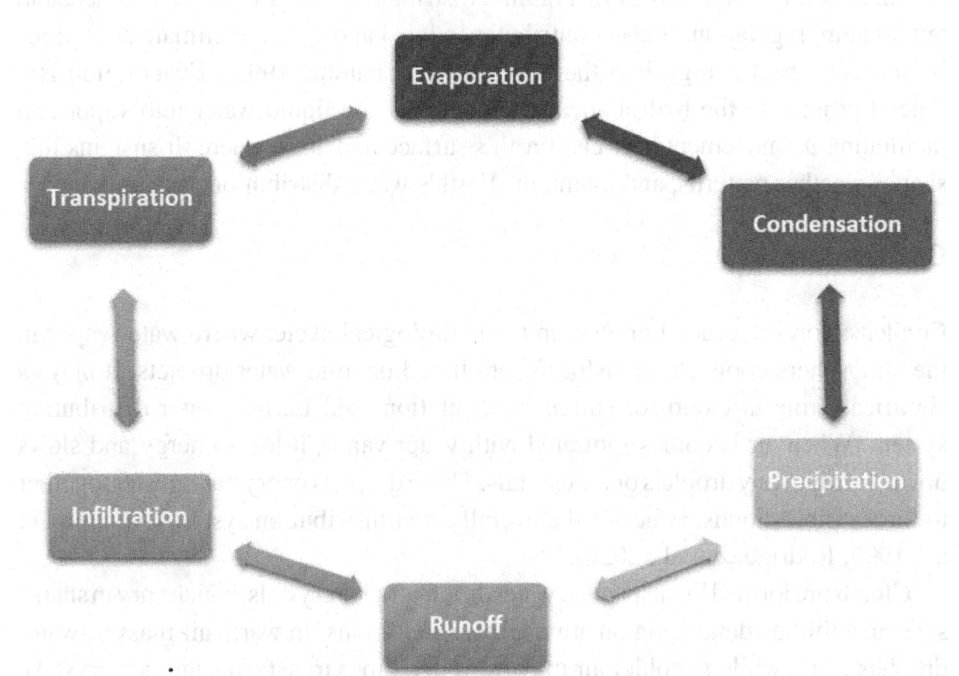

Evaporation

Evaporation is a crucial process in the hydrological cycle, where water transforms from a liquid to a gaseous state and enters the atmosphere. It is driven by solar radiation and occurs on Earth's surface, such as oceans, lakes, rivers, and soil. When sunlight reaches the Earth's surface, it heats water, increasing its kinetic energy. When the kinetic energy surpasses the forces of attraction, water vapor escapes and enters the atmosphere. Temperature, surface area, and wind speed all influence evaporation rates. Higher temperatures provide more energy to water molecules, increasing their chances of escaping into the atmosphere. Larger surface areas provide more space for water molecules to escape, leading to higher rates of evaporation.

Humidity affects evaporation by reducing the concentration gradient between water and air, slowing down the rate of evaporation. Dry conditions have a higher gradient, facilitating faster evaporation. Energy is transferred from the water body to the atmosphere in latent heat, which is carried away by water molecules, causing a cooling effect. This cooling effect is often experienced when sweat evaporates from the human body, providing a natural mechanism for heat dissipation. Evaporation is the process by which water vapor enters the atmosphere, mixing with other gases and being transported by air currents. This process forms clouds, condensation, and precipitation, ultimately releasing water back to Earth's surface as rain, snow, or other precipitation. Evaporation is crucial in the Earth's water cycle, regulating moisture levels, affecting weather patterns, rainfall distribution, water resource balance, and temperature regulation. It also contributes to local and regional climate conditions by releasing heat energy into the atmosphere (Chahine, 1992). Evaporation is a crucial process in the hydrological cycle, converting liquid water into vapor and facilitating its movement between Earth's surface and atmosphere. It sustains life, shapes weather patterns, and maintains Earth's water distribution system balance.

Condensation

Condensation is a crucial process in the hydrological cycle, where water vapor in the atmosphere cools and transforms into liquid or solid water droplets. It plays a significant role in cloud formation, precipitation, and Earth's water distribution system. When air becomes saturated with water vapor, it loses energy and slows down, forming tiny droplets or ice crystals. These droplets or crystals cluster together to form visible clouds, affecting the overall water distribution system (Akinremi et al., 1996; Rodríguez et al., 2020).

Clouds are formed by suspended water droplets or ice crystals, which vary in shape, size, and altitude depending on atmospheric conditions. In warm air masses, water droplets form, while in colder air masses, water vapor transforms into ice crystals.

Condensation is a crucial step in the hydrological cycle, as it precedes precipitation. When water droplets become too heavy, they fall to Earth's surface as precipitation, replenishing water resources, nourishing ecosystems, and sustaining life. Factors influencing cloud formation include temperature, humidity, and condensation nuclei, which provide surfaces for water vapor to condense onto.

Condensation is a crucial process in the atmosphere, regulating temperature and energy transfer. It occurs when moist air contacts a cooler surface, causing water vapor to condense. This process influences atmospheric stability, weather patterns, and the formation of convective currents. Accurate modeling of condensation processes improves weather forecasts, climate predictions, and the water cycle's dynamics. Understanding condensation is essential for meteorologists, climatologists, and scientists studying the Earth's climate system.

Condensation is the cooling of atmospheric water vapor into liquid or solid droplets, forming clouds and precipitation. This process replenishes water resources and supports life on Earth. Understanding condensation processes provides insights into weather patterns, climate dynamics, and Earth's water distribution system.

Precipitation

Precipitation is a vital part of the hydrological cycle, involving the fall of water droplets or ice crystals from the atmosphere to Earth's surface. It replenishes water resources, sustains ecosystems, and supports human activities. Rain is the most common form, consisting of liquid water droplets from clouds. Rain is essential for replenishing surface water bodies like rivers, lakes, and reservoirs, providing water for plants, animals, and human consumption. Adequate rainfall is crucial for agriculture, ensuring healthy crop growth and food production.

Snowfall, on the other hand, occurs when temperatures are below freezing, forming snowflakes. Snowfall plays a crucial role in water storage, regulating water flow, replenishing groundwater reserves, and sustaining river ecosystems. Sleet is a frozen precipitation where raindrops freeze into ice pellets before reaching the ground. This occurs when a shallow layer of freezing air below a layer of warmer air, forming small ice pellets. Sleet can create hazardous conditions, as the pellets accumulate on the ground and create slippery surfaces. Hail is a more intense form of precipitation, occurring in severe thunderstorms when strong updrafts carry raindrops upward into extremely cold regions. Hailstones can grow to various sizes, causing significant damage to crops, vehicles, and structures.

Precipitation is crucial for replenishing Earth's water resources, sustaining rivers, lakes, and groundwater reserves. It balances the water cycle by compensating for evaporation and transpiration losses. Precipitation distribution and amount affect regional water availability, agricultural productivity, and ecosystem health. However, patterns of precipitation can vary across regions and seasons, with some experiencing abundant rainfall and abundant vegetation, while others face prolonged dry spells and drought, posing challenges for water availability and agricultural activities (Hussainzada & Lee, 2021; Siddique-E-Akbor et al., 2014). Climate change impacts precipitation patterns by increasing atmospheric moisture capacity, intensifying the hydrological cycle, and increasing flooding risk. In some regions, climate change can cause precipitation shifts, causing reduced rainfall and prolonged drought. Accurate precipitation prediction is crucial for water resource management, agriculture, and disaster preparedness. Meteorological observations, remote sensing technologies, and advanced modeling techniques are used to monitor and forecast precipitation patterns at various scales.

Precipitation, including rain, snow, sleet, and hail, is essential for replenishing water resources and sustaining life on Earth. It impacts freshwater availability, agricultural productivity, and ecosystem health. Understanding precipitation patterns is crucial for effective water resource management and climate adaptation strategies.

Runoff

Runoff is the process by which excess precipitation, such as rain or snowmelt, flows over land and enters streams, rivers, lakes, and oceans. It is a crucial part of the hydrological cycle and plays a significant role in water distribution, erosion, and water bodies formation. As more water accumulates and the ground becomes saturated, excess water flows downhill due to gravity. Factors influencing runoff movement include land slope, soil type, vegetation cover, and precipitation intensity. On steeper slopes, runoff is faster and more concentrated, while on flatter terrain, it may be slower and more dispersed. Runoff collects and transports materials like sediment, organic matter, nutrients, and pollutants, which can come from natural sources or human activities. This transport can impact water quality downstream. Runoff forms channels in rivers and streams, carrying accumulated runoff towards larger bodies like lakes and oceans. This merging of runoff from various sources contributes to the formation and maintenance of river systems. Subsurface runoff, also known as groundwater runoff or base flow, also contributes to stream and river flow. Runoff plays a crucial role in the water cycle, maintaining water flow, supporting aquatic ecosystems, and replenishing water bodies like lakes and reservoirs. These freshwater resources are essential for human consumption, irrigation, industry, and recreational activities (Akinremi et al., 1996; Chahine, 1992).

Runoff impacts erosion processes, causing soil particles to dislodge, causing soil loss, impacting agricultural productivity and ecosystems. Excessive runoff and erosion also contribute to sediment deposition in water bodies, affecting water quality and biodiversity. Land use and land cover changes, like deforestation, urbanization, and natural drainage patterns, can significantly impact runoff patterns. Removing vegetation cover increases surface runoff, reducing water absorption, causing flash floods, soil erosion, and water resource loss. Effective water resource management, flood control, and erosion prevention relies on understanding and managing runoff. Hydrologists and engineers utilize tools to model and predict runoff patterns, considering factors like precipitation, land characteristics, and vegetation cover. This information helps develop strategies to mitigate the impacts of excessive runoff and ensure sustainable water management. Runoff plays a critical role in water distribution, erosion, and body formation, preserving ecosystem health and preventing adverse impacts like floods and erosion.

Infiltration

Infiltration is the process of water seeping into the ground to replenish groundwater resources, playing a crucial role in sustaining water availability, supporting ecosystems, and maintaining water balance. It occurs when precipitation falls on Earth's surface, absorbing water by soil and permeable rocks. Factors influencing infiltration rate include soil type, texture, structure, and compaction. Coarse-grained soils, like sandy ones, have higher infiltration rates compared to fine-grained ones like clay soils (Ghasemizade & Schirmer, 2013; Song et al., 2014). Vegetation cover plays a crucial role in infiltration, as plants break rainfall's impact on soil surfaces, enabling water to penetrate more effectively. Roots of plants create pathways and channels for water movement into the ground. In areas with limited vegetation, such as barren land or urban areas, infiltration may be reduced. Infiltration replenishes groundwater resources by percolating through soil and rock layers, eventually reaching the water table, which stores groundwater in the upper saturated zone.

Groundwater is essential for communities, industries, and ecosystems, providing a reliable supply during dry periods or limited surface water sources. Infiltration is crucial for sustaining groundwater availability and quality, regulating water flow in streams and rivers, and reducing surface runoff. It also supports soil moisture retention and promotes healthy plant growth, as water stored in soil pores provides a reservoir for plant roots, ensuring agricultural productivity, natural vegetation, and ecosystem health (Kraft et al., 2022; Nearing et al., 2021). Infiltration is influenced by factors like compacted soils, impermeable surfaces, land use, and land cover changes. Compacted soils in urban areas or agricultural fields can impede water movement, while impermeable surfaces like concrete or asphalt prevent water from infiltrating, causing increased runoff and reduced groundwater recharge. Deforestation, urbanization, and land cover conversion can disrupt natural infiltration, causing decreased capacity and increased surface runoff. Monitoring and managing infiltration is crucial for sustainable water management and groundwater resource protection. Techniques like soil conservation, reforestation, and permeable surfaces can improve infiltration rates and water retention. Hydrological modeling and monitoring systems assess infiltration rates, groundwater recharge, and overall water resource health.

Infiltration is essential for sustainable water management and groundwater resource protection. Techniques like soil conservation, reforestation, and permeable surfaces improve infiltration rates and soil retention. Hydrological modeling and monitoring systems assess infiltration rates, groundwater recharge, and overall water resource health. Understanding and managing infiltration is crucial for maintaining groundwater levels, mitigating flooding and erosion risks, and promoting sustainable water management practices.

Transpiration

Transpiration is the process by which plants release water vapor into the atmosphere through their leaves, playing a crucial role in the water cycle and Earth's water distribution. It occurs through respiration and photosynthesis, where plants convert carbon dioxide into glucose. Oxygen is released as a byproduct. To access carbon dioxide, plants open stomata on their leaves, allowing carbon dioxide to enter and oxygen to exit. This process results in water vapor loss through the stomata, known as transpiration (C. Liu & Zheng, 2004; Song et al., 2014). Transpiration is a crucial water cycle pathway that transports water from the soil to the atmosphere. Plants absorb water through their roots, travel through their vascular system, and eventually reach leaves. Water molecules evaporate, passing through stomata and escaping into the atmosphere. This process creates the transpiration pull, which draws more water through the plant's roots to replace the loss. This upward movement is known as the transpiration stream.

Transpiration is essential for plant nutrient uptake and temperature regulation. Roots absorb water, carrying dissolved nutrients from the soil, which are transported to leaves via xylem vessels. This distribution ensures proper growth and development. Transpiration also regulates temperature by removing heat energy from leaf surfaces, cooling plants, similar to how sweat cools our bodies. This helps maintain optimal temperature for photosynthesis and growth. Transpiration is influenced by environmental factors like temperature, humidity, wind speed, and light intensity, affecting water loss from leaves. In dry and hot conditions, rates are higher, while in cool and humid conditions, they may be lower. Transpiration impacts the water cycle and ecosystem dynamics by contributing to atmospheric moisture content, cloud formation, and precipitation. It also indirectly affects regional and global climate patterns by increasing atmospheric humidity, influencing precipitation patterns and weather systems. Additionally, the cooling effect of transpiration can impact temperature and microclimate conditions.

Transpiration is the process by which plants release water vapor through their leaves, contributing to the water cycle, nutrient uptake, temperature regulation, and ecosystem dynamics. Understanding transpiration is crucial for plant physiology, water resource management, and climate modeling.

CHALLENGES IN MODELING THE HYDROLOGICAL CYCLE

Data Scarcity and Uncertainty

Data scarcity and uncertainty in hydrology pose significant challenges in understanding and modeling the hydrological cycle. Limited availability of data, particularly in developing countries or remote regions, hinders the development of reliable models and informed decisions in water resource management. This lack of adequate monitoring infrastructure leads to data gaps and hinders our understanding of hydrological processes (Koch et al., 2013; C. Liu & Zheng, 2004; Rodríguez et al., 2020). Data collection in monitoring networks may be limited in spatial coverage and temporal resolution due to sparsely distributed stations and irregular data collection frequency. This makes it difficult to capture short-term variability and accurately represent hydrological cycle dynamics. Data reliability and quality also pose challenges in hydrological modeling, as instrumentation and measurement errors, calibration issues, and data biases introduce uncertainties. Sensor malfunctions, improper maintenance, and human errors can compromise data quality, affecting predictions and decision-making processes.

Data scarcity in hydrological modeling is primarily due to limited historical records, which can be challenging to capture long-term trends and variability. Historical records are crucial for detecting changes in hydrological patterns, assessing climate variability, and developing robust models for future projections. Climate change also introduces uncertainty to hydrological modeling, as climate variables like temperature, precipitation, and evaporation influence the cycle. Downscaling techniques are often used to bridge the gap between coarse-resolution climate models and finer resolution, but they introduce additional uncertainties into the modeling process.

To tackle data scarcity and uncertainty in hydrological modeling, strategies include expanding monitoring stations, utilizing remote sensing technologies, enhancing data quality assurance and control, and implementing data assimilation techniques. These methods improve the accuracy of hydrological models by combining observations with model simulations, ensuring proper maintenance and adherence to standardized measurement protocols. To improve historical records, digitize and integrate datasets, reconstruct past hydrological conditions, and use paleoclimate information. Utilize advanced statistical and machine learning techniques to address data scarcity, uncertainty, and improve model performance by capturing complex relationships between hydrological variables.

Data scarcity and uncertainty hinder accurate hydrological modeling due to limited availability, reliability, and historical coverage. Climate change and uncertainties hinder robust models. Expanding data collection networks, improving quality assurance, and using advanced modeling techniques can overcome these challenges and improve water resource management.

Spatial and Temporal Variability

The hydrological cycle is a dynamic system with spatial and temporal variability, involving water movement between reservoirs like the atmosphere, oceans, land surfaces, and underground aquifers. Spatial heterogeneity occurs due to variations in precipitation, evaporation rates, soil properties, topography, and vegetation cover, resulting in water availability, runoff patterns, and groundwater recharge rates. Regional variations in the hydrological cycle are influenced by climate, land use, and geological characteristics. High rainfall and dense vegetation result in higher evaporation and transpiration rates, while arid regions have limited precipitation and higher evaporation rates. Understanding the spatial distribution of water resources is crucial for sustainable water allocation, planning, and decision-making in water infrastructure development and management strategies.

Temporal variability refers to changes in hydrological processes and characteristics over time, affecting the hydrological cycle at various scales. Influenced by climate patterns, weather events, and longer-term cycles like El Niño and La Niña, seasonal patterns significantly impact the hydrological cycle's temporal variability. Changes in solar radiation, temperature, and atmospheric circulation cause wet and dry seasons, impacting water resource availability and influencing water management practices. Tropical regions experience seasonal monsoons, causing heavy rainfall and increased recharge of groundwater reservoirs. In temperate regions, precipitation is more evenly distributed, with peak flows during spring snowmelt or winter events. Interannual variability, influenced by climate phenomena like El Niño-Southern Oscillation, impacts water availability, agriculture, ecosystems, and resource planning over several years to decades.

Long-term climate cycles, like PDO and AMO, can significantly impact the hydrological cycle, causing prolonged periods of wet or dry conditions. Understanding the spatial and temporal variability is crucial for effective water resource management. This requires comprehensive monitoring networks, accurate climate data, and advanced modeling techniques. By considering spatial variations, seasonal patterns, and long-term climate cycles, decision-makers can make informed choices about water allocation, drought and flood preparedness, and sustainable water management practices.

Human Impacts

Human activities, like deforestation, urbanization, and climate change, significantly impact the water cycle's natural balance. These practices alter water resources' distribution, quality, and quantity, disrupting the hydrological cycle. Trees, particularly in forested areas, regulate the water cycle by intercepting rainfall, reducing soil erosion, and promoting groundwater recharge. Clearing forests for agriculture or urban development leads to increased surface runoff, reduced infiltration, flood risks, diminished groundwater supplies, and altered streamflow patterns.

Urbanization and infrastructure expansion impact the water cycle by replacing natural landscapes with impervious surfaces, increasing surface runoff, and causing higher peak flows and diminished groundwater recharge. This can lead to localized flooding and changes in streamflow patterns. Climate change, including rising temperatures and altered precipitation patterns, also disrupts the natural balance of the water cycle, affecting evaporation rates, soil moisture, and rainfall intensity. These changes can cause water scarcity or flooding in different regions. Climate change affects glaciers and snowpacks, which act as natural water storage systems. Rising temperatures cause glaciers to melt, affecting downstream water availability. This impacts river flows, hydropower generation, and ecosystem health. Human activities contribute to water pollution, causing unfit water bodies for various uses, disrupting aquatic ecosystems, harming human health, and limiting clean water resources availability.

To restore the water cycle and mitigate human impacts, measures like forest conservation, urban land-use planning, climate change adaptation, and water efficiency and conservation, such as water recycling and rainwater harvesting, are crucial. These measures promote groundwater recharge, reduce floods and drought risks, enhance water infiltration, and reduce runoff. Implementing green infrastructure and sustainable drainage systems can also help maintain water resources. To address water pollution, stricter regulations, environmental standards, improved wastewater treatment, responsible agricultural use, and best management practices are essential. Public awareness and education promote responsible water management, promoting sustainable practices like water conservation and watershed protection. Encouraging these practices can restore the natural balance of the water cycle.

Human activities like deforestation, urbanization, and climate change significantly impact the water cycle. Sustainable water resource management requires understanding and mitigating these impacts. Promoting conservation, adopting climate-resilient practices, and addressing water pollution can restore natural balance and ensure clean water resources for future generations.

MACHINE LEARNING AND AI IN HYDROLOGICAL MODELLING

Machine learning and AI techniques are revolutionizing hydrological modeling and prediction, improving accuracy, efficiency, and reliability. These advanced computational approaches simulate the hydrological cycle, enabling better understanding and management of water resources. Traditional models rely on mathematical equations, but machine learning and AI techniques learn patterns and relationships directly from data, enabling better analysis and identification of hidden patterns, leading to improved modeling accuracy and predictive capabilities (Kim et al., 2021; Nearing et al., 2021).

Machine learning and AI are essential in hydrology for accurate rainfall prediction, flood mitigation, and agricultural planning. By learning from historical weather data, these models capture non-linear relationships and temporal dependencies, enabling more timely and accurate predictions. Streamflow forecasting is another area where machine learning and AI techniques show promise. By analyzing historical data and climate variables, these models can capture complex relationships between rainfall, soil moisture, and river discharge, providing forecasts for future streamflow, aiding water managers in decision-making about water allocation, reservoir operations, and flood warning systems.

Flood modeling is crucial in hydrological modeling, especially in flood-prone regions. Machine learning algorithms can predict flood events using data from rainfall, river gauge readings, and topography, improving flood forecasting accuracy. Groundwater modeling also benefits from machine learning and AI techniques, integrating hydrogeological data for accurate flow models. These models aid in sustainable groundwater extraction rates, water supply management, and mitigating groundwater depletion impacts (Srivastava et al., 2013).

Machine learning and AI techniques can enhance hydrological modeling by analyzing remote sensing data and estimating parameters like soil moisture, vegetation cover, and evapotranspiration rates. These models provide valuable insights for land management and water resource planning. However, challenges include high-quality data, interpretability, and scalability. Integrating machine learning techniques with traditional physically-based models can combine the advantages of both approaches.

Machine learning and AI techniques in hydrological modeling and prediction offer significant potential for improving understanding of complex systems. These techniques enable accurate rainfall predictions, streamflow forecasting, flood modeling, and groundwater assessments, enhancing water resource management, flood resilience, and decision-making in changing hydrological environments.

Improve the Accuracy of Rainfall Prediction, Streamflow Forecasting, and Flood Modeling

Machine learning algorithms improve accuracy in rainfall prediction, streamflow forecasting, and flood modeling by capturing complex patterns and relationships that traditional models struggle to handle. By leveraging data-driven approaches, these algorithms contribute to various hydrology applications (Yang et al., 2020).

- Machine learning algorithms analyze historical weather data to develop accurate rainfall prediction models. These models capture non-linear relationships, temporal dependencies, and spatial patterns, enabling more reliable and timely predictions. This aids in water resource management, flood forecasting, and agricultural planning.
- Streamflow forecasting is crucial for water resource management, flood mitigation, and hydropower operations. Machine learning algorithms can analyze historical data, climate variables, and hydrological parameters to develop accurate models. This helps water managers make informed decisions about water allocation, reservoir operations, and flood warning systems.
- Flood modeling involves simulating and predicting flood events to assess impacts and develop mitigation strategies. Machine learning algorithms analyze data sources like rainfall, river gauge readings, and historical flood records. Integrating these algorithms improves flood predictions, enhancing preparedness, evacuation plans, and risk management.

Machine learning algorithms enhance hydrological predictions accuracy.

- Machine learning algorithms accurately represent complex hydrological processes by capturing non-linear relationships between input variables and processes, enhancing representation of rainfall, streamflow, and flood events.
- Machine learning algorithms learn from large datasets, identifying patterns and relationships, incorporating historical information for reliable predictions and overcoming human expertise challenges.
- Machine learning algorithms continuously improve predictions in dynamic hydrological systems, enabling adaptation and updating models based on new data, making them valuable in changing conditions.

- Machine learning algorithms integrate diverse data sources, improving predictive capabilities by incorporating remote sensing, weather forecasts, and ground-based observations.

Machine learning algorithms have potential for improving rainfall prediction, streamflow forecasting, and flood modeling. However, challenges include high-quality datasets, model interpretability, and considering uncertainties. Integrating machine learning techniques with physically-based models can combine physics-based accuracy with data-driven capabilities (Ardabili et al., 2020). Machine learning algorithms improve accuracy in rainfall prediction, streamflow forecasting, and flood modeling by leveraging data-driven approaches. By integrating with traditional models, they contribute to better water resource management, flood resilience, and informed decision-making in hydrology.

Challenges in Implementing Machine Learning Models for Hydrological Applications

Machine learning models in hydrological applications offer potential for improved predictions and understanding, but require careful considerations to overcome challenges and limitations (Figure 2).

Figure 2. Challenges in implementing machine learning models for hydrological applications

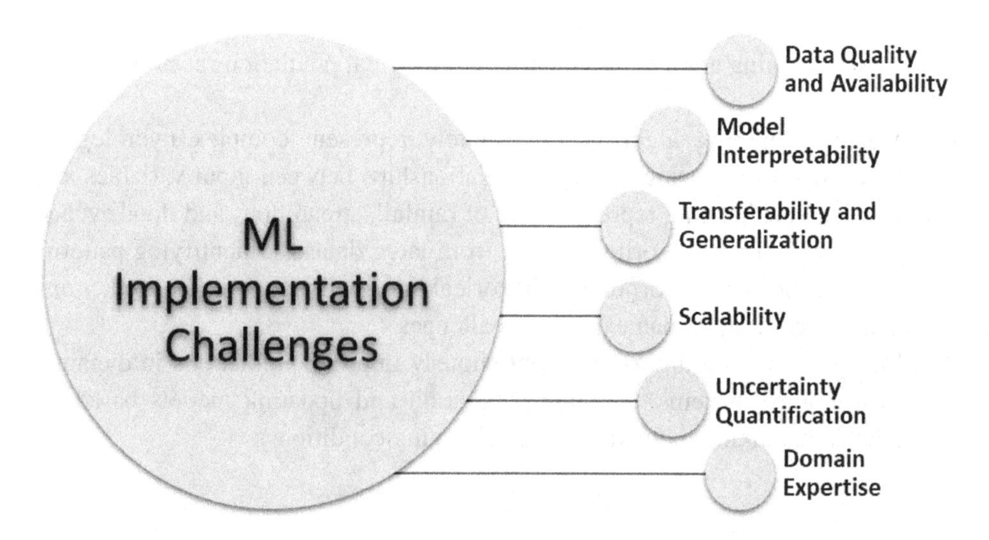

- Implementing machine learning models for hydrology faces challenges due to sparse, incomplete, or error-prone datasets. Ensuring data quality, preprocessing, and addressing missing data is crucial for robust models. Acquiring representative data across different spatiotemporal scales requires collaboration and data sharing among stakeholders.
- Machine learning models, particularly deep learning algorithms, often lack interpretability, making them difficult to understand underlying factors. This limitation is particularly important in hydrological applications, where interpretability is crucial for understanding processes, validating model behavior, and informing decision-making. Efforts are being made to develop interpretable models, such as rule-based or hybrid approaches, to address this issue.
- Scalability is a challenge in applying machine learning models to large-scale hydrological systems, as accurately capturing dynamics at larger scales can be computationally intensive. Efficient algorithms and parallel computing techniques are needed to develop scalable models that maintain accuracy and efficiency. This research area is ongoing in hydrology.
- Machine learning models trained on a single region or dataset may not perform well in different regions or hydrological conditions. This limitation requires careful consideration for long-term projections and model transferability. Techniques like transfer learning, ensemble modeling, and model adaptation are crucial for robust and reliable predictions.
- Hydrological modeling and prediction involve inherent uncertainty, making it crucial to assess and communicate these uncertainties. Machine learning models may not always provide accurate probabilistic estimates, so methods like ensemble modeling, Bayesian approaches, and Monte Carlo simulations are needed to understand model outputs and associated uncertainties.
- Machine learning models require domain expertise in hydrological processes for accurate model selection, feature engineering, and interpretation. Collaboration between hydrologists, data scientists, and experts is crucial for accurate interpretation and alignment with hydrological applications goals.
- Machine learning models in hydrology raise ethical and societal concerns, impacting communities and the environment. Ensuring transparency, fairness, and equity in model development and deployment, considering potential biases, unintended consequences, and stakeholder involvement is crucial.

Advancements in machine learning and hydrological sciences are addressing challenges and limitations. Hybrid models, improved data quality, and interdisciplinary approaches are driving progress. By collaborating and overcoming limitations, machine learning models can unlock their full potential for improved water resource management and decision support in hydrological applications.

CASE STUDY

Machine Learning for Streamflow Forecasting

One notable case study showcasing the application of machine learning in hydrological modeling is the work done by the National Center for Atmospheric Research (NCAR) and the University Corporation for Atmospheric Research (UCAR) in collaboration with the National Weather Service (NWS). They developed a machine learning-based model called the "Subseasonal Hydrologic Forecasting System" (SHyFT) for streamflow forecasting (Cheng et al., 2020; Rasouli et al., 2012).

The SHyFT model incorporates a combination of physics-based hydrological models and machine learning algorithms to improve streamflow predictions at lead times of two weeks to three months. The model integrates data from various sources, including weather forecasts, satellite observations, and ground-based measurements, to capture the complex interactions between precipitation, soil moisture, and streamflow.

Using historical streamflow and meteorological data, the SHyFT model learns the patterns and relationships between these variables and generates streamflow forecasts. Machine learning algorithms, such as artificial neural networks and support vector machines, are employed to capture the non-linear dependencies and complex interactions in the data.

The effectiveness of the SHyFT model has been demonstrated in several river basins across the United States. For instance, in the Colorado River Basin, the model provided accurate streamflow forecasts that aided in water allocation decisions and drought management. The model showed significant improvements over traditional statistical models, providing more skillful predictions even for longer lead times.

AI for Flood Monitoring and Early Warning Systems

Another noteworthy case study is the application of artificial intelligence (AI) in flood monitoring and early warning systems. The European Space Agency (ESA) and the University of Bristol collaborated on a project called "Floods from Space" to develop an AI-based system for flood detection and monitoring using satellite imagery (Rahimzad et al., 2021; Tyralis et al., 2021).

The AI system employs machine learning algorithms, such as convolutional neural networks, to analyze satellite images and automatically identify flooded areas. By training the model with labeled satellite images of both flood and non-flood scenarios, the AI system can recognize patterns and spectral characteristics associated with flooded regions.

The system can provide real-time flood maps and alerts, allowing emergency response teams and local authorities to take immediate actions to mitigate the impacts of flooding. It enables efficient resource allocation, evacuation planning, and targeted interventions to assist affected communities. The AI system has been successfully applied in various regions around the world, including the United Kingdom, Bangladesh, and Peru, to enhance flood monitoring and early warning capabilities (Boopathi, 2022b, 2022a).

The case studies demonstrate the potential of machine learning and AI in improving hydrological modeling and water resource management. The SHyFT model, a combination of physics-based hydrological models and machine learning algorithms, has shown significant improvements in streamflow predictions compared to traditional statistical models. This improved accuracy aids in water allocation decisions, drought management, and overall water resource planning. The AI system for flood monitoring and early warning uses convolutional neural networks (CNNs) trained on labeled satellite images, enabling the system to automatically detect and identify flooded areas. These technologies provide valuable insights and support informed decision-making in water-related applications.

The AI-based flood monitoring system has demonstrated effectiveness in real-time flood detection and mapping, enabling emergency response teams and local authorities to take swift action. This improves resource allocation, evacuation planning, and targeted interventions, ultimately reducing flood impact on communities and infrastructure. The integration of machine learning algorithms in hydrological modeling improves streamflow forecasting, leading to better water resource management and decision-making. The integration of AI in flood monitoring and early warning systems enhances the ability to detect and respond to floods in a timely manner, improving disaster preparedness and response efforts. These outcomes demonstrate the potential of machine learning and AI to revolutionize hydrological modeling and water resource management, enabling more accurate predictions, real-time hydrological events detection, and informed decisions for sustainable water resource management and community protection from water-related hazards.

Future Directions and Challenges

- Emerging trends involve integrating physics-based models with machine learning techniques, combining strengths for improved hydrological processes understanding and data-driven capabilities. This hybrid approach enhances model accuracy, interpretability, and uncertainty quantification.

- Transfer learning improves model generalization and robustness by transferring knowledge from one domain to another, addressing data limitations and enabling models to be applied to new contexts.
- Real-time data assimilation techniques like ensemble Kalman filtering and particle filtering improve hydrological models' accuracy and responsiveness to changing conditions by assimiling real-time observations into predictions(Jeevanantham et al., 2023).
- The integration of remote sensing data with machine learning algorithms offers new hydrological modeling and optimization opportunities. This data provides valuable information on variables like rainfall, evapotranspiration, and soil moisture. Big data analytics techniques can analyze these vast datasets, extracting meaningful patterns for improved modeling and decision-making.
- Improved uncertainty quantification in hydrological models and predictions is crucial for robust decision-making and risk management strategies. Bayesian approaches, ensemble modeling, and Monte Carlo simulations are being explored for probabilistic estimates and risk assessment.
- Machine learning and AI techniques can develop optimization models and decision support systems for water resource management, aiding in water allocation, reservoir operation, flood control, and drought mitigation. Reinforcement learning and evolutionary algorithms are being explored for intelligent decision support systems(Anitha et al., 2023; Subha et al., 2023).
- The integration of machine learning and AI in hydrological modeling requires collaborations between hydrologists, data scientists, computer scientists, and domain experts. Interdisciplinary research and knowledge exchange will develop robust, context-specific models for water resource management (Babu et al., 2023; Boopathi et al., 2023).

Challenges and Research Gaps in Machine Learning Adoption in Hydrology

- Machine learning techniques in hydrology face challenges in data quality and availability due to spatial and temporal gaps, measurement errors, and inconsistencies. Addressing these issues requires improved data collection methods, sharing protocols, quality control measures, and standardized data formats for interoperability and accessibility.
- Machine learning models, especially deep learning models, are often viewed as black boxes due to their high-dimensional parameter spaces and intricate internal workings. Enhancing interpretability and explainability is crucial for gaining trust and acceptance in the hydrological community. Further

exploration is needed in developing techniques for meaningful insights, quantifying uncertainties, and providing interpretable explanations for predictions.

- Machine learning models trained on specific hydrological settings may struggle to generalize to other regions, climates, or future time periods. To achieve model transferability and generalization, techniques must capture physical processes, domain knowledge, and account for non-stationarity and changing hydrological conditions(Boopathi, n.d.; Koshariya, n.d.; Sampath et al., 2022).
- Hydrological modeling requires scalable machine learning algorithms for large spatial domains and long time periods. Techniques like distributed computing, parallelization, and model compression improve scalability and reduce computational demands for widespread adoption.
- Machine learning techniques should integrate expert knowledge and domain-specific information for better modeling. Combining data-driven approaches with hydrology experts' insights can improve model performance, interpretability, and incorporate physical constraints. This research gap requires developing effective methodologies (J. Liu et al., 2022).
- Machine learning techniques in hydrology raise ethical and social concerns, including biases, algorithmic discrimination, and data use. Transparency, fairness, and accountability are essential for responsible and equitable water resource management.
- Machine learning models in hydrology must be robust and resilient to account for uncertainties and changing conditions. Techniques like ensemble modeling, model fusion, and uncertainty quantification can enhance robustness and resilience.

Addressing challenges and research gaps in hydrology will promote machine learning adoption by focusing on data quality, model interpretability, generalization, scalability, expert knowledge integration, ethical considerations, and robustness. This will advance water resource management, decision support, and sustainable practices.

CONCLUSION

The hydrological cycle is crucial for Earth's water distribution system, involving processes like evaporation, condensation, precipitation, runoff, infiltration, and transpiration. Understanding this cycle is essential for resource management, predicting events like floods and droughts, and ensuring sustainable water availability. However, modeling the hydrological cycle presents challenges like data scarcity, uncertainty, and human impacts. Machine learning and artificial intelligence (AI) techniques have been integrated to improve rainfall prediction, streamflow forecasting, flood modeling, and flood monitoring and early warning systems.

Machine learning models for hydrological applications face challenges like data quality, availability, interpretability, and scalability. Emerging trends include integrating physics-based models, transfer learning, real-time data assimilation, remote sensing, and big data analytics. Uncertainty quantification and risk assessment are crucial for robust decision-making. Machine learning and AI in hydrological modeling have shown benefits, such as streamflow predictions and real-time flood detection. Future directions should address challenges and research gaps, including data quality, interpretability, transferability, efficiency, expert knowledge integration, ethical considerations, and robustness. Further research, collaboration, and interdisciplinary approaches are needed to overcome challenges, improve models' capabilities, and ensure responsible, equitable use of these technologies.

ACRONYMS

- AI: Artificial Intelligence
- ML: Machine Learning
- H2O: Hydrogen Dioxide (Water)
- Evap: Evaporation
- Cond: Condensation
- Precip: Precipitation
- Runoff: Surface Runoff
- Infilt: Infiltration
- Transp: Transpiration
- Data: Datasets
- QC: Quality Control
- DL: Deep Learning
- IoT: Internet of Things
- RS: Remote Sensing
- SWMM: Storm Water Management Model
- DSS: Decision Support System
- GCM: Global Climate Model
- MLR: Multiple Linear Regression
- SVM: Support Vector Machine
- CNN: Convolutional Neural Network

REFERENCES

Akinremi, O. O., McGinn, S. M., & Barr, A. G. (1996). Simulation of soil moisture and other components of the hydrological cycle using a water budget approach. *Canadian Journal of Soil Science*, *76*(2), 133–142. doi:10.4141/cjss96-020

Alitane, A., Essahlaoui, A., Van Griensven, A., Yimer, E. A., Essahlaoui, N., Mohajane, M., Chawanda, C. J., & Van Rompaey, A. (2022). Towards a Decision-Making Approach of Sustainable Water Resources Management Based on Hydrological Modeling: A Case Study in Central Morocco. *Sustainability (Basel)*, *14*(17), 10848. doi:10.3390u141710848

Anitha, C., R, K. C., Vivekanand, C. V., Lalitha, S. D., Boopathi, S., & R, R. (2023, February). Artificial Intelligence driven security model for Internet of Medical Things ({IoMT}). *IEEE Explore*. doi:10.1109/ICIPTM57143.2023.10117713

Ardabili, S., Mosavi, A., Dehghani, M., & Várkonyi-Kóczy, A. R. (2020). Deep Learning and Machine Learning in Hydrological Processes Climate Change and Earth Systems a Systematic Review. *Lecture Notes in Networks and Systems*, *101*, 52–62. doi:10.1007/978-3-030-36841-8_5

Babu, B. S., Kamalakannan, J., Meenatchi, N., M, S. K. S., S, K., & Boopathi, S. (2023). Economic impacts and reliability evaluation of battery by adopting Electric Vehicle. *IEEE Explore*, 1–6. doi:10.1109/ICPECTS56089.2022.10046786

Boopathi, S. (2022a). Cryogenically treated and untreated stainless steel grade 317 in sustainable wire electrical discharge machining process: A comparative study. *Environmental Science and Pollution Research International*, 1–10. doi:10.100711356-022-22843-x PMID:36057706

Boopathi, S. (2022b). Experimental investigation and multi-objective optimization of cryogenic Friction-stir-welding of AA2014 and AZ31B alloys using MOORA technique. *Materials Today. Communications*, *33*, 104937. doi:10.1016/j.mtcomm.2022.104937

Boopathi, S. (2023). Deep Learning Techniques Applied for Automatic Sentence Generation. In Promoting Diversity, Equity, and Inclusion in Language Learning Environments (pp. 255-273). IGI Global. doi:10.4018/978-1-6684-3632-5.ch016

Boopathi, S., Arigela, S. H., Raman, R., Indhumathi, C., Kavitha, V., & Bhatt, B. C. (2023). Prominent Rule Control-based Internet of Things: Poultry Farm Management System. *IEEE Explore*, 1–6. doi:10.1109/ICPECTS56089.2022.10047039

Chahine, M. T. (1992). The hydrological cycle and its influence on climate. *Nature*, *359*(6394), 373–380. doi:10.1038/359373a0

Cheng, M., Fang, F., Kinouchi, T., Navon, I. M., & Pain, C. C. (2020). Long lead-time daily and monthly streamflow forecasting using machine learning methods. *Journal of Hydrology (Amsterdam)*, *590*, 125376. doi:10.1016/j.jhydrol.2020.125376

Ghasemizade, M., & Schirmer, M. (2013). Subsurface flow contribution in the hydrological cycle: Lessons learned and challenges ahead-a review. *Environmental Earth Sciences*, *69*(2), 707–718. doi:10.100712665-013-2329-8

Hussainzada, W., & Lee, H. S. (2021). Hydrological modelling for water resource management in a semi-arid mountainous region using the soil and water assessment tool: A case study in northern Afghanistan. *Hydrology*, *8*(1), 1–21. doi:10.3390/hydrology8010016

Jeevanantham, Y. A., A, S., V, V., J, S. I., Boopathi, S., & Kumar, D. P. (2023). Implementation of Internet-of Things (IoT) in Soil Irrigation System. *IEEE Explore*, 1–5. doi:10.1109/ICPECTS56089.2022.10047185

Kim, T., Yang, T., Gao, S., Zhang, L., Ding, Z., Wen, X., Gourley, J. J., & Hong, Y. (2021). Can artificial intelligence and data-driven machine learning models match or even replace process-driven hydrologic models for streamflow simulation?: A case study of four watersheds with different hydro-climatic regions across the CONUS. *Journal of Hydrology (Amsterdam)*, *598*, 126423. doi:10.1016/j.jhydrol.2021.126423

Koch, H., Liersch, S., & Hattermann, F. F. (2013). Integrating water resources management in eco-hydrological modelling. *Water Science and Technology*, *67*(7), 1525–1533. doi:10.2166/wst.2013.022 PMID:23552241

Koshariya, A. K. (n.d.). *AI-Enabled IoT and WSN-Integrated Smart*. doi:10.4018/978-1-6684-8516-3.ch011

Kraft, B., Jung, M., Körner, M., Koirala, S., & Reichstein, M. (2022). Towards hybrid modeling of the global hydrological cycle. *Hydrology and Earth System Sciences*, *26*(6), 1579–1614. doi:10.5194/hess-26-1579-2022

Liu, C., & Zheng, H. (2004). Changes in components of the hydrological cycle in the Yellow River basin during the second half of the 20th century. *Hydrological Processes*, *18*(12), 2337–2345. doi:10.1002/hyp.5534

Liu, J., Yuan, X., Zeng, J., Jiao, Y., Li, Y., Zhong, L., & Yao, L. (2022). Ensemble streamflow forecasting over a cascade reservoir catchment with integrated hydrometeorological modeling and machine learning. *Hydrology and Earth System Sciences*, *26*(2), 265–278. doi:10.5194/hess-26-265-2022

Nearing, G. S., Kratzert, F., Sampson, A. K., Pelissier, C. S., Klotz, D., Frame, J. M., Prieto, C., & Gupta, H. V. (2021). What Role Does Hydrological Science Play in the Age of Machine Learning? *Water Resources Research, 57*(3). doi:10.1029/2020WR028091

Rahimzad, M., Moghaddam Nia, A., Zolfonoon, H., Soltani, J., Danandeh Mehr, A., & Kwon, H. H. (2021). Performance Comparison of an LSTM-based Deep Learning Model versus Conventional Machine Learning Algorithms for Streamflow Forecasting. *Water Resources Management*, *35*(12), 4167–4187. doi:10.100711269-021-02937-w

Rasouli, K., Hsieh, W. W., & Cannon, A. J. (2012). Daily streamflow forecasting by machine learning methods with weather and climate inputs. *Journal of Hydrology (Amsterdam)*, *414–415*, 284–293. doi:10.1016/j.jhydrol.2011.10.039

Rodríguez, E., Sánchez, I., Duque, N., Arboleda, P., Vega, C., Zamora, D., López, P., Kaune, A., Werner, M., García, C., & Burke, S. (2020). Combined Use of Local and Global Hydro Meteorological Data with Hydrological Models for Water Resources Management in the Magdalena - Cauca Macro Basin – Colombia. *Water Resources Management*, *34*(7), 2179–2199. doi:10.100711269-019-02236-5

Sampath, B. C. S., & Myilsamy, S. (2022). Application of TOPSIS Optimization Technique in the Micro-Machining Process. In Trends, Paradigms, and Advances in Mechatronics Engineering (pp. 162–187). IGI Global. doi:10.4018/978-1-6684-5887-7.ch009

Siddique-E-Akbor, A. H. M., Hossain, F., Sikder, S., Shum, C. K., Tseng, S., Yi, Y., Turk, F. J., & Limaye, A. (2014). Satellite Precipitation Data–Driven Hydrological Modeling for Water Resources Management in the Ganges, Brahmaputra, and Meghna Basins. *Earth Interactions*, *18*(17), 1–25. doi:10.1175/EI-D-14-0017.1

Song, X., Zhang, J., Wang, G., He, R., & Wang, X. (2014). Development and challenges of urban hydrology in a changing environment: II: Urban stormwater modeling and management. *Shuikexue Jinzhan. Shui Kexue Jinzhan*, *25*(5), 752–764.

Srivastava, P. K., Han, D., Ramirez, M. R., & Islam, T. (2013). Machine Learning Techniques for Downscaling SMOS Satellite Soil Moisture Using MODIS Land Surface Temperature for Hydrological Application. *Water Resources Management*, *27*(8), 3127–3144. doi:10.100711269-013-0337-9

Subha, S., Inbamalar, T. M., R, K. C., Suresh, L. R., Boopathi, S., & Alaskar, K. (2023, February). A Remote Health Care Monitoring system using internet of medical things ({IoMT}). *IEEE Explore*. doi:10.1109/ICIPTM57143.2023.10118103

Tyralis, H., Papacharalampous, G., & Langousis, A. (2021). Super ensemble learning for daily streamflow forecasting: Large-scale demonstration and comparison with multiple machine learning algorithms. *Neural Computing & Applications*, *33*(8), 3053–3068. doi:10.100700521-020-05172-3

Yang, S., Yang, D., Chen, J., Santisirisomboon, J., Lu, W., & Zhao, B. (2020). A physical process and machine learning combined hydrological model for daily streamflow simulations of large watersheds with limited observation data. *Journal of Hydrology (Amsterdam)*, *590*, 125206. doi:10.1016/j.jhydrol.2020.125206

Chapter 4
Artificial Intelligence in Water Treatments and Water Resource Assessments

K. Gunasekaran
Department of Mechanical Engineering, Muthayammal Engineering College, Namakkal, India

Sampath Boopathi
https://orcid.org/0000-0002-2065-6539
Department of Mechanical Engineering, Muthayammal Engineering College, Namakkal, India

ABSTRACT

This chapter explores the use of AI in water treatment, evaporation management, and water resource management. It begins with an introduction, highlighting AI's motivation and objectives. The chapter then discusses AI applications, challenges, and opportunities in their implementation. It compares traditional approaches and AI-driven solutions for evaporation control and optimization and presents case studies and applications to demonstrate real-world examples. The chapter also discusses water resource management challenges, data-driven modeling, forecasting, optimization, and decision support systems. It also discusses the benefits and limitations of AI, interdisciplinary collaboration, ethical considerations, and policy frameworks for responsible AI implementation. The chapter also provides recommendations for future research to advance AI in water treatment and resource management.

DOI: 10.4018/978-1-6684-6791-6.ch004

INTRODUCTION

Water treatment is a crucial process involving purifying and managing water resources to ensure safe and sustainable access to clean water. With the growing global population, urbanization, and industrialization, demand for clean water is increasing, and resources are becoming scarcer and more contaminated. Traditional methods face limitations in efficiency, resource optimization, and decision-making due to the complexity of systems, variability of water quality, and dynamic nature of resources. Artificial Intelligence (AI) is a transformative technology that has the potential to revolutionize water treatment. It uses machine learning, neural networks, and deep learning techniques to learn from data, identify patterns, and make predictions without explicit programming. AI can analyze vast amounts of data from sensors, monitoring systems, and historical records, providing insights into water quality, usage patterns, and resource availability. This data-driven approach enables accurate prediction of treatment outcomes, optimization of treatment processes, and proactive resource management (Nourani et al., 2018).

AI can improve water treatment efficiency and automation by enabling real-time monitoring and control of treatment systems, minimizing energy consumption and facilitating timely adjustments. AI-driven decision support systems aid operators in making informed choices about treatment strategies, resource allocation, and maintenance scheduling. AI techniques can optimize evaporation processes, reducing water loss by analyzing weather data and humidity levels, resulting in improved overall efficiency. AI in water treatment presents challenges such as data availability, quality, compatibility, interpretability, and interdisciplinary collaboration. However, it holds immense promise in addressing traditional approaches and promoting efficient, sustainable, and data-driven water resource management. Further research, development, and implementation of AI-driven solutions are crucial to unlock the full potential of this technology in water treatment. AI is gaining attention in water treatment as a powerful tool for improving efficiency, resource optimization, and decision-making. By integrating AI techniques, traditional approaches face challenges in water quality, energy consumption, and resource management (S. et al., 2022; Vanitha et al., 2023). AI can enhance efficiency, accuracy, and automation, leading to improved water quality, reduced energy consumption, and optimized resource management. By leveraging machine learning algorithms, neural networks, and deep learning models, AI can analyze complex data sets, predict outcomes, and optimize operations in real-time. AI has great potential in evaporation processes and water resource management, reducing water wastage and improving efficiency. It can provide accurate assessments, forecast future water availability, and support decision-making processes for optimal allocation and utilization of resources. This chapter explores the applications of AI in evaporation processes and water resource management within water treatment, discussing benefits, limitations, and challenges associated with integrating AI. Key research findings, future directions, and the potential impact of AI revolutionize water treatment processes. By harnessing AI's power, the field of water treatment can make significant strides towards efficient and sustainable water resource management, ensuring clean water availability and mitigating challenges posed by population growth, climate change, and pollution (Alam et al., 2022; Fan et al., 2018).

Evaporation and water resource management are crucial aspects of water treatment, ensuring clean and sustainable water resources. Evaporation, the conversion of water into vapor and loss to the atmosphere, is a major challenge in treatment due to water loss and its impact on efficiency. Water resource management involves assessing, allocating, and utilizing resources to meet various sectors' demands, considering factors like availability, quality, and environmental sustainability. Evaporation accounts for a significant portion of water loss in reservoirs, lakes, and other water bodies, exacerbating water scarcity challenges. Effective management of water resources is essential for meeting diverse needs in domestic, industrial, and agricultural sectors while ensuring long-term sustainability. AI techniques can significantly improve water treatment efficiency, accuracy, and sustainability by enhancing evaporation control and water resource management. By utilizing advanced algorithms and data analytics, AI can analyze complex datasets, identify patterns, and make informed predictions. This approach optimizes evaporation control, enhances water resource allocation, and enables proactive management strategies. Evaporation is a significant water loss factor, and conventional methods often rely on simple empirical formulas or physical barriers. AI techniques, such as machine learning algorithms, neural networks, and deep learning models, can analyze various factors influencing evaporation, such as weather conditions, humidity levels, wind speed, and solar radiation. By integrating AI into evaporation control, water treatment systems can accurately predict evaporation rates, optimize water storage and release strategies, and minimize water loss (Lafta & Amori, 2022; Li et al., 2021).

Water resource management involves assessing water availability, forecasting future conditions, and allocating resources efficiently. AI can revolutionize water resource management by analyzing diverse datasets like historical records, meteorological data, and satellite imagery. By leveraging predictive modeling, optimization algorithms, and decision support systems, water managers can make informed decisions about water allocation, demand management, and infrastructure planning. AI can also enable real-time monitoring of water resources, enabling early detection of anomalies, pollution events, and unsustainable usage patterns. However, challenges such as data availability, quality, compatibility, interpretability, and ethical considerations must be addressed to effectively integrate AI in evaporation control and water resource management (S. et al., 2022; Vanitha et al., 2023).

This chapter explores the use of AI techniques in evaporation control and water resource management in water treatment. It discusses the benefits, limitations, and challenges of integrating AI in these areas. Case studies and real-world applications are presented to illustrate practical implementation. Key research findings and future directions are discussed, with the potential impact on revolutionizing water treatment processes, improving sustainability, and ensuring clean water resources for present and future generations.

EVAPORATION IN WATER TREATMENT

Evaporation is a natural process that converts water into vapor and releases it into the atmosphere. It is a significant concern in water treatment due to water loss, especially in water bodies like reservoirs, lakes, and open channels. Evaporation rate is influenced by factors like temperature, humidity, wind speed, solar radiation, and water surface area. Understanding evaporation mechanisms is crucial for effective management in water treatment. The energy balance between the water surface and the atmosphere influences evaporation, with higher humidity reducing the rate (Dind & Schmid, 1978; Heins & Peterson, 2018; Yuan et al., 2021).

Challenges in Evaporation Management

Evaporation challenges water treatment, requiring optimal resource utilization and efficiency to overcome.

Water Loss

Evaporation leads to the loss of significant volumes of water from reservoirs and other water bodies. This water loss exacerbates the challenges of water scarcity, especially in arid and water-stressed regions. Efficient evaporation management strategies are necessary to minimize water wastage and preserve valuable water resources.

2.2.2 Impact on Water Quality

Evaporation can contribute to the concentration of dissolved solids and other impurities in the remaining water. As water evaporates, the dissolved solids become more concentrated, potentially leading to issues such as scaling, salinization, and increased treatment requirements. Effective management of evaporation can help mitigate these water quality concerns.

Energy Consumption

Evaporation requires energy in the form of heat to convert liquid water into vapor. This energy consumption contributes to the overall operational costs of water treatment processes. Managing evaporation effectively can help optimize energy usage and reduce associated expenses.

Environmental Factors

Evaporation management must consider the impact on the local environment and ecosystems. Altering evaporation rates through artificial means may have unintended consequences on the surrounding flora and fauna. Balancing evaporation control strategies with environmental sustainability is crucial.

Complex and Dynamic Nature

Evaporation rates are influenced by multiple factors, including meteorological conditions, water temperature, wind speed, and humidity. These factors are subject to continuous change, making evaporation management a complex and dynamic task. Traditional approaches based on empirical formulas or physical barriers may not adequately account for these variations, necessitating more sophisticated and adaptive solutions.

Advanced techniques like Artificial Intelligence (AI) are crucial for evaporation management challenges. AI can analyze relationships between factors affecting evaporation rates and provide real-time insights for decision-making. Water treatment systems can optimize storage, release strategies, and evaporation control measures, minimizing water loss and improving efficiency. This chapter explores AI algorithms, machine learning techniques, and case studies demonstrating their application in optimizing evaporation processes in water treatment.

Traditional Approaches vs. AI-Driven Solutions

Traditionally, evaporation management in water treatment has relied on conventional methods like empirical formulas and physical barriers. These methods offer some evaporation reduction but lack precision, adaptability, and efficiency. AI-driven solutions offer advantages and improvements in evaporation management. Comparing traditional approaches with AI-driven solutions is essential for understanding their effectiveness in managing water treatment (Sharshir et al., 2020).

- Precision and Accuracy: Traditional evaporation methods use simplified formulas based on factors like temperature and wind speed, which may not fully capture the complexity of evaporation dynamics. AI-driven solutions use advanced algorithms and machine learning to analyze multiple factors, historical data, and real-time information, enabling more precise predictions and improved evaporation control.
- Adaptability and Flexibility: Traditional evaporation control methods often use fixed strategies or physical barriers, which may not adapt well to changing environmental conditions or water bodies. AI-driven solutions can continuously analyze and adapt to evolving conditions, using machine learning algorithms to make real-time adjustments based on weather forecasts, humidity levels, and other relevant variables.
- Optimization and Resource Efficiency: AI-driven solutions can optimize evaporation management by considering multiple factors, analyzing large datasets, identifying complex patterns, and optimizing control strategies. This dynamic adjustment ensures optimal water resource utilization, reducing energy consumption and operational costs.
- Real-Time Monitoring and Decision Support: Traditional approaches often lack real-time monitoring and decision support capabilities. In contrast, AI-driven solutions can incorporate real-time data from sensors, weather forecasts, and other sources to continuously monitor evaporation rates and provide timely insights. This enables operators to make informed decisions and take proactive actions to manage evaporation effectively.
- Scalability and Adaptation to Various Water Bodies: Traditional approaches may be limited in their applicability to different types of water bodies or scales of operation. AI-driven solutions, with their ability to analyze diverse datasets and adapt to different environments, offer greater scalability. They can be applied to a wide range of water bodies, including reservoirs, lakes, ponds, and open channels, providing tailored evaporation management solutions.
- Integration with Overall Water Treatment Systems: AI-driven solutions can be integrated with other components of water treatment systems, such as water quality monitoring, demand forecasting, and resource allocation. This integration enables a holistic approach to water treatment, where evaporation management is optimized in conjunction with other processes, leading to enhanced system efficiency and performance.

AI-driven solutions offer superior precision, adaptability, optimization, and integration with water treatment systems, surpassing traditional approaches in evaporation reduction. This application holds promise for improving water resource utilization, reducing water loss, and enhancing efficiency in water treatment processes.

AI Techniques for Evaporation Control and Optimization

Machine Learning Algorithms

Machine learning algorithms are essential for evaporation control and optimization in water treatment. They analyze historical data, real-time sensor readings, and environmental factors to make accurate predictions and decisions on evaporation rates. Commonly used algorithms include:

- Linear Regression: Linear regression models can establish relationships between evaporation and relevant factors such as temperature, wind speed, humidity, and solar radiation. By fitting a linear equation to the data, these models can predict evaporation rates based on the input variables.
- Support Vector Machines (SVM): SVM algorithms are effective for both classification and regression tasks. In the context of evaporation control, SVM can be used to predict evaporation rates based on multiple factors. SVM algorithms can handle complex relationships and adapt to nonlinear patterns in the data.
- Random Forests: Random Forest models are an ensemble learning technique that combines multiple decision trees to make predictions. Random forests can handle high-dimensional data and capture nonlinear relationships between input variables and evaporation rates. These models can also assess the importance of different factors in predicting evaporation, aiding in decision-making.
- Gradient Boosting: Gradient boosting algorithms, such as XGBoost and LightGBM, iteratively build a series of weak predictive models to make accurate predictions. These algorithms can handle large datasets, capture complex interactions between variables, and provide robust predictions for evaporation rates.

Machine learning algorithms can predict evaporation rates in real-time using historical data and environmental factors, offering valuable insights for evaporation control and optimization strategies.

Neural Networks

Neural networks, a subfield of machine learning, excel at analyzing complex relationships and patterns in evaporation data. These interconnected nodes, organized in layers, learn from training data and adjust connection weights for predictions(Anitha et al., 2023; Jeevanantham et al., 2023).

- Feedforward Neural Networks: Feedforward neural networks consist of an input layer, one or more hidden layers, and an output layer. They can capture complex nonlinear relationships between input variables and evaporation rates. By training these networks on historical data, they can provide accurate predictions of evaporation rates based on current environmental conditions.
- Recurrent Neural Networks (RNN): RNNs are suitable for modeling time series data, making them applicable to evaporation prediction. RNNs have recurrent connections that allow information to persist through time, enabling them to capture temporal dependencies in evaporation data. Long Short-Term Memory (LSTM) networks, a type of RNN, are particularly effective in capturing long-term dependencies and making accurate predictions for evaporation rates.
- Convolutional Neural Networks (CNN): While commonly used in image analysis, CNNs can also be applied to evaporation prediction by treating the input data as a spatial grid. CNNs can identify spatial patterns in environmental factors that influence evaporation rates. By training these networks on historical data, they can learn to make predictions based on the spatial characteristics of the input variables.

Deep Learning Models

Deep learning models, like deep neural networks and convolutional neural networks, improve evaporation control and optimization by capturing intricate relationships and patterns in data. They automatically learn features and representations, making accurate predictions and optimizing evaporation control strategies.

- Autoencoders: Autoencoders are unsupervised learning models that can learn efficient representations of the input data. These models consist of an encoder network that compresses the input data into a lower-dimensional representation and a decoder network that reconstructs the original data from the compressed representation. By training autoencoders on historical evaporation data, they can capture the underlying patterns and relationships, aiding in evaporation prediction and control.

- Generative Adversarial Networks (GANs): GANs are composed of a generator network and a discriminator network that play a "game" against each other. The generator network generates synthetic data samples that resemble the real data, while the discriminator network tries to distinguish between real and synthetic samples. GANs can be used to generate synthetic evaporation data, facilitating data augmentation and improving the robustness of evaporation models.

Deep learning models analyze complex relationships, making accurate predictions for evaporation rates, enabling advanced control and optimization in water treatment processes, despite computationally intensive training and large data requirements.

This chapter explores the practical application of AI techniques like machine learning algorithms, neural networks, and deep learning models in evaporation control and optimization. Case studies and real-world examples demonstrate their effectiveness in water treatment.

CASE STUDIES AND APPLICATIONS

Case Study 1: Reservoir Evaporation Control In a large reservoir management project, AI techniques were employed to optimize evaporation control strategies. Historical data on evaporation rates, temperature, wind speed, humidity, and solar radiation were collected and used to train machine learning models. These models accurately predicted evaporation rates based on real-time environmental conditions. By integrating these models with the reservoir management system, operators were able to make informed decisions on water release schedules, storage capacity utilization, and the use of evaporation mitigation techniques such as covers or sprays. This resulted in significant reduction in water loss due to evaporation, improved water resource utilization, and cost savings.

Case Study 2: Urban Water Storage Evaporation Reduction In an urban water storage facility, evaporation reduction was achieved using AI-driven optimization techniques. Real-time sensor data on temperature, wind speed, and humidity were continuously monitored and fed into a neural network model. The model learned the complex relationships between these variables and evaporation rates, allowing for accurate predictions. Based on the predictions, an adaptive control algorithm adjusted the water levels and implemented evaporation mitigation strategies such as water misting systems. This dynamic approach resulted in substantial evaporation reduction, improved water availability, and reduced energy consumption.

Case Study 3: Agriculture Water Management In agricultural water management, AI techniques were applied to optimize irrigation scheduling and minimize evaporation losses. Machine learning algorithms were trained using historical data on weather conditions, soil moisture levels, and crop water requirements. These algorithms provided recommendations on optimal irrigation schedules based on real-time weather forecasts and soil moisture monitoring. By fine-tuning irrigation practices, the amount of water exposed to evaporation was minimized, leading to water conservation and increased crop yield.

Benefits and Limitations of AI in Evaporation Management

1. Enhanced Precision: AI techniques, such as machine learning algorithms and neural networks, can provide accurate predictions of evaporation rates by considering multiple factors and complex relationships. This precision allows for more effective evaporation control strategies.
2. Real-Time Monitoring and Decision Support: AI-driven systems can continuously monitor environmental conditions, process real-time data, and provide timely insights for decision-making. This enables proactive evaporation management and optimized resource allocation.
3. Improved Efficiency and Resource Utilization: AI techniques can optimize evaporation control strategies, leading to reduced water loss, improved water resource utilization, and energy savings. By dynamically adjusting water storage, release schedules, and evaporation mitigation techniques, resources can be utilized more efficiently.
4. Adaptability to Changing Conditions: AI models can adapt to changing environmental conditions, allowing for real-time adjustments in evaporation management strategies. This adaptability ensures that evaporation control measures remain effective even in dynamic situations.
5. Integration with Overall Water Treatment Systems: AI-driven evaporation management can be seamlessly integrated with other components of water treatment systems, such as water quality monitoring and demand forecasting. This integration allows for a holistic approach to water management, improving overall system efficiency.

Limitations of AI in Evaporation Management

1. Data Requirements: AI techniques, especially deep learning models, require large amounts of high-quality data for training. Obtaining comprehensive and reliable historical data for evaporation rates and environmental factors can be challenging in some cases.
2. Computational Complexity: Some AI models, particularly deep learning models, can be computationally intensive and require significant computing

resources. Implementing and running these models may pose challenges in terms of computational power and infrastructure.

3. Interpretability: Some AI models, such as deep neural networks, are often considered black boxes, making it difficult to interpret the reasoning behind their predictions. This lack of interpretability may limit the understanding of the underlying mechanisms driving evaporation control.
4. Implementation and Operational Costs: Deploying and maintaining AI-driven evaporation management systems may require initial investment and ongoing operational costs. These costs include data collection, model development, integration with existing infrastructure, and system maintenance.

Despite these limitations, the benefits offered by AI in evaporation management outweigh the challenges. The advancements in AI techniques and their successful application in various case studies demonstrate the potential for improved evaporation control, water resource management, and overall efficiency in water treatment processes.

WATER RESOURCE MANAGEMENT

Water Resource Challenges

Water resource management is crucial for ensuring sustainable water supply and addressing the challenges associated with water scarcity, population growth, climate change, and increasing water demand (Diop et al., 2020; Shivaprakash et al., 2022; Xiang et al., 2021).

1. Water Scarcity: Many regions face water scarcity due to limited freshwater availability and increasing demand from various sectors such as agriculture, industry, and domestic use.
2. Climate Change Impacts: Climate change is altering precipitation patterns, leading to increased frequency and intensity of droughts and floods. These changes pose significant challenges for managing water resources effectively.
3. Population Growth and Urbanization: Rapid population growth and urbanization put additional pressure on water resources, leading to increased demand for drinking water, sanitation, and industrial needs.
4. Inefficient Water Use: Inefficient water use practices, such as outdated irrigation methods and water losses from leakages in distribution systems, result in the wastage of valuable water resources.

5. Lack of Data and Monitoring: Inadequate data collection and monitoring systems make it challenging to assess water availability, usage patterns, and make informed decisions in water resource management.

Role of AI in Water Resource Management: Artificial intelligence (AI) technologies offer significant potential in addressing the challenges of water resource management. AI can enhance data analysis, prediction, and decision-making processes, leading to more efficient and sustainable water resource management practices. Here are some key roles of AI in water resource management:

1. Data Analysis and Modeling: AI techniques, such as machine learning and data analytics, can analyze large volumes of data from various sources, including remote sensing, weather forecasts, and sensor networks. These techniques can identify patterns, correlations, and trends in water availability, demand, and quality. AI models can also develop predictive models for water availability, streamflow, and groundwater levels, aiding in long-term planning and decision-making.
2. Water Demand Forecasting: AI algorithms can analyze historical data, population trends, and socioeconomic factors to forecast water demand accurately. This helps in optimizing water allocation, infrastructure planning, and developing strategies to meet future water needs.
3. Leakage Detection and Infrastructure Maintenance: AI-based systems can analyze sensor data and patterns of water consumption to detect leaks and anomalies in water distribution networks. This allows for early identification of leaks and proactive maintenance, reducing water losses and improving the efficiency of water supply systems.
4. Smart Irrigation and Agriculture: AI-driven irrigation systems can optimize water use in agriculture by considering factors such as soil moisture levels, weather conditions, and crop water requirements. These systems can automatically adjust irrigation schedules and water application rates, minimizing water wastage and improving crop yield.
5. Integrated Water Management: AI can facilitate the integration of various components in water resource management, including water supply, wastewater treatment, and stormwater management. AI-driven systems can optimize the coordination and operation of these systems, leading to improved efficiency, water reuse, and overall water management.
6. Decision Support Systems: AI can assist decision-making processes by providing real-time data, predictive models, and scenario analysis. Decision support systems powered by AI can consider multiple factors and trade-offs

to help stakeholders make informed decisions regarding water allocation, infrastructure investments, and policy development.

By leveraging AI technologies, water resource management can become more efficient, adaptive, and sustainable. AI-driven approaches enable better understanding of water systems, precise predictions, optimized resource allocation, and improved decision-making, contributing to the effective management of water resources in the face of growing challenges.

Figure 1. AI techniques for water resource assessment

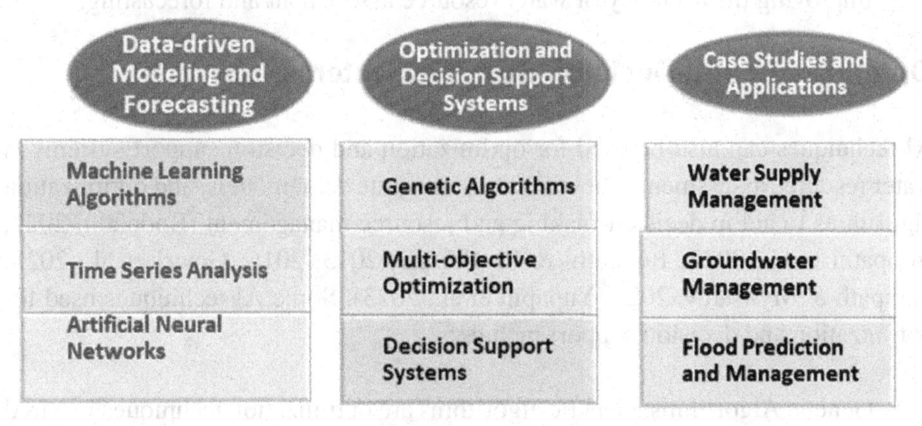

AI TECHNIQUES FOR WATER RESOURCE ASSESSMENT

Data-driven Modeling and Forecasting

AI techniques can be employed for data-driven modeling and forecasting in water resource assessment. By analyzing historical data and incorporating various environmental and socio-economic factors, AI models can provide insights into water availability, streamflow, and groundwater levels (Abba et al., 2020; Dharmaraj & Vijayanand, 2018; Zhao et al., 2020). AI Techniques for Water Resource Assessment is illustrated in Figure 1.

- Machine Learning Algorithms: Machine learning algorithms, such as regression models, support vector machines, random forests, and gradient boosting, can be trained on historical data to predict water resource variables. These models can capture complex relationships between input variables (e.g., precipitation, temperature, land use) and water resource parameters, enabling accurate forecasting.

- Time Series Analysis: Time series analysis techniques, including autoregressive integrated moving average (ARIMA), seasonal decomposition of time series (STL), and state space models, can analyze historical water data to identify patterns, trends, and seasonal variations. These techniques can provide short-term and long-term forecasts of water availability and demand.
- Artificial Neural Networks: Neural networks, such as feedforward neural networks and recurrent neural networks (e.g., LSTM), can handle complex nonlinear relationships in water resource data. These models can learn from historical data to make predictions and capture temporal dependencies, improving the accuracy of water resource assessment and forecasting.

Optimization and Decision Support Systems

AI techniques can also be used for optimization and decision support systems in water resource assessment. These systems integrate data, models, and optimization algorithms to aid in decision-making and resource management (Boopathi, 2022; Boopathi et al., 2022; Boopathi & Sivakumar, 2013, 2016; Gowri et al., 2023; Sampath & Myilsamy, 2021; Yupapin et al., 2023). Some AI techniques used for optimization and decision support include:

- Genetic Algorithms: Genetic algorithms are optimization techniques inspired by the principles of natural selection. These algorithms can be applied to optimize water allocation, reservoir operation, and infrastructure planning. By considering multiple constraints and objectives, genetic algorithms can identify optimal solutions for water resource management.
- Multi-objective Optimization: Multi-objective optimization approaches aim to find a set of optimal solutions that balance multiple conflicting objectives, such as maximizing water supply while minimizing environmental impact. Evolutionary algorithms, such as NSGA-II (Non-dominated Sorting Genetic Algorithm II) and SPEA2 (Strength Pareto Evolutionary Algorithm 2), can efficiently solve multi-objective optimization problems in water resource management.
- Decision Support Systems: AI-driven decision support systems integrate data, models, and visualization tools to assist stakeholders in making informed decisions. These systems provide real-time data, scenario analysis, and visualization of potential impacts, helping stakeholders evaluate alternative management strategies and select the most suitable options for water resource assessment and management.

By leveraging AI techniques for data-driven modeling, forecasting, optimization, and decision support, water resource assessment can be enhanced. AI-driven approaches enable more accurate predictions, optimize resource allocation, consider multiple objectives, and support decision-making processes, leading to effective and sustainable water resource management.

Case Studies and Applications

Case Study 1: Water Supply Management In a water-stressed region, AI techniques were employed to optimize water supply management. Historical data on water demand, supply, and weather conditions were collected and used to develop machine learning models. These models accurately predicted water demand based on various factors such as population, weather patterns, and economic indicators. By integrating these models into a decision support system, water authorities were able to optimize water allocation, prioritize supply to critical areas, and implement demand management strategies. This resulted in improved water availability, reduced water wastage, and better response to water scarcity events.

Case Study 2: Groundwater Management In a region heavily reliant on groundwater resources, AI techniques were used to assess and manage groundwater levels. Historical data on groundwater levels, rainfall, and pumping rates were analyzed using neural networks. The models developed were able to predict groundwater levels accurately, considering the effects of climate variability and pumping activities. These predictions helped in optimizing pumping rates, identifying areas prone to groundwater depletion, and implementing sustainable pumping strategies. This case study demonstrated the effectiveness of AI in supporting informed decision-making for sustainable groundwater management.

Case Study 3: Flood Prediction and Management AI techniques were utilized in a flood-prone area to improve flood prediction and management. Machine learning algorithms were trained on historical rainfall data, river levels, and flood occurrence records. These models were able to forecast flood events with high accuracy, providing early warning to authorities and communities. The predictions facilitated proactive measures such as timely evacuation, deployment of resources, and implementation of flood control strategies. By leveraging AI, the region was able to minimize flood damages, enhance public safety, and improve emergency response(Alam et al., 2022; Chau, 2006; Lai et al., 2008).

Benefits and Limitations of AI in Water Resource Management

1. Improved Decision-Making: AI techniques provide accurate and timely insights into water availability, demand, and usage patterns. This information enables

stakeholders to make data-driven decisions and implement effective water resource management strategies.

2. Enhanced Efficiency: AI-driven optimization algorithms help in optimizing water allocation, infrastructure planning, and operational decisions. By considering multiple objectives, constraints, and scenarios, AI can improve the efficiency of water resource management, leading to reduced water wastage and improved resource utilization.
3. Adaptive and Real-Time Management: AI models can adapt to changing conditions and provide real-time monitoring and forecasting. This allows for proactive management of water resources, including timely response to water scarcity or flood events.
4. Integrated Approach: AI techniques enable the integration of various components in water resource management, such as water supply, demand, and infrastructure. This integrated approach helps in addressing complex challenges and optimizing the overall water system.
5. Scalability and Replicability: AI models and algorithms can be scaled and replicated across different regions and contexts, allowing for the transfer of knowledge and best practices in water resource management.

Limitations of AI in Water Resource Management

1. Data Availability and Quality: AI techniques rely on data for training and analysis. Limited data availability, particularly in certain regions or for specific variables, can pose challenges in implementing AI-driven approaches. Additionally, data quality issues, such as inaccuracies or missing values, can impact the performance of AI models.
2. Interpretability: Some AI models, such as neural networks, are often considered black boxes, making it difficult to interpret the underlying reasoning behind their predictions. This lack of interpretability can hinder stakeholders' understanding and trust in the AI-driven systems.
3. Implementation Challenges: Implementing AI in water resource management requires expertise in data collection, model development, and integration with existing systems. It may also require investment in computational resources and infrastructure.
4. Uncertainty and Future Projections: AI techniques can provide accurate predictions based on historical data, but uncertainties in future climate conditions and socio-economic factors can affect the reliability of long-term projections.
5. Ethical Considerations: The use of AI in water resource management raises ethical considerations, such as data privacy, equity in decision-making, and

Figure 2. Integration of AI in water treatment systems

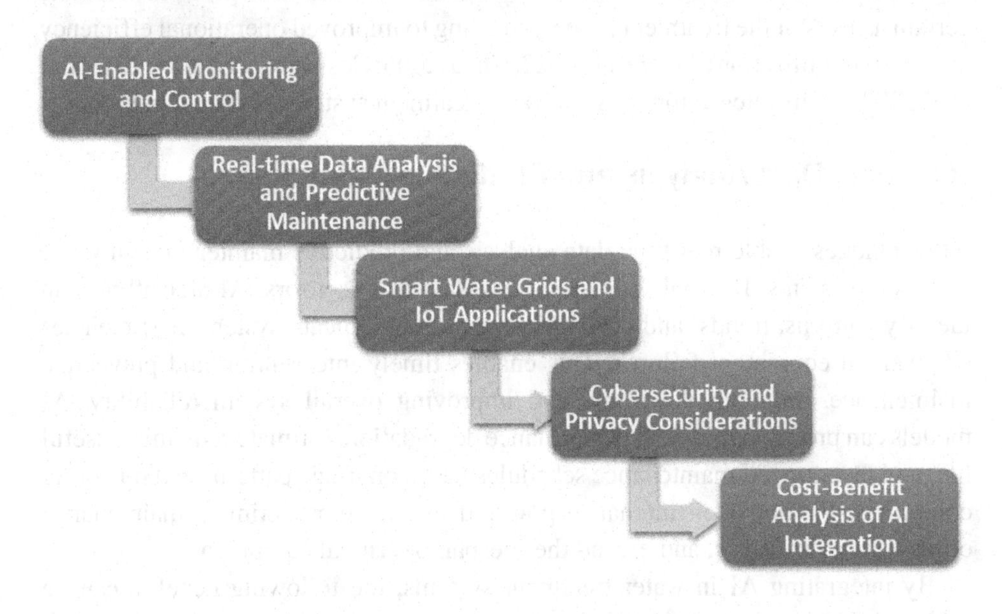

potential biases in the algorithms. These ethical concerns need to be addressed to ensure fair and equitable water resource management practices.

While AI offers significant benefits in water resource management, it is important to consider these limitations and address them through appropriate data governance, transparency, and stakeholder engagement. By leveraging the strengths of AI and addressing its limitations, water resource management can benefit from improved efficiency, resilience, and sustainability.

INTEGRATION OF AI IN WATER TREATMENT SYSTEMS

AI-Enabled Monitoring and Control

AI can be integrated into water treatment systems to enable advanced monitoring and control capabilities. By leveraging real-time data from sensors and other monitoring devices, AI algorithms can continuously monitor water quality parameters, process variables, and system performance. This allows for early detection of anomalies, deviations from optimal conditions, and potential equipment failures. AI-enabled monitoring systems can provide alerts and notifications to operators, enabling proactive actions to maintain water quality and system efficiency. Moreover, AI algorithms can learn from historical data to optimize control strategies and automate certain aspects of the treatment process, leading to improved operational efficiency and resource utilization (Alam et al., 2022; Chau, 2006; Cosgrove & Loucks, 2015; Lai et al., 2008). The integration of AI in water treatment systems is shown in Figure 2.

Real-time Data Analysis and Predictive Maintenance

AI techniques enable real-time data analysis and predictive maintenance in water treatment systems. By analyzing streaming data from sensors, AI algorithms can identify patterns, trends, and correlations that may indicate system inefficiencies or potential equipment failures. This enables timely intervention and preventive maintenance, reducing downtime and improving overall system reliability. AI models can predict equipment performance degradation, estimate remaining useful life, and recommend maintenance schedules based on usage patterns and operating conditions. Predictive maintenance powered by AI helps optimize maintenance efforts, minimize costs, and extend the lifespan of critical equipment.

By integrating AI in water treatment systems, the following benefits can be achieved:

- Enhanced Operational Efficiency: AI-enabled monitoring and control systems optimize process parameters in real-time, leading to improved treatment efficiency, reduced energy consumption, and optimized chemical dosing. This results in cost savings and improved overall system performance.
- Improved Water Quality: AI algorithms can continuously monitor water quality parameters and make real-time adjustments to treatment processes. This ensures that water meets the required quality standards and reduces the risk of waterborne diseases.
- Early Detection of Anomalies: AI algorithms can identify anomalies and deviations from normal operating conditions, allowing operators to take corrective actions promptly. This prevents process upsets, minimizes production losses, and ensures consistent water quality.
- Cost Savings: AI-driven predictive maintenance helps avoid unexpected equipment failures and costly repairs. By scheduling maintenance activities based on actual equipment condition, resources can be allocated efficiently, reducing downtime and repair costs.
- Data-Driven Decision Making: AI techniques provide data-driven insights and recommendations to operators and managers. Real-time data analysis, predictive modeling, and scenario analysis support informed decision-making, enabling proactive responses to changing operating conditions and optimizing system performance.

However, it is important to consider the limitations of AI integration in water treatment systems, such as initial setup costs, data availability, system complexity, and the need for continuous monitoring and maintenance of AI algorithms. These challenges can be addressed through proper planning, data management, and training of personnel to ensure successful integration and sustainable operation of AI-driven water treatment systems.

Smart Water Grids and IoT Applications

The integration of AI in water treatment systems can be further enhanced through the deployment of smart water grids and Internet of Things (IoT) applications. Smart water grids combine advanced sensors, communication technologies, and data analytics to enable real-time monitoring and management of water distribution networks. AI algorithms can analyze data collected from IoT devices, such as smart meters and pressure sensors, to optimize water distribution, detect leaks, and reduce non-revenue water.

IoT applications in water treatment systems can provide valuable data for AI-driven decision-making. For example, IoT devices can collect data on water quality, flow rates, energy consumption, and equipment performance. This data can be analyzed by AI algorithms to optimize treatment processes, improve energy efficiency, and enable predictive maintenance.

Cybersecurity and Privacy Considerations

As AI and IoT technologies are integrated into water treatment systems, it is crucial to address cybersecurity and privacy considerations. The interconnectedness of devices and systems in smart water grids increases the vulnerability to cyber threats and unauthorized access. Measures should be taken to secure data transmission, protect against cyberattacks, and ensure the integrity and confidentiality of sensitive information.

Additionally, privacy concerns arise from the collection and use of personal data in IoT applications. Proper data governance practices, including anonymization and data encryption, should be implemented to safeguard individual privacy and comply with relevant regulations.

Water utilities and system operators need to invest in robust cybersecurity measures, conduct regular risk assessments, and implement appropriate security protocols to protect AI-driven water treatment systems and IoT devices from potential threats.

By addressing cybersecurity and privacy considerations, water utilities can confidently leverage AI, IoT, and smart water grid technologies to enhance water treatment processes, improve operational efficiency, and provide sustainable and reliable water services to communities.

Cost-Benefit Analysis of AI Integration

Integrating AI into water treatment systems involves upfront costs and ongoing expenses. However, it is essential to conduct a cost-benefit analysis to evaluate the long-term advantages and return on investment (Boopathi et al., 2022; S. et al., 2022). Here are some key factors to consider in the cost-benefit analysis of AI integration:

- Implementation Costs: This includes the costs associated with acquiring and installing AI infrastructure, such as sensors, data storage and processing systems, and software development. It also includes the expenses for training personnel, integrating AI algorithms into existing systems, and conducting pilot studies.
- Operational Costs: Ongoing operational costs include maintenance, upgrades, and personnel training to ensure the smooth functioning of AI-driven systems. Additionally, costs associated with data storage, processing, and energy consumption should be considered.
- Efficiency Gains: AI integration can lead to improved operational efficiency, reduced energy consumption, and optimized resource utilization. These efficiency gains can result in cost savings, such as reduced chemical usage, energy bills, and maintenance costs.
- Improved Decision-Making: AI-driven systems provide real-time data analysis, predictive insights, and decision support. Enhanced decision-making can lead to better resource allocation, optimized treatment processes, and reduced operational errors, which contribute to cost savings and improved overall performance.
- Increased System Reliability: AI-enabled monitoring, control, and predictive maintenance can minimize equipment failures, prevent process upsets, and reduce downtime. This results in enhanced system reliability, reduced emergency repairs, and improved service continuity.
- Enhanced Water Quality and Customer Satisfaction: AI integration can help maintain consistent water quality, detect anomalies, and respond quickly to quality issues. Improved water quality leads to increased customer satisfaction, reduced customer complaints, and potential savings on remediation efforts.
- Risk Mitigation: AI-driven systems can help in early detection of potential risks, such as equipment failures, water contamination, and leaks. Timely intervention and preventive measures can minimize risks, avoid costly damages, and reduce the impact on operations.

- Regulatory Compliance: AI integration can facilitate compliance with regulatory requirements and water quality standards. Avoiding penalties, fines, and legal actions associated with non-compliance can contribute to cost savings.

It is important to quantify both the tangible and intangible benefits of AI integration and compare them against the implementation and operational costs. Conducting a thorough cost-benefit analysis will provide insights into the financial viability and potential returns of integrating AI into water treatment systems. Moreover, considering the long-term benefits and sustainability aspects is crucial for making informed decisions regarding AI integration.

CHALLENGES AND FUTURE DIRECTIONS

Data Availability and Quality

One of the primary challenges in implementing AI in water treatment and resource management is the availability and quality of data. AI algorithms rely on large and diverse datasets for training and analysis. However, water-related data, such as historical water quality records, real-time monitoring data, and operational data, may be limited in some regions or fragmented across different organizations. Furthermore, data quality issues, such as missing values, data gaps, and measurement errors, can affect the performance and reliability of AI models. Addressing these challenges requires efforts in data collection, standardization, integration, and validation. Collaboration among stakeholders, including water utilities, regulatory agencies, and research institutions, is crucial to ensure the availability of high-quality data for AI-driven applications (Cosgrove & Loucks, 2015; Gill et al., 2022; Hallaji et al., 2022; Khor et al., 2014).

Explainability and Transparency of AI Models

Another challenge is the explainability and transparency of AI models used in water treatment and resource management. Many AI techniques, such as neural networks and deep learning models, are often considered black boxes, making it difficult to understand the underlying decision-making process. Explainability is important for stakeholders to trust and accept AI-driven systems, particularly in critical domains like water management. Researchers are working on developing interpretable AI models and techniques that provide insights into how decisions are made. Explainable AI can help in understanding the reasoning behind AI predictions and recommendations, enabling informed decision-making and fostering public trust.

Ethical Considerations and Societal Implications

The integration of AI in water treatment and resource management raises ethical considerations and societal implications. For instance, the collection and use of personal data in IoT applications and AI-driven systems need to comply with privacy regulations and ensure data protection. Moreover, AI algorithms should be designed and implemented with fairness and equity in mind, considering potential biases in data or models that may disproportionately affect certain communities or groups. It is essential to establish ethical guidelines and frameworks to address these concerns and ensure that AI is deployed in a responsible and inclusive manner.

Interdisciplinary Collaboration and Knowledge Sharing

Addressing the complex challenges in water treatment and resource management requires interdisciplinary collaboration and knowledge sharing. Bringing together experts from diverse fields, such as water engineering, data science, environmental science, and social sciences, can foster innovation and holistic approaches to tackle the multifaceted issues in water management. Collaborative efforts can include sharing data, tools, and best practices, as well as fostering dialogue among stakeholders to exchange knowledge and experiences. Such collaborations can drive advancements in AI techniques, develop context-specific solutions, and promote sustainable water management practices.

Emerging Technologies and Innovations

The future of AI in water treatment and resource management lies in the exploration of emerging technologies and innovations. Researchers are continuously developing new AI techniques, such as reinforcement learning and hybrid models, to address the unique challenges in water management. Additionally, the integration of AI with other emerging technologies, including blockchain, Internet of Things (IoT), and remote sensing, holds great potential for improving data collection, enhancing system monitoring, and optimizing resource allocation. These innovations can further enhance the capabilities of AI in addressing water-related challenges and achieving sustainable water management goals.

Policy and Regulation Frameworks

Policy and regulation frameworks play a crucial role in shaping the adoption and implementation of AI in water treatment and resource management. Governments and regulatory bodies need to establish guidelines, standards, and frameworks that promote responsible AI use, protect data privacy, and ensure ethical considerations are addressed. These frameworks should also encourage data sharing, collaboration, and transparency among stakeholders. Additionally, policies that incentivize the integration of AI in water management and provide support for research and development can foster innovation and accelerate the adoption of AI-driven solutions.

AI has the potential to revolutionize water treatment and resource management, but challenges must be addressed. These include data availability, quality, transparency, ethical considerations, interdisciplinary collaboration, and knowledge sharing. Staying updated on emerging technologies and establishing supportive policy frameworks are crucial for shaping AI's future in the water sector. By addressing these challenges and leveraging AI opportunities, we can achieve more efficient, sustainable, and resilient water management practices for communities and the environment.

CONCLUSION

In conclusion, the integration of artificial intelligence (AI) in water treatment and resource management has the potential to revolutionize the way we address challenges in the water sector. This chapter has explored various aspects of AI applications in water treatment, evaporation management, water resource assessment, and overall water resource management.

- AI can enhance evaporation management by optimizing evaporation processes, predicting evaporation rates, and improving water resource utilization.
- AI techniques such as machine learning, neural networks, and deep learning models can be effectively used for evaporation control and optimization.
- Water resource management can benefit from AI through data-driven modeling, optimization, and decision support systems.
- AI-driven systems enable real-time data analysis, predictive insights, and proactive decision-making in water treatment and resource management.
- Challenges such as data availability and quality, explainability of AI models, ethical considerations, and interdisciplinary collaboration need to be addressed for successful implementation of AI in the water sector.

Implications and Potential Impact of AI in Water Treatment

The implications and potential impact of AI in water treatment are significant. By leveraging AI technologies, water treatment processes can be optimized, leading to improved operational efficiency, reduced energy consumption, and better water quality. Real-time monitoring and control systems driven by AI can enhance system performance, detect anomalies, and enable proactive responses. Predictive maintenance can prevent equipment failures, minimize downtime, and extend the lifespan of critical assets. Overall, AI has the potential to transform water treatment systems into more efficient, reliable, and sustainable infrastructure.

Recommendations for Future Research

To further advance the field of AI in water treatment and resource management, the following recommendations for future research can be made:

- Develop robust AI models and algorithms that provide explainability and transparency. This will enable stakeholders to understand the decision-making process and foster trust in AI-driven systems.
- Address the challenges related to data availability, quality, and interoperability. Efforts should be made to standardize data formats, promote data sharing, and develop data quality assurance frameworks.
- Explore interdisciplinary collaborations and knowledge sharing platforms to foster innovation and holistic approaches to water management challenges.
- Conduct comprehensive assessments of the ethical considerations and societal implications of AI in water treatment. Develop guidelines and frameworks that ensure fairness, equity, and privacy in AI-driven water management systems.
- Investigate the potential of emerging technologies and innovations, such as blockchain, IoT, and remote sensing, in conjunction with AI, to further enhance water treatment and resource management practices.
- Establish supportive policy and regulation frameworks that encourage responsible AI adoption in the water sector and provide incentives for research and development.

By addressing these research recommendations, we can advance the field of AI in water treatment and resource management, leading to more efficient, sustainable, and resilient water management practices.

AI integration in water treatment and resource management has potential to improve operational efficiency, optimize resource utilization, enhance water quality, and support decision-making processes. However, ethical implications and challenges must be addressed to ensure responsible implementation. Collaborative efforts and research can unlock AI's full potential, transforming water resource management and conserving resources for a sustainable future.

REFERENCES

Abba, S. I., Pham, Q. B., Usman, A. G., Linh, N. T. T., Aliyu, D. S., Nguyen, Q., & Bach, Q. V. (2020). Emerging evolutionary algorithm integrated with kernel principal component analysis for modeling the performance of a water treatment plant. *Journal of Water Process Engineering*, *33*, 101081. doi:10.1016/j.jwpe.2019.101081

Alam, G., Ihsanullah, I., Naushad, M., & Sillanpää, M. (2022). Applications of artificial intelligence in water treatment for optimization and automation of adsorption processes: Recent advances and prospects. *Chemical Engineering Journal*, *427*, 130011. doi:10.1016/j.cej.2021.130011

Anitha, C., R, K. C., Vivekanand, C. V., Lalitha, S. D., Boopathi, S., & R, R. (2023, February). Artificial Intelligence driven security model for Internet of Medical Things ({IoMT}). *IEEE Explore*. doi:10.1109/ICIPTM57143.2023.10117713

Boopathi, S. (2022). Experimental investigation and multi-objective optimization of cryogenic Friction-stir-welding of AA2014 and AZ31B alloys using MOORA technique. *Materials Today. Communications*, *33*, 104937. doi:10.1016/j.mtcomm.2022.104937

Boopathi, S., & Sivakumar, K. (2013). Experimental investigation and parameter optimization of near-dry wire-cut electrical discharge machining using multi-objective evolutionary algorithm. *International Journal of Advanced Manufacturing Technology*, *67*(9–12), 2639–2655. doi:10.100700170-012-4680-4

Boopathi, S., & Sivakumar, K. (2016). Optimal parameter prediction of oxygen-mist near-dry Wire-cut EDM. *International Journal of Manufacturing Technology and Management*, *30*(3–4), 164–178. doi:10.1504/IJMTM.2016.077812

Boopathi, S., Sureskumar, M., Jeyakumar, M., Sanjeev Kumar, R., & Subbiah, R. (2022). Influences of Fabrication Parameters on Natural Fiber Reinforced Polymer Composite (NFRPC) Material: A Review. *Materials Science Forum*, *1075*, 115–124. doi:10.4028/p-095f0t

Chau, K. (2006). A review on integration of artificial intelligence into water quality modelling. *Marine Pollution Bulletin*, *52*(7), 726–733. doi:10.1016/j.marpolbul.2006.04.003 PMID:16764895

Cosgrove, W. J., & Loucks, D. P. (2015). Water management: Current and future challenges and research directions. *Water Resources Research*, *51*(6), 4823–4839. doi:10.1002/2014WR016869

Dharmaraj, V., & Vijayanand, C. (2018). Artificial Intelligence (AI) in Agriculture. *International Journal of Current Microbiology and Applied Sciences*, *7*(12), 2122–2128. doi:10.20546/ijcmas.2018.712.241

Dind, P., & Schmid, H. (1978). Application of solar evaporation to waste water treatment in galvanoplasty. *Solar Energy*, *20*(3), 205–211. doi:10.1016/0038-092X(78)90098-1

Diop, L., Samadianfard, S., Bodian, A., Yaseen, Z. M., Ghorbani, M. A., & Salimi, H. (2020). Annual Rainfall Forecasting Using Hybrid Artificial Intelligence Model: Integration of Multilayer Perceptron with Whale Optimization Algorithm. *Water Resources Management*, *34*(2), 733–746. doi:10.100711269-019-02473-8

Fan, M., Hu, J., Cao, R., Ruan, W., & Wei, X. (2018). A review on experimental design for pollutants removal in water treatment with the aid of artificial intelligence. *Chemosphere*, *200*, 330–343. doi:10.1016/j.chemosphere.2018.02.111 PMID:29494914

Gill, S. S., Xu, M., Ottaviani, C., Patros, P., Bahsoon, R., Shaghaghi, A., Golec, M., Stankovski, V., Wu, H., Abraham, A., Singh, M., Mehta, H., Ghosh, S. K., Baker, T., Parlikad, A. K., Lutfiyya, H., Kanhere, S. S., Sakellariou, R., Dustdar, S., ... Uhlig, S. (2022). AI for next generation computing: Emerging trends and future directions. *Internet of Things (Netherlands)*, *19*, 100514. doi:10.1016/j.iot.2022.100514

Gowri, N. V., Dwivedi, J. N., Krishnaveni, K., Boopathi, S., Palaniappan, M., & Medikondu, N. R. (2023). Experimental investigation and multi-objective optimization of eco-friendly near-dry electrical discharge machining of shape memory alloy using Cu/SiC/Gr composite electrode. *Environmental Science and Pollution Research International*, 1–19. doi:10.100711356-023-26983-6 PMID:37126160

Hallaji, S. M., Fang, Y., & Winfrey, B. K. (2022). Predictive maintenance of pumps in civil infrastructure: State-of-the-art, challenges and future directions. *Automation in Construction*, *134*, 104049. doi:10.1016/j.autcon.2021.104049

Heins, W., & Peterson, D. (2018). Use of evaporation for heavy oil produced water treatment. *Canadian International Petroleum Conference 2003, CIPC 2003, 44*(1). 10.2118/2003-178

Jeevanantham, Y. A., A, S., V, V., J, S. I., Boopathi, S., & Kumar, D. P. (2023). Implementation of Internet-of Things (IoT) in Soil Irrigation System. *IEEE Explore*, 1–5. doi:10.1109/ICPECTS56089.2022.10047185

Khor, C. S., Chachuat, B., & Shah, N. (2014). Optimization of water network synthesis for single-site and continuous processes: Milestones, challenges, and future directions. *Industrial & Engineering Chemistry Research, 53*(25), 10257–10275. doi:10.1021/ie4039482

Lafta, A. M., & Amori, K. E. (2022). Hydrogel materials as absorber for improving water evaporation with solar still, desalination and wastewater treatment. *Materials Today: Proceedings, 60*, 1548–1553. doi:10.1016/j.matpr.2021.12.061

Lai, E., Lundie, S., & Ashbolt, N. J. (2008). Review of multi-criteria decision aid for integrated sustainability assessment of urban water systems. *Urban Water Journal, 5*(4), 315–327. doi:10.1080/15730620802041038

Li, L., Rong, S., Wang, R., & Yu, S. (2021). Recent advances in artificial intelligence and machine learning for nonlinear relationship analysis and process control in drinking water treatment: A review. *Chemical Engineering Journal, 405*, 126673. doi:10.1016/j.cej.2020.126673

Nourani, V., Elkiran, G., & Abba, S. I. (2018). Wastewater treatment plant performance analysis using artificial intelligence - An ensemble approach. *Water Science and Technology, 78*(10), 2064–2076. doi:10.2166/wst.2018.477 PMID:30629534

S., P. K., Sampath, B., R., S. K., Babu, B. H., & N., A. (2022). Hydroponics, Aeroponics, and Aquaponics Technologies in Modern Agricultural Cultivation. In *Trends, Paradigms, and Advances in Mechatronics Engineering* (pp. 223–241). IGI Global. doi:10.4018/978-1-6684-5887-7.ch012

Sampath, B., & Myilsamy, S. (2021). Experimental investigation of a cryogenically cooled oxygen-mist near-dry wire-cut electrical discharge machining process. *Strojniski Vestnik. Jixie Gongcheng Xuebao, 67*(6), 322–330. doi:10.5545v-jme.2021.7161

Sharshir, S. W., Algazzar, A. M., Elmaadawy, K. A., Kandeal, A. W., Elkadeem, M. R., Arunkumar, T., Zang, J., & Yang, N. (2020). New hydrogel materials for improving solar water evaporation, desalination and wastewater treatment: A review. *Desalination, 491*, 114564. doi:10.1016/j.desal.2020.114564

Shivaprakash, K. N., Swami, N., Mysorekar, S., Arora, R., Gangadharan, A., Vohra, K., Jadeyegowda, M., & Kiesecker, J. M. (2022). Potential for Artificial Intelligence (AI) and Machine Learning (ML) Applications in Biodiversity Conservation, Managing Forests, and Related Services in India. *Sustainability (Basel)*, *14*(12), 7154. doi:10.3390u14127154

Vanitha, S. K. R., & Boopathi, S. (2023). Artificial Intelligence Techniques in Water Purification and Utilization. In *Human Agro-Energy Optimization for Business and Industry* (pp. 202–218). IGI Global. doi:10.4018/978-1-6684-4118-3.ch010

Xiang, X., Li, Q., Khan, S., & Khalaf, O. I. (2021). Urban water resource management for sustainable environment planning using artificial intelligence techniques. *Environmental Impact Assessment Review*, *86*, 106515. doi:10.1016/j.eiar.2020.106515

Yuan, B., Zhang, C., Liang, Y., Yang, L., Yang, H., Bai, L., Wei, D., Wang, W., Wang, Q., & Chen, H. (2021). A Low-Cost 3D Spherical Evaporator with Unique Surface Topology and Inner Structure for Solar Water Evaporation-Assisted Dye Wastewater Treatment. *Advanced Sustainable Systems*, *5*(3), 2000245. doi:10.1002/adsu.202000245

Yupapin, P., Trabelsi, Y., Nattappan, A., & Boopathi, S. (2023). Performance Improvement of Wire-Cut Electrical Discharge Machining Process Using Cryogenically Treated Super-Conductive State of Monel-K500 Alloy. *Iranian Journal of Science and Technology. Transaction of Mechanical Engineering*, *47*(1), 267–283. doi:10.100740997-022-00513-0

Zhao, L., Dai, T., Qiao, Z., Sun, P., Hao, J., & Yang, Y. (2020). Application of artificial intelligence to wastewater treatment: A bibliometric analysis and systematic review of technology, economy, management, and wastewater reuse. *Process Safety and Environmental Protection*, *133*, 169–182. doi:10.1016/j.psep.2019.11.014

Chapter 5
The Use and Awareness of ICT to Facilitate the Adoption of Artificial Intelligence in Agriculture

Mushtaq Ahmad Shah
Lovely Professional University, India

Mihir Aggarwal
Lovely Professional University, India

ABSTRACT

Agricultural practices are changing drastically with the incorporation of artificial intelligence (AI). These innovations have the potential to cause a sea change in farming by increasing production and solving issues of sustainability. More than 58% of rural households in India rely on agriculture as their main source of income, which contributes 18% to India's GDP. India's agricultural output is much lower compared to that of China, Brazil, and the United States. There is growing empirical evidence that adopting the most cutting-edge technologies, such as AI, improves farmers' economic situations and production. This chapter explores the use and awareness of ICT in facilitating the adoption of AI in agriculture and the barriers to accessing information sources. A questionnaire survey was administered to farmers to understand their experiences and perspectives on the use of ICT in agriculture and to facilitate the adoption of AI in agriculture.

DOI: 10.4018/978-1-6684-6791-6.ch005

1. INTRODUCTION

Artificial intelligence (AI) is revolutionising various industries, and agriculture is no exception. With the world's population projected to reach 9.7 billion by 2050, the demand for food is expected to increase by at least 70%, which means that agriculture must become more efficient, productive, and sustainable. In India, agriculture accounts for approximately 18% of GDP and employs 58% of the total labour force. Agriculture is an important sector in most developing countries, with the majority of the rural population relying on it (Stienen, Bruinsma, and Neuman 2007). Green et al. (2005) concluded that agriculture has been shown to significantly contribute to economic growth, food security, poverty alleviation, livelihoods, food security, and ecological sustainability. India is the largest producer of pulses, rice, wheat, and spices in the world. However, in terms of marketing facilities and technological know-how, Indian agriculture continues to face challenges such as a lack of business sector harmonisation and integration, as well as farmers' provision of trustworthy and accessible relevant information, cold storage facilities, and technical know-how (Shah, 2015).

The adoption of AI in agriculture requires the use of information and communications technology (ICT). ICT provides a framework for integrating various technologies and systems, making it possible to develop and deploy AI solutions that can help farmers optimize production processes, reduce costs, and increase yields. As such, the adoption of AI in agriculture requires continued investment in ICT infrastructure, connectivity, and automation systems. In recent years, the farming sector has become increasingly information-dependent, requiring a wide range of technical and scientific data for the farming community to make informed decisions (Cash, 2001). However, large-scale knowledge asymmetry exists in practically all levels of the agricultural supply chain in underdeveloped countries, which leads to victimization of the farming community and redundancies across the chain (Eggleston, Jensen & Zeckhauser, 2001; Ravallion, 1986). This is one of the most serious issues confronting farmers in emerging nations, including India (Gollakota, 2008). Information and awareness are critical variables in speeding agricultural growth by implementing suitable production planning, adopting improved farming techniques, and implementing efficient post-harvest management and marketing (Bertolini, 2004; Kizilaslan, 2006). India's agriculture systems are immensely complex and dispersed, the demand for information and expertise varies greatly throughout various stages of the supply chain (Adhiguru & Mruthyunjaya, 2004; Rao, 2006). Other aspects, such as land, labour, money, and management skill, can be enhanced by relevant, trustworthy, and usable information provided by extension services, research institutes, and other agricultural organisations to assist farmers in making better decisions. Farmers become more aware of agricultural technology adoption as a result of agricultural information. As a result, it is critical for every country, particularly emerging countries, to have a solid agricultural information system that supports agricultural growth.

In the present era, effective agricultural growth necessitates access to and utilisation of ICT tools that underpin modern information systems. ICTs simply make it easier to create, manage, store, retrieve, and disseminate any relevant data, knowledge, and information that has already been processed and altered (Bachelor 2002; Chapman and Slay maker 2002; Rao 2007; Heeks 2002). ICT in agriculture is a new field in India that focuses on improving agriculture and rural development. ICT can give farmers with reliable information, resulting in higher agricultural output.

Based on the foregoing discussion on ICT's significance in agriculture, no one can argue that farmers in India and other developing nations should not benefit from current ICT tools and applications. Various government programmes for agricultural development are being implemented, including public and private sector participation. However, in India, where ICT is still in its infancy and growing as a trend, the benefits of ICT have yet to reach all farmers. To facilitate the integration of AI technology into agriculture, it requires awareness and adoption of ICT (information and communication technology) among farmers. Many farmers may not obtain enough information and services as a result of poor economic conditions and societal constraints. This paper examines the adoption and use of ICT tools by farmers in India, focusing on the state of Punjab. The paper is particularly interested in identifying barriers to farmers' effective use of ICT in farming in the study region.

2. REVIEW OF LITERATURE

The potential benefits of AI in agriculture are enormous, and it is essential to continue investing in this technology to maximise its potential for the benefit of farmers and consumers alike. The present section provides an overview of a few studies relevant to the potential use of AI in agriculture. Bhatt & Muduli (2022) explores the potential of AI technology in agriculture by analysing existing literature on AI-based applications in agriculture. The authors discuss the importance of crop monitoring, yield prediction, disease detection, and weed control in modern agriculture and highlight the advantages of using AI in these areas. They also discuss the challenges in adopting AI-based solutions, including the high cost of technology and the need for specialized knowledge to use it effectively. Bhat & Huang (2021) provides an in-depth analysis of the potential benefits and challenges of AI in agriculture. The authors discuss the advantages of using AI in crop management, livestock monitoring, and soil analysis, and highlight the need for better data management and analysis to support these applications. They also discuss the social implications of AI adoption, including the impact on employment and the need for training and education to ensure that farmers can effectively use AI technology.

Saranya, Sridevi & Anbananthen (2023) focuses on the potential of AI to improve crop management. The authors discuss the importance of precision agriculture, soil fertility analysis, and crop disease detection, and highlight the advantages of using AI in these areas. They also discuss the challenges in adopting AI-based solutions, including the need for better data management and the high cost of technology. Singh & Singh (2020) also provides a comprehensive overview of the applications of AI in different stages of agriculture, such as pre-planting, planting, and post-harvest. The authors discuss the importance of using AI to optimize crop management, soil analysis, and water management, and highlight the advantages of using AI in these areas. They also discuss the challenges in adopting AI-based solutions, including the need for better data management and the high cost of technology. Kakani et al. (2020) provides a detailed analysis of the various AI techniques used in agriculture, such as computer vision, machine learning, and deep learning, and their applications in crop management, soil analysis, and livestock monitoring. The authors discuss the importance of using AI to optimize crop yields, reduce waste, and enhance sustainability, and highlight the advantages of using AI in these areas. They also discuss the challenges in adopting AI-based solutions, including the need for better data management and the high cost of technology

Overall, these reviews highlight the potential of AI and information and technology to significantly improve agricultural productivity, reduce waste, and enhance sustainability. They also discuss the challenges in adopting AI-based IT solutions and the need for better data management and specialized knowledge to use AI technology effectively. From the literature review, it is clear most of the studies regarding the role of AI in agriculture are done in advanced economies. However, from the Indian perspective, farmers are not well versed in technology. To facilitate the integration of AI technology into agriculture, it requires awareness and adoption of ICT among farmers.

3. RESEARCH METHODOLOGY

The Indian economy has been predominantly agricultural. Adopting innovative agricultural methods, such as artificial intelligence and information and communication technology, is critical for long-term growth and increased output. ICT tools such as mobile phones, the internet, and other communication devices are significant mediums for obtaining and delivering information. Information technologies are a key term in the agriculture sector, with access to market prices, weather updates, technical specifications, improved agricultural technologies, finance and insurance, and on-demand information. Recognising the significance of ICT in agriculture and the impediments to its adoption, the study's major aims are as follows:

To understand the demographic characteristics of farmers and determine the use and accessibility of key ICT tools by farmers to facilitate the adoption of AI in agriculture practises. To achieve the study objectives, both primary and secondary data were collected and analysed. Primary data was obtained from 186 farmers from several villages in Punjab using a scheduled schedule and face-to-face interaction. Data was collected using simple random and purposive sampling approaches. Secondary data was gathered from journals, periodicals, newspapers, and books, among other sources. Both descriptive and inferential statistical approaches were used to analyse the data. For data analysis, statistical approaches such as frequency and percentage analysis were applied. The quantitative data was analysed using a Microsoft Excel Spread Sheet, and the qualitative data was analysed using a content analysis process.

4. RESULTS AND DISCUSSION

Demographic features of the farmer community revealed that 14 percent and 19 percent of respondents were between the ages of 25 - 35, and 36 - 45, respectively (Table 1). Only 6% of respondents were under the age of 25, with a maximum of 62 percent being beyond the age of 45. Scholars believe that age has a significant impact in the dissemination, acceptance, and digital revolution, which are seen to be positively connected with age. Younger farmers are acknowledged to be less reluctant to change than older farmers, and they rapidly embrace and adapt innovations and new techniques, resulting in a faster diffusion rate (Crusan et al. 1982 and Habib et al 2007). According to the data gathered, 10% of respondents were illiterate, while 20% and 45% had primary and middle school education, respectively. Only 23% of respondents had a secondary school education, and only 2 percent had a higher-level degree. The study area's total literacy rate of 90% was highly encouraging. When opposed to illiterate individuals, educated people are likely to have more favourable attitudes regarding agricultural capabilities, information, and knowledge (Hassan, 1991; Habib et al., 2007).

According to the calculated statistics, 73 percent of respondents were landowners working on their farms. Similarly, tenants made up 20% of the total, while owner cum tenants made up 7%. The size of land status indicated that the majority of the farmers (186 respondents) held marginal land. Only 110 (59%) had 1-5 Acres of land, while 22 percent had 05-10 Acres. The majority of respondents, 76 percent, were solely involved in agricultural operations, while the remaining 24 were involved in other occupations such as tailoring, street vending businesses, and daily-wage

Table 1. Demographic data of the farmers

Age	Number	Percent	Education	Number	Percent
Below 25	11	6%	illiterate	19	10%
25-35	26	14%	Primary Education	37	20%
36-45	35	19%	Middle Education	84	45%
Above 45	115	62%	Secondary Education	43	23%
Total	186	100%	Above Secondary Education	4	2%
Tenancy status	**Number**	**Percent**	Total	186	100
Owner	136	73%	**Size of land Holdings Acres**	**Number**	**Percent**
Tenant	37	20%	01-05 Acres	110	59%
Owner cum Tenant	13	7%	05-10 Acres	41	22%
Total	186	100%	11-15 Acres	22	12%
Other Occupation	**Number**	**Percent**	Above 15	13	7%
No	45	24%		186	100
Yes	141	76%	**Average Monthly Income**	**Number**	**Percent**
Total	186	100	1000-5000	97	52%
Farming Experience	**Number**	**Percent**	6000-10000	50	27%
01-05 Year	7	4%	11000-15000	30	16%
06-10 Year	11	6%	Above 15000	9	5%
11-15 year	30	16%	Total	186	100
Above 15	138	74%			
Total	186	100			

Source: primary data collection

jobs. More land means more opportunity to boost production and efficiency by implementing current technology. The extent of landholding is significant in the spread and acceptance of contemporary agricultural methods among farmers. Income is directly or indirectly related to landholding size and also has a direct link with the use of contemporary technology. According to research findings, the majority of respondents had 11 to more than 15 years of agricultural experience, with 33 percent having 11–15 years and 29.5 percent having more than 15 years (Table 1).

According to findings, 52 percent of participants in the sample survey receive average monthly revenues ranging from 1000 to 5000 thousand per month from agricultural produce. 27 percent of respondents earn between 6,000 to 10,000 per month, while 16 percent make between 11,000 to 15,000. Another key element influencing respondents' attitudes about the acquisition of enhanced agricultural technology and its use for higher agricultural productivity is their income.

5. AGRICULTURAL INFORMATION SOURCES FOR FARMERS

Farmers' information sources differ depending on the type of activity and services they do. Information sources are tools or communication carriers that assist extension professionals in meeting their information needs. Several studies have been undertaken to evaluate the sources of information used by extension workers. According to Alfred and Odefadehan (2007) extension workers' information sources include organisations, interpersonal associates, regional, national, and worldwide conferences, training, physical and digital communication, telecommunications, and internet service. Koyenikan (2011) bifurcates the above source of information accessed by farmers formal and informal, national radio stations, local and international print media (such as newspapers, bulletins, and journals), and conferences as official sources, while farmers, family friends, and personal evaluations and judgement are cited as informal sources. Farooq et al. (2010) conducted similar research that particularly emphasized the role of Agriculture research Institutes and Agricultural Executives as information sources, whereas Rama and Joan (1996) recognised agents in the headquarters, representatives in other regions, extension specialists, direct supervisor, media organizations, state/federal agencies, teachers, and administrators as important information sources for agricultural extension workers. Mugwisi, Ocholla, and Mostert (2012) argued that farmers preferred face-to-face interaction over libraries, the internet, colleagues, personal and departmental collections, workshops, and seminars, after emphasising the position of libraries, the internet, colleagues, personal and departmental collections, workshops, and seminars.

Through an analysis, it was observed that farmers obtain agricultural information from a variety of sources, as indicated in Table 2. The majority of farmers (171) (92%) get agricultural information from mobile phones, followed by television 167 (90%), radio 166 (89%), fellow farmers 160 (86%), YouTube 158 (85%), and other sources such as Kisan helpline, E-magazine/E-newspaper, social media, and agricultural books. Other information sources and services, such as conferences and seminars, information kiosks/Common Service Centres, CD/DVD, E-books, and E-mails, are scored low, implying that they are insufficiently available to farmers in the region.

Farmers get access to information from a variety of sources, including seed and pesticide company experts, cooperative advisory council office-bearers, distributors and retailers, government agricultural support officers, market-commission agents/traders, veterinary specialists, and so on, using their mobile phones. Farmers can not only easily access information, but they can also reduce transaction costs and

Table 2. Agricultural information sources

S.No.	Statements	Number	Percent	Rank
1	Fellow Farmers	160	86%	4
2	Community Radio/AI Radio	166	89%	3
3	Television	167	90%	2
4	Information kiosk/Common Service Centres	125	67%	12
5	Mobile Phones	171	92%	1
6	Personal Computers/Laptop	140	75%	9
7	Internet	141	76%	8
8	WhatsApp	140	75%	9
9	Facebook	130	70%	10
10	YouTube	158	85%	5
11	Conference and Workshops	126	68%	11
12	E-mail	56	30%	15
13	E-Books	87	47%	14
14	CD/DVD	112	60%	13
14	E-magazine/ E-newspaper	149	80%	7
15	Kisan help line	153	82%	6

Source: primary data collection

increase the incentives on their agriculture produce. They emphasised the need of precise information arriving at the appropriate time in order to decrease waste and hence maximize productivity.

6. CONSTRAINTS IN THE USAGE AND ACCESS OF ICT IN AGRICULTURAL

The ranking technique was used to analyse the hurdles to farmers' in adoption of ICTs tools . Knowing the major barriers to ICT tool adoption is critical for reaching out to farmers' voices and problems so that planners, administrators, development workers, and policymakers can implement developmental programmes and interventions that can cater to farmers' needs and benefit them more effectively. Table 3 displays the results, which are further discussed below:

The main barriers in accessing agricultural information that farmers face are listed in Table 3 above. The key obstacles reported by the farmers are as follows: 'Lack of extension officers', 'lack of confidence in operating ICTs', primarily

Table 3. Challenges in accessing and using information sources

S.No	Statements	Number	Percent	Rank
1	No network connectivity	93	50%	10
2	A weak response from the agriculture department	112	60%	8
3	Lack of financial resources	141	76%	5
4	Inadequate Training	86	46%	12
5	High illiteracy	130	70%	6
6	Poor electricity	149	80%	4
7	Do not understand information delivered	67	36%	14
8	Format and language problem	117	63%	7
9	Lack of confidence in operating ICTs	167	90%	2
10	Lack of Agricultural Extension Officers	173	93%	1
11	Lack good leadership	56	30%	15
12	Lack of seminars, workshops and training programme	160	86%	3
13	Lack of agricultural demonstrations	74	40%	13
14	Growers get information untimely	99	53%	9
15	Unavailability of the local information centre	90	48%	11

Source: primary data collection

mobile phone applications as a result of limited exposure and understanding of how to effectively use ICTs to reap their benefits. Other issues that farmers encounter include 'inadequate training programmes', 'seminars, and workshops', delayed access to information, minimal electricity to recharge their smart phones, and the inability to view farming-related TV programmes owing to irregular and inconstant power supply. Furthermore, it was discovered that the lack of a local information centre, a slow response from the agriculture department, and misinterpretation or misunderstanding of information supplied are all significant barriers to access and use of farming information.

It is critical for the farming community to have the correct information at the right time and place, delivered through the relevant channels, in order to make informed decisions. For a long time, extension has played an essential role in agricultural growth, particularly during the early stages of India's first Green Revolution (Babu et al. 2013). It has principally contributed to the dissemination of agricultural technologies and management methods, thereby significantly increasing agricultural expansion and rural development. With changes in government regulations, technological demand and supply characteristics, and marketing reforms, the agriculture extension system is confronting greater opportunities as well as problems. access to and use of agricultural information

The findings of this study are consistent with those of Chavula (2014; Eucharia et al. 2016), they found that smartphones are the most commonly utilised ICT tools among fish farmers due to their convenience, extensive coverage, and relatively inexpensive. Some farmers expressed concern about a lack of training and practical knowledge how to use mobile phone applications and the internet, and they believed that they required some training to understand how to use ICTs that could improve agriculture and rural development. Farmers also complained about the high expense of fixing ICTs such as mobile phones and television sets. This, however, has prevented them from using ICTs on occasion because electronic devices are frequently broken. Some of the farmers struggled to understand the English language.

This is because most smartphones have English language menus, and some survey participants, particularly older farmers, do not understand how to get relevant data from cell phones. Due to illiteracy and a lack of expertise in using it, they also don't know how to use the primary features of mobile phones, such as SMS, access to agricultural apps, YouTube, and so on. The results also supported Rebekka Syiem and Saravanan Raj's (2015) study lack of training and awareness programmes, seminars, and workshops was also seen as a significant challenge; without any training programme, they were unaware of how to successfully use ICTs to their benefit. With the exception of personal interaction, they are uninformed of the socioeconomic benefits and improvements that ICTs may bring to their lives.

7. CONCLUSION

Agriculture has always been an important sector for economic growth and social development. However, with the growing population, food security has become a major challenge. This has led to the need for advanced technologies such as Artificial Intelligence (AI) to increase agricultural productivity and efficiency. The adoption of AI in agriculture can lead to a significant increase in crop yields and decrease in production costs. For instance, AI can be used to predict weather patterns, optimize irrigation and fertilizer usage, and detect crop diseases early on. This can help farmers make informed decisions and improve crop quality and yield.

However, the adoption of AI in agriculture requires awareness and effective use of Information and Communication Technologies (ICT). ICT can play a critical role in facilitating the integration of AI into agricultural practices. For instance, farmers can use ICT to access real-time weather data and make informed decisions on when to plant or harvest their crops. They can also use mobile applications to monitor crop growth and health, detect pests and diseases, and optimize irrigation and fertilization. The convergence of smart agriculture technology and contemporary data know-hows allows seed planting to be tailored to a specialized industry in order to ensure an effective production process. This study provides the insights on the use of ICT technologies for agricultural practise and growth in Punjab, as well as impediments farmers experienced in accessing modern ICT tools. The findings demonstrated that farmers in India commonly utilised mobile phones, radios, and television to share agricultural information. These technologies innovations have the ability to significantly improve farmers' capabilities, transforming agriculture and providing new opportunities. However, the findings revealed that farmers encountered several challenges in accessing and using information, such as a 'lack of extension officers', 'lack of confidence in operating ICTs', particularly mobile phone applications, inconsistent electrical supply and poor network connectivity. This situation can have a negative impact on the timely sharing of information related to farming technologies or innovations, including artificial intelligence, and, consequently, can cause poor farming productivity in the country.

The finding suggests that the pertinent ministry, through their extension officers, should encourage as well as organise training sessions to the agricultural community about use of ICTs tools (mobile phones, radios, and television) in sharing and exchanging information on agriculture. The government, non-governmental organisations (NGOs), and other private institutions should invest more in smart infrastructure to improve mobile phone, radio, and television reach and network connectivity to rural India as well. Furthermore, the government should make low-cost agriculture related digital infrastructure so that more farm owners can own and use it. The government should promote media owners to telecast more agricultural programmes related to agricultural production on both radio and television. A holistic ICT strategy has thus been devised, not just to reach out to growers in a more efficient and effective manner, but also for the planning and management of schemes, allowing policy decisions to be made more instantly and farmers to receive support more quickly. ICT can also help farmers access market information and connect with buyers, thereby increasing their profitability. Additionally, it can help in the collection and analysis of data, which can be used to develop AI models that are specific to local conditions.

the adoption of AI in agriculture can revolutionize the sector and contribute to global food security. However, this requires the effective use of ICT to facilitate the integration of AI into agricultural practices. Governments, organizations, and farmers need to work together to increase awareness, promote the use of ICT, and provide training and support to ensure the successful adoption of AI in agriculture.

REFERENCES

Adhiguru, P., & Mruthyunjaya. (2004). *Institutional innovations for using information and communication technology in agriculture.* Policy Brief 18. New Delhi: National Centre for Agricultural Economics and Policy Research.

Aina, I. O., Kaniki, A. M., & Ojiambo, J. B. (1995). *Agricultural information in Africa.* Third World Information Services.

Akullo, W. N., & Mulumba, O. (2016). *Making ICTs relevant to rural farmers in Uganda: a case of Kamuli district.* Available at: https://library.ifla.org/1488/1/110-akullo-en.pdf

Babu, S. C., Joshi, P. K., Glendenning, C. J., Asenso-Okyere, K., & Sulaiman, V. R. (2013). The State of Agricultural Extension Reforms in India: Strategic Priorities and Policy Options §. *Agricultural Economics Research Review*, *26*(2), 159–172.

Batchelor, S. (2002). *Using ICTs to Generate Development Content. IICD Research Report 10.* International Institute for Communication and Development.

Benard, R., Dulle, F., & Lamtane, H. (2018). The influence of ICTs usage in sharing information on fish farming productivity in the Southern Highlands of Tanzania. *International Journal of Science and Technoledge*, *6*(2), 67.

Bertolini, R. (2004). *Making information and communication technologies work for food security in Africa.* 2020 Africa Conference Brief 11. International Food Policy Research Institute.

Bhalekar, P., Ingle, S., & Pathak, K. (2015). The study of some ICTs projects in agriculture for rural development of India. *Asian Journal of Computer Science and Information Technology*, *5*(1), 5–7.

Bhat, S. A., & Huang, N. F. (2021). Big data and ai revolution in precision agriculture: Survey and challenges. *IEEE Access : Practical Innovations, Open Solutions*, *9*, 110209–110222. doi:10.1109/ACCESS.2021.3102227

Bhatt, P., & Muduli, A. (2022). Artificial intelligence in learning and development: A systematic literature review. *European Journal of Training and Development.*

Cash, D. W. (2001). In Order to Aid in Diffusing Useful and Practical Information: Agricultural Extension and Boundary Organizations. *Science, Technology & Human Values*, *26*(4), 431–453. doi:10.1177/016224390102600403

Chapman, R., & Slaymaker, T. (2002). *ICTs and Rural Development: Review of the Literature, Current Interventions, and Opportunities for Action*. ODI Working Paper 192. London: Overseas Development Institute.

Chavula, H. K. (2014). The role of ICTs in agricultural production in Africa. *Journal of Development and Agricultural Economics*, *6*(7), 279–289. doi:10.5897/JDAE2013.0517

Crusan. (1982). agricultural research committees: Complementary platforms for integrated decision-making in sustainable agriculture Network Paper. *Agric. Res. and Ext. Network.*, *23*, 105–111.

Davis, F. D. (1989). Perceived usefulness, perceived ease of use, and use acceptance of information technology. *Management Information Systems Quarterly*, *13*(3), 319–339. doi:10.2307/249008

Eggleston, K., Jensen, R., & Zeckhauser. (2001). Information and Communication Technologies, Markets and Economic Development. In *The Global Competitiveness Report, World Economic Forum & Centre for International Development*. Harvard Business School.

Eucharia, E.-O., Ubochioma, N., Chikaire, J., Ifeanyi, O. E., & Patience, C. N. (2016). Roles of information and communications technologies in improving fish farming and production in Rivers state, Nigeria. *Library Philosophy and Practice*. Available at: www. ejournalofscience.org/1445

Gollakota, K. (2008). ICT use by businesses in rural India: The case of EID Parry's Indiagriline. *International Journal of Information Management*, *28*(4), 336–341. doi:10.1016/j.ijinfomgt.2008.04.003

Green, D., Lee, B., Morrison, J., & Werth, A. (2005). Sustainable development, poverty and agricultural trade reform. *Commodities, Trade and Sustainable Development, 15*.

Habib, M., Khan, Z., Iqbal, M., Nawab, M., & Ali, S. (2007). Role of farmer field school on sugarcane productivity in Malakand Pakistan. *African Crop Science Conference Proceedings*, 1443-1446

Hassan, J. (1991). Influence of NPK fertilizer on the technological qualities of plant cane Varietty CB. *International Sugar Journal*, *84*, 76–82.

Kakani, V., Nguyen, V. H., Kumar, B. P., Kim, H., & Pasupuleti, V. R. (2020). A critical review on computer vision and artificial intelligence in food industry. *Journal of Agriculture and Food Research*, *2*, 100033. doi:10.1016/j.jafr.2020.100033

Kizilaslan, N. (2006). Agricultural information systems: A national case study. *Library Review*, *55*(8), 497–50. doi:10.1108/00242530610689347

Koyenikan, M.J. (2011). Extension Workers' Access to Climate Information and Sources in Edo State. *Nigeria Scholars Research Library Archives of Applied Science Research, 3*(4), 11-20. Retrieved 6/10/2012 from http://scholarsresearchlibrary/archieve.html

Ravallion, M. (1986). Testing market integration. *American Journal of Agricultural Economics*, *68*(1), 102–109. doi:10.2307/1241654

Rebekka, S., & Sravanan, R. (2015). Access and usage of ICTs for agriculture and rural development by the tribal farmers in Meghalaya state of North-East India. *J Agric. Informatics. Nigeria. Agricultural Information Worldwide*, *6*(1), 18–24.

Saranya, T., Deisy, C., Sridevi, S., & Anbananthen, K. S. M. (2023). A comparative study of deep learning and Internet of Things for precision agriculture. *Engineering Applications of Artificial Intelligence*, *122*, 106034. doi:10.1016/j.engappai.2023.106034

Shah, M. A. (2015). Accelerating Public Private Partnership in Agricultural Storage Infrastructure in India. *Global Journal of Management and Business Research*, *15*(A13), 23–30.

Singh, P., & Singh, N. (2020). Blockchain with IoT and AI: A review of agriculture and healthcare. *International Journal of Applied Evolutionary Computation*, *11*(4), 13–27. doi:10.4018/IJAEC.2020100102

Stienen, J., Bruinsma, W. & Neuman, F. (2007). *How ICT can make a difference in agricultural livelihoods. The commonwealth ministers book-2007*. International Institute for Communication and Development.

Chapter 6
Artificial Intelligence, Internet of Things, and Machine-Learning:
To Smart Irrigation and Precision Agriculture

Fatima-Zahra Akensous
https://orcid.org/0000-0001-9518-8146
Faculty of Science Semlalia, Cadi Ayyad University, Marrakesh, Morocco

Naira Sbbar
Faculty of Science Semlalia, Cadi Ayyad University, Marrakesh, Morocco

Lahoucine Ech-chatir
https://orcid.org/0009-0005-0729-0789
Faculty of Science and Technology Guéliz, Cadi Ayyad University, Marrakesh, Morocco

Abdelilah Meddich
https://orcid.org/0000-0001-9590-4405
Faculty of Science Semlalia, Cadi Ayyad University, Marrakesh, Morocco

ABSTRACT

Water scarcity has been escalating both in terms of frequency and severity, owing to climate change and global warming. Furthermore, water is a vital source that is at the core of crucial sectors like agriculture. Yet, this source is labeled scarce, and its distribution is uneven globally. For the aforementioned reasons, achieving a rational use of water is of utmost importance. In this framework, computational

DOI: 10.4018/978-1-6684-6791-6.ch006

intelligence like artificial intelligence (AI), the internet of things (IoT), and machine-learning, has been gaining momentum for implementing smart irrigation and precision agriculture. Thus, the chapter surveys a selection of recent studies, corroborating AI, IoT, and machine-learning as promising approaches to advance agriculture. The chapter also sheds light on the notion of virtual water, proposes strategies to deal with water scarcity, and highlights the essential components to achieve effective smart irrigation, thereby switching toward an innovative model of sustainable digital agriculture.

1. INTRODUCTION

Water scarcity has evolved into one of the major and critical issues of the 21st century, owing to an insanely increasing water demand versus water supply (Veldkamp et al., 2017). Water unavailability affects more geographical regions than others, owing to the variability in temperature patterns and uneven distribution of precipitation (van Vliet et al., 2017; Coffel et al., 2019). Climate change and global warming are only worsening the case, since statistical and predictive models indicate extreme water-related events for the next one hundred years to come (Brown et al., 2019). Water is at the core of requisite fields like agriculture (Markland et al., 2018). Seventy percent of all water withdrawals are consumed by agriculture, which is justified toward securing food and crop productivity, for instance (Mabhaudhi et al.,

Figure 1. The disposition of water on Earth
Source: Adapted from Shiklomanov (1991)

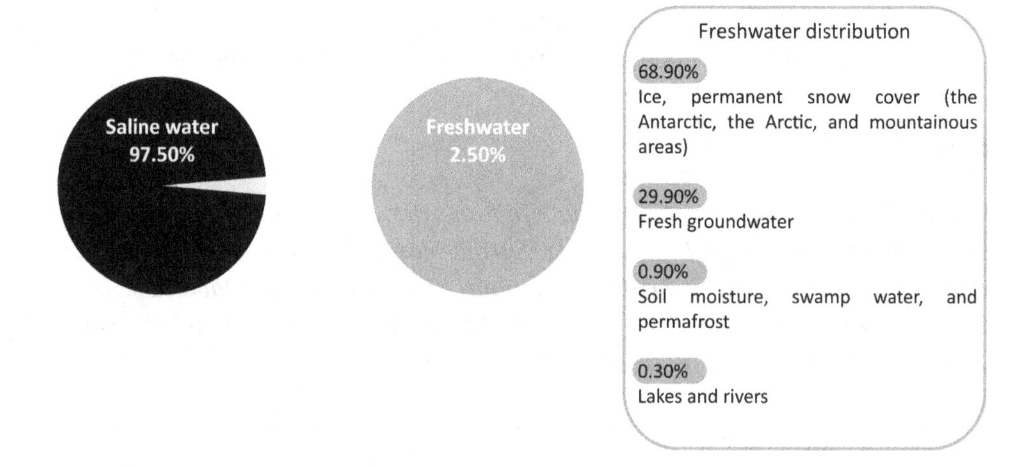

2018). However, only 2.5% represents the globe's freshwater, two-thirds of which is inaccessible (Qadri et al., 2020) (Figure 1). For the above-mentioned reasons, managing water use and assuring sustainable agriculture is of the highest importance (Hoekstra, 2017).

During the past few decades, advanced computational tools, such as Artificial Intelligence (AI), the Internet of Things (IoT), and Machine-Learning have been gaining momentum (Liakos et al., 2018; Jha et al., 2019; Antony et al., 2020). In recent times, advances regarding *in situ* features have revolutionized our understanding of complicated issues like those related to water (Doorn, 2021). These powerful human-set tools have been reported for their time-efficient, data-generation, and decision-making roles (Tien, 2017). AI relies on simulating human intelligence on an automated basis. Thus, AI as a branch of computer science that can recognize its environment should be able to contribute towards improving machine labor and the development of more advanced technologies, including Convolutional Neural Network (CNN), Artificial Neural Network (ANN), and Machine-Learning (Currie and Rohren, 2021). IoT is a network of physical items with integrated technology that enable communication and internal sensing (Antony et al., 2020). It represents an ensemble of physical items that can link and share data like software, hardware, and sensors (Sadique et al., 2018). Thus, the complete automated scheme acts apprehensively and reacts in accordance. Since there is a limited supply of fresh water on Earth, using it reasonably is the main purpose (Kamienski et al., 2019). Thus, AI, IoT, and Machine-Learning can automate practices related to irrigation and optimizing crop growth and productivity. Nevertheless, challenging aspects that hinder switching to precision agriculture are the need for trained agriculture stakeholders like farmers, crop consultants, and advisors (Patrício and Rieder, 2018). Another issue is the non-uniformity of agricultural lands and crop types, which makes coming up with practical smart irrigation setups a challenging mission (Kassim, 2020). Therefore, the present book chapter surveys a selection of the recent advances in technologies powered by AI, IoT, and machine-based, as well as their applications for modern irrigation and precision agriculture. Furthermore, research gaps and challenging issues related to these technologies are addressed.

The book chapter is structured as follows: (i) section one highlights the scarcity of water resources regarding agriculture, and how computational intelligence deals with managing hydrological resources, (ii) section two brings about precision agriculture, in addition to why and how computational intelligence can help switch to smart and sustainable agricultural systems, (iii) section three corroborates the underscored points at the level of section two, hence showcases how precision agriculture market is gradually expanding. Finally, (iv) section four addresses limiting factors and hurdles to be surmounted to improve the performance and widen the scope of dealt-with computational intelligence tools. The chapter concludes with the tremendous roles played by AI, IoT, and Machine-Learning toward smart irrigation and precision agriculture.

The recency of case studies indexed in Scopus, Web of Science, and Google Scholar, is *mostly and not exclusively* set up in the last five years (2018-2023).

2. WATER RESOURCES: ENDANGERED ECOSYSTEMS

The gap flanked by water supply and water demand is anticipated to attain approximately 40% within less than a decade, owing to the population's speedily escalating growth tendency (Islam and Karim, 2020). Hence, scarce water can imply numerous consequences, including ecological imbalance, hampered economic growth, and hydrological conflicts (Distefano and Kelly, 2017). Water resources occur on Earth in vapor, liquid, and solid forms, and are most abundant in oceans but less available at the level of glaciers (Castellazzi et al., 2019). However, climatological and anthropogenic factors have been leading to tremendous pressure on freshwater resources (Shrestha et al., 2017). According to a 2022 IPCC report, the temperature of water at the level of rivers and lakes has escalated (up to 1°C and 0.45 °C/decade, respectively), whereas the ice cover extent has decreased by 25% due to climate change and global warming significant impact (Parmesan et al., 2022). Furthermore, growing water scarcity and biodiversity-related risks are primarily triggered by the degradation and exhaustion of freshwater ecosystems, which emphasizes the need for coordinated approaches to assure integrated and inclusive water resource management resolutions (Vollmer et al., 2018; Coffel et al., 2019; Slimani et al., 2023).

3. WATER RESOURCES CLASSIFICATION AND APPLICATIONS IN AGRICULTURE

The scarcity of water resources is becoming a serious global issue (van Vliet et al., 2021). Every year, approximately 510,000 km^3 of water is precipitated worldwide in the form of rain, snow, or sleet. Of this amount, 400,000 km^3 falls into the oceans and 110,000 km^3 on the land. However, the geographical precipitation distribution is uneven, which severely affects sectors like agriculture (Ruane et al., 2008; Fishman, 2016; Mardero et al., 2020). In this regard, water can be classified into three categories (Figure 2): green water, which is the soil moisture generated by precipitation and available right to the plant root proximity; blue water, which is the stored runoff from precipitation in lakes, rivers, dams, and aquifers; and grey water, which is an already used water and has the potential to hold impure bodies (Rakesh et al.,

Figure 2. Categories of virtual water

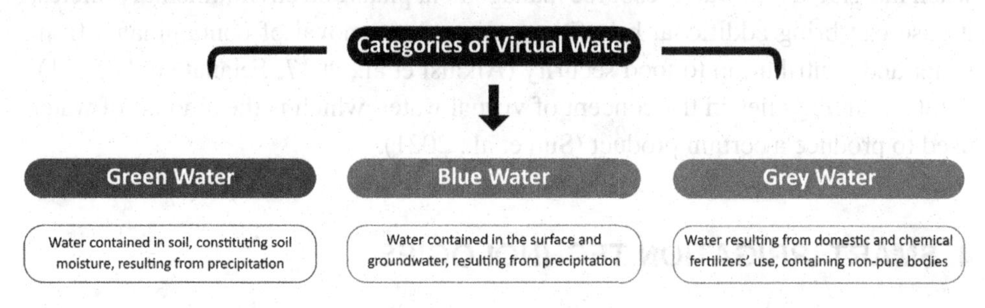

2020). Green water is the main resource for rain-fed agriculture and blue water is dedicated to irrigated agriculture (Ruane et al., 2008; Zhuo and Hoekstra, 2017).

To preserve this source, the means to reduce the risks of water scarcity are varied and depend on the risk's type and scale (Figure 3).

The FAO highlights a series of key strategies to address global water scarcity and increase its availability, including desalination of saline water, reuse of wastewater, virtual trading of water and food, increasing agricultural yields, and improving water use efficiency in agriculture through biotechnology advancement (Falkenmark, 2013). Desalination is mainly used to assure drinking water supply, especially in water-scarce regions, whereas wastewater is utilized for irrigation and aquaculture. Desalination processes require large amounts of energy. Nevertheless, the cost of desalination of brackish water and seawater has been a major constraint hindering their use (Alkaisi et al., 2017). Wastewater treatment is an expensive process, but

Figure 3. Common water scarcity reduction strategies

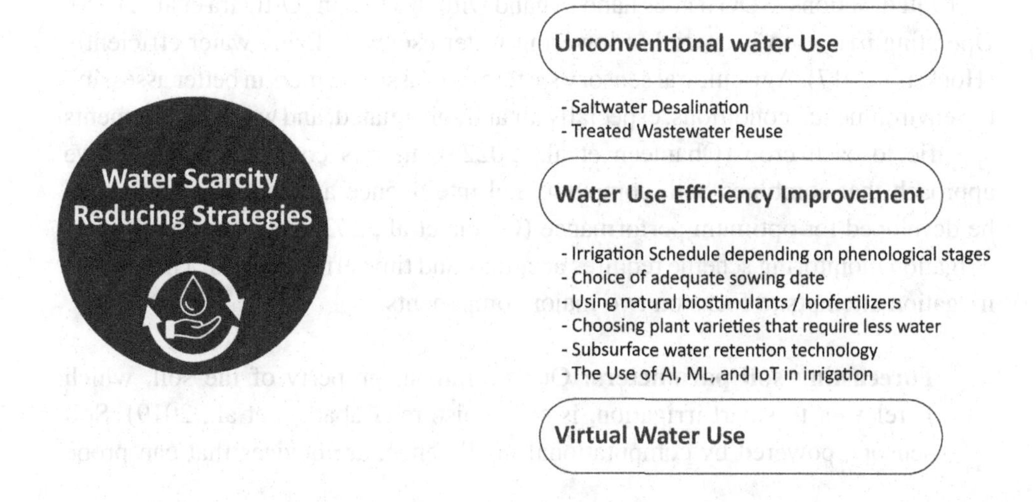

when integrated into water resource management plans and environmental policies, its use can bring additional benefits, such as the removal of contaminants from water and contribution to food security (Alkaisi et al., 2017; Fetanat et al., 2021). Another strategy lies in the concept of virtual water, which is the amount of water used to produce a certain product (Sun et al., 2021).

4. SMART IRRIGATION TECHNOLOGIES

To assure effectiveness and sustainability in irrigation systems, it is important to implement AI, IoT, and Machine-Learning technologies (Figure 4).

Figure 4. Computational intelligence for smart irrigation

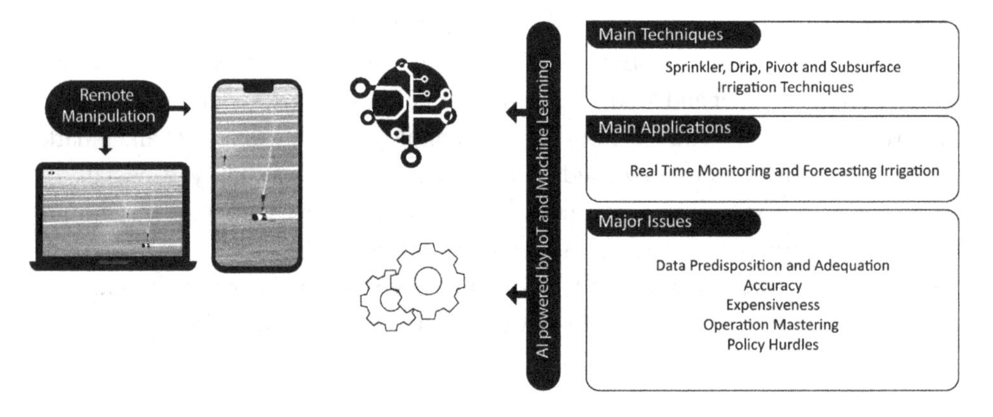

United Nations' SDG 6 goes hand in hand with this vision (Ortigara et al., 2018). Operating irrigation is crucial to lowering water use by utilizing water efficiently (Hoekstra, 2017). Agricultural sensory systems can assist farmers in better assessing the environmental conditions, especially air and soil-related, and water requirements specific to each crop (Obaideen et al., 2022a). In this context, an integrative approach that combines both computational intelligence and sensor systems can be developed for optimum performance (García et al., 2020). An operative water irrigation monitoring scheme requires accuracy and time effectiveness. Thus, smart irrigation attempts to focus on two major components:

- **Forecasting soil parameters:** One prominent property of the soil, which is relevant to smart irrigation, is soil moisture (Babaeian et al., 2019). Soil sensors, powered by computational intelligence, are devices that can probe soil's gravimetric and volumetric-like areas. The gradual depth capacity that they offer permits exploring soil moisture at a comprehensive level (Leh et al., 2019). Soil moisture depends on prime parameters, such as water content and water potential. Soil water content expresses water available per soil unit, whereas soil water potential indicates the potential energy of water contained in the soil. A wireless sensor network, a set of sensors spatially distant that can register environmental and physical properties, can track and monitor soil moisture (Thakur et al., 2019).

- **Forecasting climate-plant evolution:** Climate variation-specific sensors can measure evapotranspiration (ET), based on prior information and data, mostly regarding the agricultural site and crop in question (Touil et al., 2022). The sensors that monitor ET can modify schedules related to irrigation, based on prior site and weather data. To gather ET information on a daily and/or monthly basis, parameters like humidity and temperature should be forecasted, which can be enabled thanks to smart irrigation (Yin et al., 2020).

4.1. Artificial Intelligence and Its Applications

The intersection of AI and irrigation in agriculture represents one of the most effective and suitable combinations that can advance smart irrigation (Ramdinthara et al., 2022). Irrigating following each crop water requirement is a major step toward better use of water resources (Talaviya et al., 2020). AI can collect and generate information regarding each crop (e.g., growth evolution) and soil (e.g., soil moisture), thanks to a plethora of applications (Sanikhani et al., 2019). When it comes to irrigation, AI can play a central role in managing water. Table 1 describes different applications of AI for smart irrigation in different parts of the world and varied crops. According to a study carried out by Katimbo et al. (2023), a computerized Irrigation-Decision Support System/IDSS contributed to estimating crop evapotranspiration/ETc as well as crop water stress index/CWSI in maize. The authors pointed out the promising role of the used IDSS in optimizing water use efficiency, in comparison to conventional irrigation methods. In another study carried out by Alves et al. (2023), the authors applied the digital twin computer program for evaluating varied irrigation methods before their implementation in the field. The authors approved of the applied computer program, stating that, based on prior weather and soil probing, the digital twin can assist farmers in better water utilization in the future. In an attempt to allow judicious irrigation of local dry areas at the level of an agricultural field, Jalajamony et al. (2023) suggested a LoRa-WAN-based drone-assisted smart irrigation system, powered by a Thermal-Infrared/TIR camera and a GPS component to localize the dry areas. One particular highlighted point by the authors was the possible option of scalability of the smart irrigation system, covering a wide range of crops and geographical localities. Based on such findings, AI has assets to revolutionize conventional agricultural systems, especially regarding irrigation and water optimization.

4.2. Internet of Things (IoT) and Its Applications

The IoT relies on connecting devices, such as sensors and software, through the Internet or any other networking form, which serves as a medium service, especially when it comes to geographically remote areas (Antony et al., 2020). In this regard, Table 1 describes different applications of IoT for smart irrigation in different parts of the world for different crops. The minimal human interference alternative is possible via IoT, which is an advantage of this technology that can register, forecast, and regulate every interaction involving the connected devices (Vij et al., 2020). As illustrated in Figure 5, using this technology, real-time data can be collected from multiple types of sensors (e.g., soil moisture sensor) using microcontrollers like Arduino (Satish et al., 2017) and Raspberry Pi (García et al., 2020), that are then sent to the cloud for processing. Once the data reaches the end user for visualization and decision-making, the farmer can choose to turn the pump on or off and control the flow and volume of irrigation water, as well as the time and areas to be irrigated. Depending

Figure 5. Components of an IoT-based smart irrigation system

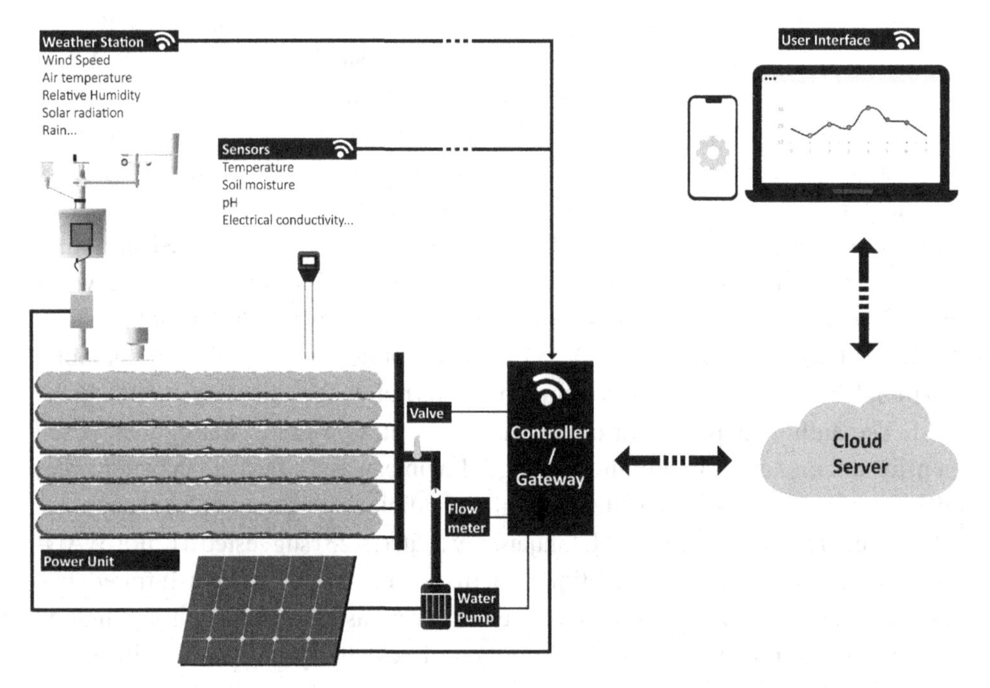

Table 1. AI and IoT applications in smart irrigation

Locality	Technological Feature	Objective	Tested Crop(s)	Observations	Reference
Santomera, Murcia, Spain	Horti-Control-Expert software	Assessment of irrigation frequency management in a greenhouse environment and optimization of water use efficiency for soil-exempted cultures	Tomato	Irrigation and frequency management	Rodriguez-Ortega et al., 2017
Canal del Zújar Irrigation District, southwest Spain	Genetic fuzzy systems (GFS) and Adaptive neuro-fuzzy inference system (ANFIS)	Testing GFS and ANFIS combination in predicting the depth of daily irrigation applied by every farmer	Rice, maize, tomato	Farming practices differed regarding each crop though the irrigation system was the same	González Perea et al., 2018
Australian cotton farm, Australia	Hardware and software specific for bay irrigation	Monitored furrow irrigation system testing	Cotton	Better and time-efficient irrigation performance	Uddin et al., 2018
Unspecified	Fuzzy inference system	Development of a smart fuzzy inference structure built upon smart irrigation information for central-pivot irrigation	Corn, soy	Fast control of maps	Mendes et al., 2019
Indian smart farming as a model, India	Neural Network (NN)	Presentation of a developed forecasting and an automatic system of irrigation IoT-based	Spinach	Irrigation scheduling, remote data observation, and neural network assessment Up to 67% of water saving	Nawandar & Satpute, 2019
Central Java, Indonesia	Real-Time Clock (RTC) DS-1302, Arduino systems	Design of criteria for time-efficient irrigation specific to sandy soil systems Evaluation of the functioning of automated irrigation	Onion	Efficient time setting up for irrigation	Sudarmaji et al., 2019
Rwanda, Central Africa	A cost-effective IoT system	Proposing a cost-effective smart irrigation system adaptable to the Rwandan rice setting	Rice	Management of irrigation The capacity of decision-making regarding irrigation	Bamurigire et al., 2020
Mozambique, Tanzania, and Zimbabwe, southern Africa	Agricultural Innovation Platforms (AIPs)	Analysis of the introduction of the AIPs into three small-scaled irrigation systems	Unspecified	Providing information to farmers to enhance the effectiveness of the agronomic system, farmers' knowledge as well as decreasing conflict	Bjornlund et al., 2020
Unspecified	Fuzzy logic, powered by IoT	Conceiving an integrated approach enabling smart irrigation and crop productivity	Unspecified	Cost-effectiveness and smart water use/ irrigation	Krishnan et al., 2020
North-eastern Australia	Cybernetic closed-loop irrigation system	Enhancing irrigation management by integrating an automated irrigation system with an irrigation decision support tool	Sugarcane	Effective irrigation scheduling management	Wang et al., 2020

continued on following page

Table 1. continued

Locality	Technological Feature	Objective	Tested Crop(s)	Observations	Reference
Afghanistan	Networks of wireless sensors, powered by IoT	Determination of IoT-based wireless sensors' applications in precision agriculture	Saffron, wheat	Checking-in soil parameters' evolution, crops performance, and automating irrigation	Rasooli et al., 2020
Universiti Teknologi, Malaysia	Internet Cloud, Discrete-time Model Predictive Control (DMPC)	Presentation of a precision irrigation procedure built upon DMPC	Cantaloup	Up to three times reduction in computational complexity compared to existing procedures, thus time-efficient Less consumption of water	Abioye et al., 2021
Date Palm Research Center of Excellence, King Faisal University, Al-Ahsa, Saudi Arabia	Automated Controlled Subsurface Irrigation System (CSIS), based on cloud IoT solutions	Design, construction, and validation of a completely automated subsurface irrigation scheme to control water requirements	Date palm	Positive effect on irrigation water control	Mohammed et al., 2021
Mekong Delta, Vietnam	Alternate wetting-drying (AWD) technology, powered by IoT	Investigation of IoT's inputs to advancing the AWD technology as well as smallholder farmers' conceiving of the technology	Rice	Reduction of water used for irrigation Enhancement of water use efficiency	Pham et al., 2021
Barani Agricultural Research Institute Chakwal, Punjab, Pakistan	Scientific irrigation scheduling (SIS), IoT based	Investigation of an SIS toward a reasonable olive grove productivity	Olive groves	Contribution to the economy in terms of water use, thus optimization of water use efficiency	Aziz et al., 2022
India	A brainy system designed to track and schedule irrigation accurately, relying in part on IoT	Introduction of a smart irrigation system to schedule irrigation and use water more effectively	Tomato, eggplant, banana, rice	Precision in terms of providing the crops with their water requirements Optimum irrigation	Prasanna Lakshmi et al., 2023

on the system components, it could also be automatically responsive to water levels through algorithms that adjust irrigation requirements (Kumar et al., 2017).

Hence, IoT offers several advantageous points, such as effective monitoring of crop growth, yield, conservation of nutrients and the structure of the soil, as well as irrigation water optimization (Obaideen et al., 2022b).

Since agriculture consumes the lion's share of water resources worldwide, optimizing irrigation to achieve sustainable agricultural systems is necessary (Ramachandran et al., 2022). In this sense, Veerachamy et al. (2022) suggested a methodology based on a modified IoT platform to allow a smart irrigation monitoring system. The system was able to sense a set of parameters, including soil moisture, humidity, temperature, and rainfall. The authors stated that the employed methodology permitted effective remote monitoring of the agricultural area, more specifically the irrigation aspect. In another study conducted by Muley and Bhonge

(2019), the authors proposed a Wireless-Sensor-Network (WSN) and IoT-based irrigation tracking system. According to this study, the system was able to forecast soil moisture, temperature, and humidity. Hence, the system could effectively control the irrigation of diverse crops like cereals (e.g., wheat and maize), legumes (e.g., beans), and fruit trees (e.g., apricots). Hence, IoT represents a promising technology for the effective monitoring of irrigation and optimum water use.

4.3. Machine-Learning and Its Applications

A division of AI, Machine-Learning usage depends on algorithms that allow training data and is widely applied in agriculture. Figure 6 describes how a Machine-Learning based smart irrigation scheme works. The system operates by gathering data concerning climate and soil moisture through sensors and remote sensing imagery. Then, a cloud server analyzes and filters the gathered data, which permits the prediction of the right water irrigation requirements as well as the decision to take.

The particularity of Machine-Learning lies in its usage where it is tough or non-practical to employ conventional algorithms (Liakos et al., 2018; S. Ray, 2019). Therefore, Machine-Learning constitutes a crucial passage toward achieving smart irrigation and optimal water use. In this regard, Table 2 highlights the accuracy of Machine-Learning-based schemes in tracking water-related parameters. In this regard, Chandrappa et al. (2023) proposed a Spatiotemporal modeling method for soil moisture, which features Machine-Learning algorithms, specifically conceived to

Figure 6. Components of a machine-learning-based smart irrigation system

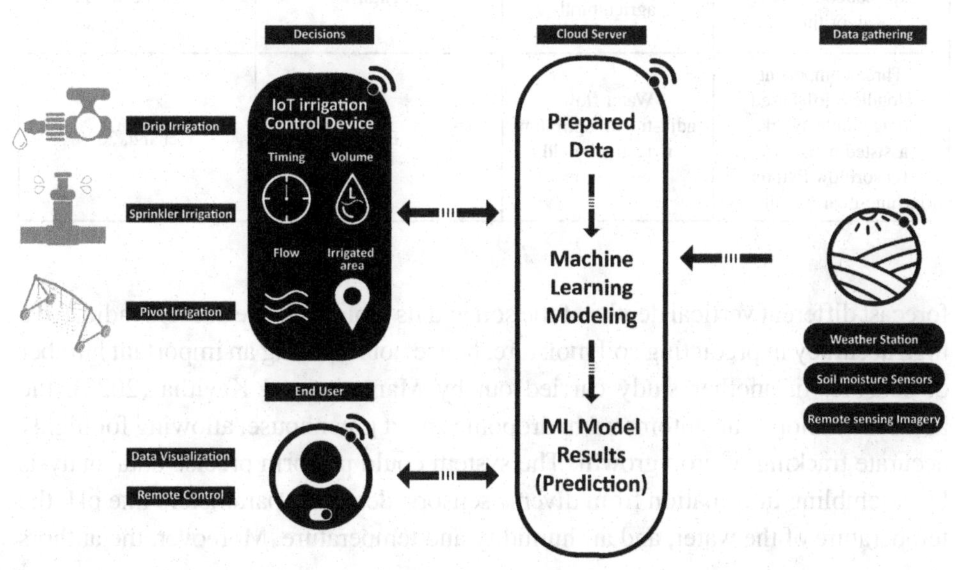

Table 2. Machine-learning applications in smart irrigation

Technique	Tested Parameters	Accuracy	Reference
A new algorithm, powered by supervised and unsupervised Machine-Learning	Soil moisture, air temperature, air relative humidity, soil temperature, radiation	High	Goap et al., 2018
Time Division Multiple Access (TDMA)-based Medium Access Control (MAC) protocol (LoRa P2P)	Soil moisture, soil temperature, soil humidity, air temperature, air humidity, light intensity	High	Chang et al., 2019
Weather-aware Runoff-Prevention-Irrigation Control (WaRPIC), powered by Machine-Learning	Maximum Allowable Runtime (MAR) relative to each sprinkler, irrigation scheduling	High	Murthy et al., 2019
Machine-Learning algorithms as irrigation forecasting support schemes	Soil moisture, temperature, humidity, water level	High	Vij et al., 2020
Machine-Learning algorithms, in particular, Light Gradient Boosting Machine (LightGBM)	Soil moisture, soil matric potential	High	Togneri et al., 2022
DHT22, BH1750, YL-69 sensors, powered by Machine-Learning algorithms	Humidity, light intensity, temperature, soil moisture	High	Abdurahman, 2022
A novel sequential gap-reduction (SGR) algorithm	Optimum dispersal of soil sensors over an agricultural field	High	Goodrich et al., 2023
Three-component cloudless IoT-based irrigation network, assisted by Google TensorFlow Python Machine-Learning library	Water flow adjustment, optimum irrigation, yield estimation	High	Tschand, 2023

forecast different vertical depths of the soil and its moisture. The authors indicated a high accuracy in predicting soil moisture, hence not requiring an important number of sensors. In another study carried out by Mamatha and Kavitha (2023), the authors developed an automated hydroponic smart greenhouse, allowing for highly accurate tracking of crop growth. The system could perform precise data analysis by assembling information from diverse sensors, detecting parameters like pH, the temperature of the water, and air humidity and temperature. Moreover, the authors emphasized that the Machine-Learning-based system assisted the hydroponic crops' growth improvement, enhancing their production and resilience to biotic factors (e.g., pests and diseases). According to such recorded results, Machine-Learning can help optimize water use and crops' productivity.

5. PRECISION AGRICULTURE: THE FUTURE BLUEPRINT

The practice of exploiting data to monitor as well as ameliorate agricultural systems and productivity is referred to as precision agriculture (Lowenberg-DeBoer and Erickson, 2019; Singh et al., 2020). Forecasting the tiniest signs of an anomaly (e.g., plant hormones signaling in response to drought) within crops' behavior is key (Ait-El-Mokhtar et al., 2022; Vrchota et al., 2022). Precision agriculture constitutes a farming managerial concept, built upon observation, measurements, and responses to sensible variations that occur within the plant and soil. The particularity of precision agriculture is that it allows crop and soil anomaly assumption and adequate decision-making, via sophisticated technological tools and sensors, such as IoT. To build an effective model of precision agriculture, an important amount of data related to plant and soil health statuses should be acquired and processed. In this sense, a varied range of factors can be analyzed like the level of water, temperature nuances, soil nutrient content, and the presence of pests and weeds. Thus, information on nutrient content can indicate whether or not a certain dose of a given fertilizer should be applied. Therefore, a timely decision can be made, which plays a key role in optimizing agricultural systems (Shafi et al., 2019). In this sense, digital agriculture can and is made possible thanks to computational intelligence, which allows appropriate agricultural field-tracking (Tantalaki et al., 2019) (Figure 7). Computational intelligence is provided by AI and can be defined

Figure 7. Computational intelligence for precision agriculture

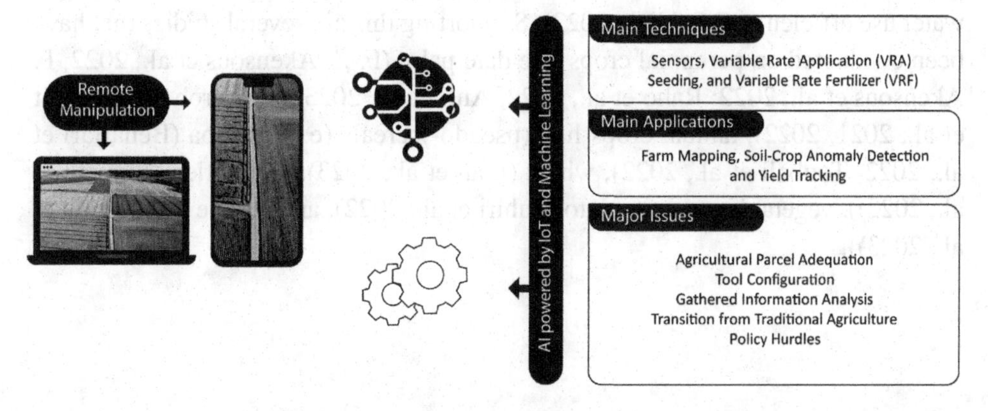

as the capacity of a computer system to acquire a precise task using data, relying on experimental observation. Image processing of remote sensing is an area that applies computational intelligence regarding precision agriculture. Images provided by satellite systems are extensively used in precision agriculture, to track fields and optimize crop productivity (Caballero et al., 2019; Sharma et al., 2021). Water use, in addition to the type and suitable application of fertilizers, is crucial to switch to precision agriculture, especially since these two components can be critical under extreme environmental constraints (Singh et al., 2020; Akensous et al., 2022). Another attribute of precision agriculture is that it can permit farmers to optimize water meant for irrigation, therefore produce more by using less (Sharma et al., 2021).

Due to an unparalleled decrease in terms of the frequency and number of wells (pumping), which are already running dry, pumping water from the underground is insufficient, and preserving it for agricultural purposes easily stands out as a challenging hurdle (LaVanchy, 2017). On the one hand, computational intelligence can help sustain the most precious source on Earth by rationally regulating water usage, as supported by several studies (S. S. Ray et al., 2023; Syrmos et al., 2023; Zanfei et al., 2023). Agriculture is yet to prevent the wasteful usage of water. It is, therefore, necessary to improve water use efficiency for reasonable and sustainable irrigated agriculture. Precision agriculture is promising in this regard since it can assure smart irrigation systems, wireless communication technologies, monitoring systems, and advanced AI-assisted control strategies, which are crucial for optimal irrigation scheduling and monitoring (Bwambale et al., 2022a). On the other hand, managing the soil component in terms of soil moisture, beneficial microorganisms, and overall health status can also contribute to optimizing the water use efficiency of crops. In this sense, upgrading the soil's health profile begins with enriching it with biostimulants that can boost its humus, which can retain water (Anli et al., 2020; Ben-Laouane et al., 2023). Moreover, the application of biostimulants, which are substances, amendment, and/or microorganisms with the capacity of increasing nutrient assimilation by uptaking water and improving the water-plant relations even under drastic environmental constraints, plays a key role in enhancing crops' water use efficiency (Meddich, 2022). Supporting this are several studies that have been conducted on perennial crops like date palm (F.-Z. Akensous et al., 2022; F. Akensous et al., 2022; Raho et al., 2022; Anli et al., 2023) and carob (Boutasknit et al., 2021, 2022), annual crops like (pseudo-)cereals (e.g., quinoa (Benaffari et al., 2022; Toubali et al., 2022), wheat (Ikan et al., 2023), and barley (Slimani et al., 2022)), vegetables (e.g., tomato (Tahiri et al., 2022), and lettuce (Ouhaddou et al., 2023)).

6. PRECISION AGRICULTURE AND SUSTAINABILITY

Smartly managing both crop productivity and animal husbandry come under the umbrella of precision agriculture. However, the focus is on crop productivity at the level of this chapter. By recovering information concerning the climate, surrounding soil environment, and crops via sensors and machine networks, time-efficient and opportune decision-making are made possible (Shafi et al., 2019). Precision agriculture considers implementing sustainability in every aspect of it. The chief objective of precision agriculture is to boost yield production and optimize the reasonable usage of resources, particularly water, whilst also helping to decide when to apply biostimulants, for instance. In this regard, precision agriculture can contribute to quantifying resources, thus preventing wasteful practices and providing adequate information for better resource management (Shafi et al., 2019). In this regard, Lahbouki et al. (2022) experimented with the Subsurface Water Retention Technology/SWRT in a field experiment to explore its impact on the resilience of prickly pear to drought. The authors reported improved growth, development, and yield of the plants. SWRT is a technique designed to conserve water at the level of sandy soils utilizing water-saving membranes (Roy et al., 2019). SWRT relies on the installation of an impermeable water-retaining, polyethylene-based membrane that can lessen the loss of water and nutrients by percolation. The procedure consists of installing the membrane below the plant's root system area, to preserve water and nutrients from percolating. The application of this technology can enhance water and nutrient uptake by plants, contributing to less leaching and percolation, thereby improving water and nutrient use efficiency, especially under drought stress (Kavdir et al., 2014). Precision agriculture is often regarded as agriculture that uses data for both the characterization of heterogeneity and the modulation of spatialized technical itineraries (McBratney et al., 2005; Lowenberg-DeBoer and Erickson, 2019). It is at the center of efforts to provide solutions to the main problems of agriculture (Bwambale et al., 2022b). Thus, utilizing these technologies could help mitigate the environmental repercussions of agricultural practices (Moreiro, 2017). Precision agriculture can contribute in many ways to the long-term sustainability of agrosystems' productivity, confirming that it should reduce the environmental load by applying mainly biostimulants and nano-fertilizers where and when needed (Mahapatra et al., 2022). Precision agriculture can ameliorate the sustainable quality of agricultural products through eco-friendly practices, such as fertilizing agricultural soils and boosting plants' water and nutrient uptake as well as resilience, using biostimulants. Moreover, precision agriculture can upgrade the output gained from renewable energies under uncertain weather conditions. Thus, smart farming has the potential to empower agricultural systems, offering quality products (Sharma et al., 2021). Not only precision agriculture can accurately predict yield but also estimate agricultural residues, which are by-products of farming. These materials can include stalks from cereal crops, such as corn, rice, and wheat, and other crops like nut shells, among others. The potential to increase the world's energy production capacity through agricultural residues is largely untapped. Agricultural residues contain abundant energy that can be treated and contribute to renewable energy production, such as biofuels, which rhymes with the notion of a circular economy. Biofuels are considered to be a safer option for the environment, due to their lower emissions (Rodionova et al., 2017). By harnessing renewable energy from agricultural residues and waste, a significant contribution to mitigating the effects of climate change and global warming can be executed (Odara et al., 2015). Energy management can also be improved, which will better synchronize demand and supply. Moreover, providing consumers with generated power, even in remote areas, has the potential to reduce transmission losses (Pham et al., 2021).

7. PRECISION AGRICULTURE AND COMPUTATIONAL INTELLIGENCE

Conventional agriculture has the challenge of assuring food security globally. However, the process of feeding the globe can take considerable time (Meemken and Qaim, 2018; Rao, 2018). In this context, technological advancements, such as AI, IoT, and Machine-Learning can lead to time-effectiveness, accuracy, and better yielding (Tantalaki et al., 2019). Moreover, automated systems, powered by sophisticated algorithms, can help assess crops' quantitative and qualitative traits efficiently (Sinha et al., 2019). Precision Agriculture works through the collection of enormous data regarding the soil's overall sanitary status, including harmful agents (e.g., pests and diseases), weeds, and heavy metal(loid)s, beginning with climatological traits (Shafi et al., 2019; Raklami et al., 2022). The massive data is then examined and filtered to generate the right agricultural decision. Smart irrigation, the pillar of precision agriculture, is effectuated based on the phenological stage of the plant, for instance (Campos et al., 2019). Therefore, computational intelligence can help conventional farmers proceed like strategists, by permitting the usage of sophisticated technological tools and their implementation into and on their fields and crops, thus implicating the right solution at the opportune moment.

8. PAVING THE PATTERN TOWARDS DIGITAL AGRICULTURE

Precision agriculture holds tremendous potential for sustaining and affording better agricultural systems (Ben Ayed and Hanana, 2021). Furthermore, digital agriculture can withstand and deal with unprecedented climate variations (Lindblom et al., 2017). AI, IoT, and Machine-Learning applications, either applied separately or altogether, can both process and generate important data, hence providing information of utmost interest to the farmers (Pathan et al., 2020; Ramdinthara et al., 2022). Digital agriculture relies heavily on technological features like AI, IoT, and Machine-Learning, intended to assemble, process, analyze, and emit data on agricultural characteristics. Its highest advantages are a rise in agricultural practices' efficacity and sustainability. The digital tools that can track and monitor a given agricultural system can give insightful output on the nature of the decision to be taken. Thus, farmers of the 21st century can retrieve critical details on crops' overall performance status, in addition to surrounding conditions (Sishodia et al., 2020). The worldwide market of precision agriculture valuation was 5.49 billion USD in 2021 and is expected to attain 19.24 billion USD by the end of the current decade (Straits-Research, 2022). Simply put, technological advancement of pondering computing set-ups is progressively, yet drastically, transforming agriculture (Ciruela-Lorenzo et al., 2020).

9. CHALLENGES AND LIMITATIONS

Though intelligent computing systems like AI, IoT, and Machine-Learning offer a plethora of potential concerning the water-agriculture scheme, several issues deserve specific attention and to be acted upon efficiently.

AI-Related

Among major issues that surround AI applications, denial of service attacks emerges as the most worrying. The governmental implication to deal with this escalating issue is needed constantly. Moreover, setting up policies that should allow the approval of AI as a technological feature to advance small farmers' as well as stakeholders' agricultural systems on a comprehensive level is much needed.

IoT-Related

IoT is strongly associated with the Internet. Hence, Internet access is primordial. Consequently, the question that arises is whether every agricultural system can be facilitated with a suitable connection.

Machine-Learning-Related

Machine-Learning requires a thorough learning process that begins with assessing algorithms and, eventually, training data and addressing the right architecture associated with agricultural complicated problems, such as scarce water resources' efficient use and enhancing productivity, both of which are contradictory, yet requires specific attention.

On the one hand, the above-mentioned issues hindering a better performance of these advanced tools can be intersected. On the other hand, wells are running dry, which is contributing to the water scarcity problem. In addition, frequency and well numbers are declining. Another point is the use of watering mode by water-saving tools with computerized monitoring (drip, sprinkler, misting...) can only be completed by the management of soil as a vital component, including its humus which retains water. Opting for plant species requiring less water is another practical strategy. Therefore, it has become crucial to combine soil management, plant choice, and water-saving schemes with digitalization for smart irrigation and precision agriculture efficiency.

10. CONCLUSIONS AND FUTURE OUTLOOKS

Optimizing water use begins with monitoring and forecasting water and water-related parameters. In this regard, AI, IoT, and Machine-Learning constitute a successful combination, with tremendous potential for achieving this goal. In this regard, the present chapter underscores the varied roles played by these technological features in optimizing as well as preserving water at many levels. Water is a vital source that is at the core of every sector, yet immense pressure is put on freshwater resources that are already scarce. In this regard, computational intelligence has been revolutionizing the understanding of crucial field sectors like agriculture, in terms of water requirements. Water is at the heart of agriculture. Hence, rational irrigation is critical to both preserve water, an already scarce source, and optimize agrosystems' production. The application of AI, IoT, and Machine-Learning are some of the computational methods used to achieve the smart irrigation-precision agriculture approach, which relies on features like specified algorithms and software. Consequently, effective methods are required to monitor soils' moisture levels, improve water use, and boost yields. This is how the present chapter discusses many cutting-edge methods incorporating computational intelligence into agricultural systems. Though several experiments have been successful, completely switching to precision agriculture is yet to be embraced by farmers and stakeholders on a broader level, owing to issues to be surmounted like setting up adequate policies, assuring extensive training, and further improving computational intelligence scope. Moreover, improving the physicochemical and biological soil properties, in addition to plant species choice, water-wise, represents a strategy to behold.

CONFLICT OF INTEREST

The authors declare no conflict of interest.

REFERENCES

Abdurahman, R. (2022). *IoT-Based Smart Farming Using Machine Learning For Red Spinach*. Academic Press.

Abioye, E. A., Abidin, M. S. Z., Aman, M. N., Mahmud, M. S. A., & Buyamin, S. (2021). A model predictive controller for precision irrigation using discrete Lagurre networks. *Computers and Electronics in Agriculture, 181*, 105953. doi:10.1016/j.compag.2020.105953

Ait-El-Mokhtar, M. (2022). Cereals and Phytohormones Under Drought Stress. *Sustainable Remedies for Abiotic Stress in Cereals*. https://doi.org/https://doi.org/10.1007/978-981-19-5121-3_13

Ait Rahou, Y., Ait-El-Mokhtar, M., Anli, M., Boutasknit, A., Ben-Laouane, R., Douira, A., Benkirane, R., El Modafar, C., & Meddich, A. (2021). Use of mycorrhizal fungi and compost for improving the growth and yield of tomato and its resistance to *Verticillium dahliae. Archiv für Phytopathologie und Pflanzenschutz, 54*(13–14), 665–690. doi:10.1080/03235408.2020.1854938

Akensous, F., Anli, M., Boutasknit, A., Ben-Laouane, R., Ait-Rahou, Y., Ahmed, H., Nasri, N., Hafidi, M., & Meddich, A. (2022). Boosting Date Palm (*Phoenix dactylifera* L.) Growth under Drought Stress: Effects of Innovative Biostimulants. *Gesunde Pflanzen, 74*(4), 961–982. doi:10.100710343-022-00651-0

Akensous, F.-Z., Anli, M., & Meddich, A. (2022). Biostimulants as Innovative Tools to Boost Date Palm (*Phoenix dactylifera* L.) Performance under Drought, Salinity, and Heavy Metal(Oid)s′ Stresses: A Concise Review. *Sustainability (Basel), 14*(23), 15984. doi:10.3390u142315984

Alkaisi, A., Mossad, R., & Sharifian-Barforoush, A. (2017). A Review of the Water Desalination Systems Integrated with Renewable Energy. *Energy Procedia, 110*, 268–274. doi:10.1016/j.egypro.2017.03.138

Alves, R. G., Maia, R. F., & Lima, F. (2023). Development of a Digital Twin for smart farming: Irrigation management system for water saving. *Journal of Cleaner Production, 388*, 135920. doi:10.1016/j.jclepro.2023.135920

Anli, M., Ait-El-Mokhtar, M., Akensous, F.-Z., Boutasknit, A., Ben-Laouane, R., Fakhech, A., Ouhaddou, R., Raho, O., & Meddich, A. (2023). Biofertilizers in Date Palm Cultivation. In *Date Palm* (pp. 266–296). CABI. doi:10.1079/9781800620209.0009

Anli, M., Baslam, M., Tahiri, A., Raklami, A., Symanczik, S., Boutasknit, A., Ait-El-Mokhtar, M., Ben-Laouane, R., Toubali, S., Ait Rahou, Y., Ait Chitt, M., Oufdou, K., Mitsui, T., Hafidi, M., & Meddich, A. (2020). Biofertilizers as Strategies to Improve Photosynthetic Apparatus, Growth, and Drought Stress Tolerance in the Date Palm. *Frontiers in Plant Science, 11*, 1–22. doi:10.3389/fpls.2020.516818 PMID:33193464

Antony, A. P., Leith, K., Jolley, C., Lu, J., & Sweeney, D. J. (2020). A review of practice and implementation of the internet of things (IoT) for smallholder agriculture. *Sustainability (Basel), 12*(9), 1–19. doi:10.3390u12093750

Aziz, M., Khan, M., Anjum, N., Sultan, M., Shamshiri, R. R., Ibrahim, S. M., Balasundram, S. K., & Aleem, M. (2022). Scientific Irrigation Scheduling for Sustainable Production in Olive Groves. *Agriculture, 12*(4), 564. doi:10.3390/agriculture12040564

Babaeian, E., Sadeghi, M., Jones, S. B., Montzka, C., Vereecken, H., & Tuller, M. (2019). Ground, Proximal, and Satellite Remote Sensing of Soil Moisture. *Reviews of Geophysics, 57*(2), 530–616. doi:10.1029/2018RG000618

Bamurigire, P., Vodacek, A., Valko, A., & Rutabayiro Ngoga, S. (2020). Simulation of Internet of Things Water Management for Efficient Rice Irrigation in Rwanda. *Agriculture, 10*(10), 431. doi:10.3390/agriculture10100431

Ben Ayed, R., & Hanana, M. (2021). Artificial Intelligence to Improve the Food and Agriculture Sector. *Journal of Food Quality, 2021*(Ml), 1–7. doi:10.1155/2021/5584754

Ben-Laouane, R., Ait-El-Mokhtar, M., Anli, M., Boutasknit, A., Ait-Rahou, Y., Oufdou, K., ... & Meddich, A. (2023). Potential of Biofertilizers for Soil Enhancement: Study on Growth, Physiological, and Biochemical Traits of Medicalo Sativa. *Bioremediation And Phytoremediation Technologies in Sustainable Soil Management*, 75-97.

Benaffari, W., Boutasknit, A., Anli, M., Ait-El-Mokhtar, M., Ait-Rahou, Y., Ben-Laouane, R., Ben Ahmed, H., Mitsui, T., Baslam, M., & Meddich, A. (2022). The Native Arbuscular Mycorrhizal Fungi and Vermicompost-Based Organic Amendments Enhance Soil Fertility, Growth Performance, and the Drought Stress Tolerance of Quinoa. *Plants*, *11*(3), 393. doi:10.3390/plants11030393 PMID:35161374

Bjornlund, H., van Rooyen, A., Pittock, J., Parry, K., Moyo, M., Mdemu, M., & de Sousa, W. (2020). Institutional innovation and smart water management technologies in small-scale irrigation schemes in southern Africa. *Water International*, *45*(6), 621–650. doi:10.1080/02508060.2020.1804715

Boutasknit, A., Baslam, M., Ait-El-Mokhtar, M., Anli, M., Ben-Laouane, R., Ait-Rahou, Y., Mitsui, T., Douira, A., El Modafar, C., Wahbi, S., & Meddich, A. (2021). Assemblage of indigenous arbuscular mycorrhizal fungi and green waste compost enhance drought stress tolerance in carob (*Ceratonia siliqua* L.) trees. *Scientific Reports*, *11*(1), 22835. doi:10.103841598-021-02018-3 PMID:34819547

Boutasknit, A., Baslam, M., Anli, M., Ait-El-Mokhtar, M., Ben-Laouane, R., Ait-Rahou, Y., El Modafar, C., Douira, A., Wahbi, S., & Meddich, A. (2022). Impact of arbuscular mycorrhizal fungi and compost on the growth, water status, and photosynthesis of carob (*Ceratonia siliqua*) under drought stress and recovery. *Plant Biosystems - An International Journal Dealing with All Aspects of Plant Biology*, *156*(4), 994–1010. doi:10.1080/11263504.2021.1985006

Brown, T. C., Mahat, V., & Ramirez, J. A. (2019). Adaptation to Future Water Shortages in the United States Caused by Population Growth and Climate Change. *Earth's Future*, *7*(3), 219–234. doi:10.1029/2018EF001091

Bwambale, E., Abagale, F. K., & Anornu, G. K. (2022a). Smart irrigation monitoring and control strategies for improving water use efficiency in precision agriculture: A review. *Agricultural Water Management*, *260*, 107324. doi:10.1016/j.agwat.2021.107324

Bwambale, E., Abagale, F. K., & Anornu, G. K. (2022b). Smart irrigation monitoring and control strategies for improving water use efficiency in precision agriculture: A review. *Agricultural Water Management*, *260*, 107324. doi:10.1016/j.agwat.2021.107324

Caballero, D., Calvini, R., & Amigo, J. M. (2019). Hyperspectral imaging in crop fields: precision agriculture. In Data Handling in Science and Technology (Vol. 32, pp. 453–473). doi:10.1016/B978-0-444-63977-6.00018-3

Campos, N., Rocha, A. R., Gondim, R., Coelho da Silva, T. L., & Gomes, D. G. (2019). Smart & Green: An Internet-of-Things Framework for Smart Irrigation. *Sensors (Basel), 20*(1), 190. doi:10.339020010190 PMID:31905749

Castellazzi, P., Burgess, D., Rivera, A., Huang, J., Longuevergne, L., & Demuth, M. N. (2019). Glacial Melt and Potential Impacts on Water Resources in the Canadian Rocky Mountains. *Water Resources Research, 55*(12), 10191–10217. doi:10.1029/2018WR024295

Chandrappa, V. Y., Ray, B., Ashwatha, N., & Shrestha, P. (2023). Spatiotemporal modeling to predict soil moisture for sustainable smart irrigation. *Internet of Things, 21*, 100671. doi:10.1016/j.iot.2022.100671

Chang, Y.-C., Huang, T.-W., & Huang, N.-F. (2019). A Machine Learning Based Smart Irrigation System with LoRa P2P Networks. *2019 20th Asia-Pacific Network Operations and Management Symposium (APNOMS)*, 1–4. 10.23919/APNOMS.2019.8893034

Ciruela-Lorenzo, A. M., Del-Aguila-Obra, A. R., Padilla-Meléndez, A., & Plaza-Angulo, J. J. (2020). Digitalization of Agri-Cooperatives in the Smart Agriculture Context. Proposal of a Digital Diagnosis Tool. *Sustainability (Basel), 12*(4), 1325. doi:10.3390u12041325

Coffel, E. D., Keith, B., Lesk, C., Horton, R. M., Bower, E., Lee, J., & Mankin, J. S. (2019). Future Hot and Dry Years Worsen Nile Basin Water Scarcity Despite Projected Precipitation Increases. *Earth's Future, 7*(8), 967–977. doi:10.1029/2019EF001247

Currie, G., & Rohren, E. (2021). Intelligent Imaging in Nuclear Medicine: The Principles of Artificial Intelligence, Machine Learning and Deep Learning. *Seminars in Nuclear Medicine, 51*(2), 102–111. doi:10.1053/j.semnuclmed.2020.08.002 PMID:33509366

Distefano, T., & Kelly, S. (2017). Are we in deep water? Water scarcity and its limits to economic growth. *Ecological Economics, 142*, 130–147. doi:10.1016/j.ecolecon.2017.06.019

Doorn, N. (2021). Artificial intelligence in the water domain: Opportunities for responsible use. *The Science of the Total Environment, 755*, 142561. doi:10.1016/j.scitotenv.2020.142561 PMID:33039891

Falkenmark, M. (2013). Growing water scarcity in agriculture: future challenge to global water security. *Philosophical Transactions of the Royal Society A: Mathematical, Physical and Engineering Sciences, 371*(2002), 20120410. doi:10.1098/rsta.2012.0410

Fetanat, A., Tayebi, M., & Mofid, H. (2021). Water-energy-food security nexus based selection of energy recovery from wastewater treatment technologies: An extended decision making framework under intuitionistic fuzzy environment. *Sustainable Energy Technologies and Assessments, 43*, 100937. doi:10.1016/j.seta.2020.100937

Fishman, R. (2016). More uneven distributions overturn benefits of higher precipitation for crop yields. *Environmental Research Letters, 11*(2), 024004. doi:10.1088/1748-9326/11/2/024004

García, L., Parra, L., Jimenez, J. M., Lloret, J., & Lorenz, P. (2020). IoT-Based Smart Irrigation Systems: An Overview on the Recent Trends on Sensors and IoT Systems for Irrigation in Precision Agriculture. *Sensors (Basel), 20*(4), 1042. doi:10.339020041042 PMID:32075172

Goap, A., Sharma, D., Shukla, A. K., & Rama Krishna, C. (2018). An IoT based smart irrigation management system using Machine learning and open source technologies. *Computers and Electronics in Agriculture, 155*(May), 41–49. doi:10.1016/j.compag.2018.09.040

González Perea, R., Camacho Poyato, E., Montesinos, P., & Rodríguez Díaz, J. A. (2018). Prediction of applied irrigation depths at farm level using artificial intelligence techniques. *Agricultural Water Management, 206*(May), 229–240. doi:10.1016/j.agwat.2018.05.019

Goodrich, P., Betancourt, O., Arias, A. C., & Zohdi, T. (2023). Placement and drone flight path mapping of agricultural soil sensors using machine learning. *Computers and Electronics in Agriculture, 205*, 107591. doi:10.1016/j.compag.2022.107591

Hoekstra, A. Y. (2017). Water Footprint Assessment: Evolvement of a New Research Field. *Water Resources Management, 31*(10), 3061–3081. doi:10.100711269-017-1618-5

Ikan, C., Ben-Laouane, R., Ouhaddou, R., Ghoulam, C., & Meddich, A. (2023). Co-inoculation of arbuscular mycorrhizal fungi and plant growth-promoting rhizobacteria can mitigate the effects of drought in wheat plants (*Triticum durum*). *Plant Biosystems - An International Journal Dealing with All Aspects of Plant Biology*, 1–13. doi:10.1080/11263504.2023.2229856

Jalajamony, H. M., Nair, M., Mead, P. F., & Fernandez, R. E. (2023). Drone Aided Thermal Mapping for Selective Irrigation of Localized Dry Spots. *IEEE Access, 11*, 7320–7335. doi:10.1109/ACCESS.2023.3237546

Jha, K., Doshi, A., Patel, P., & Shah, M. (2019). A comprehensive review on automation in agriculture using artificial intelligence. *Artificial Intelligence in Agriculture*, *2*, 1–12. doi:10.1016/j.aiia.2019.05.004

Kamienski, C., Soininen, J.-P., Taumberger, M., Dantas, R., Toscano, A., Salmon Cinotti, T., Filev Maia, R., & Torre Neto, A. (2019). Smart Water Management Platform: IoT-Based Precision Irrigation for Agriculture. *Sensors (Basel)*, *19*(2), 276. doi:10.339019020276 PMID:30641960

Kassim, M. R. M. (2020). IoT Applications in Smart Agriculture: Issues and Challenges. *2020 IEEE Conference on Open Systems (ICOS)*, 19–24. 10.1109/ICOS50156.2020.9293672

Katimbo, A., Rudnick, D. R., Zhang, J., Ge, Y., DeJonge, K. C., Franz, T. E., Shi, Y., Liang, W., Qiao, X., Heeren, D. M., Kabenge, I., Nakabuye, H. N., & Duan, J. (2023). Evaluation of artificial intelligence algorithms with sensor data assimilation in estimating crop evapotranspiration and crop water stress index for irrigation water management. *Smart Agricultural Technology, 4*, 100176. doi:10.1016/j.atech.2023.100176

Kavdir, Y., Zhang, W., Basso, B., & Smucker, A. J. M. (2014). Development of a new long-term drought resilient soil water retention technology. *Journal of Soil and Water Conservation*, *69*(5), 154A–160A. doi:10.2489/jswc.69.5.154A

Krishnan, R. S., Julie, E. G., Robinson, Y. H., Raja, S., Kumar, R., Thong, P. H., & Son, L. H. (2020). Fuzzy Logic based Smart Irrigation System using Internet of Things. *Journal of Cleaner Production*, *252*, 119902. doi:10.1016/j.jclepro.2019.119902

Kumar, A., Surendra, A., Mohan, H., Valliappan, K. M., & Kirthika, N. (2017). Internet of things based smart irrigation using regression algorithm. *2017 International Conference on Intelligent Computing, Instrumentation and Control Technologies (ICICICT),* 1652–1657. 10.1109/ICICICT1.2017.8342819

Lahbouki, S., Ech-chatir, L., Er-Raki, S., Outzourhit, A., & Meddich, A. (2022). Improving drought tolerance of *Opuntia ficus-indica* under field using subsurface water retention technology: Changes in physiological and biochemical parameters. *Canadian Journal of Soil Science*, *102*(4), 888–898. doi:10.1139/cjss-2022-0022

LaVanchy, G. T. (2017). When wells run dry: Water and tourism in Nicaragua. *Annals of Tourism Research*, *64*, 37–50. doi:10.1016/j.annals.2017.02.006

Leh, N. A. M., Kamaldin, M. S. A. M., Muhammad, Z., & Kamarzaman, N. A. (2019). Smart Irrigation System Using Internet of Things. *2019 IEEE 9th International Conference on System Engineering and Technology (ICSET),* 96–101. 10.1109/ICSEngT.2019.8906497

Liakos, K. G., Busato, P., Moshou, D., Pearson, S., & Bochtis, D. (2018). Machine learning in agriculture: A review. *Sensors (Basel)*, *18*(8), 1–29. doi:10.339018082674 PMID:30110960

Lindblom, J., Lundström, C., Ljung, M., & Jonsson, A. (2017). Promoting sustainable intensification in precision agriculture: Review of decision support systems development and strategies. *Precision Agriculture*, *18*(3), 309–331. doi:10.100711119-016-9491-4

Lowenberg-DeBoer, J., & Erickson, B. (2019). Setting the Record Straight on Precision Agriculture Adoption. *Agronomy Journal*, *111*(4), 1552–1569. doi:10.2134/agronj2018.12.0779

Mabhaudhi, T., Mpandeli, S., Nhamo, L., Chimonyo, V. G. P., Nhemachena, C., Senzanje, A., Naidoo, D., & Modi, A. T. (2018). Prospects for improving irrigated agriculture in Southern Africa: Linking water, energy and food. *Water (Basel)*, *10*(12), 1–16. doi:10.3390/w10121881

Mahapatra, D. M., Satapathy, K. C., & Panda, B. (2022). Biofertilizers and nanofertilizers for sustainable agriculture: Phycoprospects and challenges. *The Science of the Total Environment*, *803*, 149990. doi:10.1016/j.scitotenv.2021.149990 PMID:34492488

Mamatha, V., & Kavitha, J. C. (2023). Machine learning based crop growth management in greenhouse environment using hydroponics farming techniques. *Measurement: Sensors, 25*, 100665. doi:10.1016/j.measen.2023.100665

Mardero, S., Schmook, B., Christman, Z., Metcalfe, S. E., & De la Barreda-Bautista, B. (2020). Recent disruptions in the timing and intensity of precipitation in Calakmul, Mexico. *Theoretical and Applied Climatology*, *140*(1–2), 129–144. doi:10.100700704-019-03068-4

Markland, S. M., Ingram, D., Kniel, K. E., & Sharma, M. (2018). Water for Agriculture: the Convergence of Sustainability and Safety. In Preharvest Food Safety (pp. 143–157). ASM Press. doi:10.1128/9781555819644.ch8

McBratney, A., Whelan, B., Ancev, T., & Bouma, J. (2005). Future Directions of Precision Agriculture. *Precision Agriculture*, *6*(1), 7–23. doi:10.100711119-005-0681-8

Meddich, A. (2022). Biostimulants for Resilient Agriculture—Improving Plant Tolerance to Abiotic Stress: A Concise Review. *Gesunde Pflanzen*, 1–19. doi:10.100710343-022-00784-2

Meemken, E.-M., & Qaim, M. (2018). Organic Agriculture, Food Security, and the Environment. *Annual Review of Resource Economics*, *10*(1), 39–63. doi:10.1146/annurev-resource-100517-023252

Mendes, W. R., Araújo, F. M. U., Dutta, R., & Heeren, D. M. (2019). Fuzzy control system for variable rate irrigation using remote sensing. *Expert Systems with Applications*, *124*, 13–24. doi:10.1016/j.eswa.2019.01.043

Mohammad Fakhrul Islam, S., & Karim, Z. (2020). World's Demand for Food and Water: The Consequences of Climate Change. In Desalination - Challenges and Opportunities (pp. 1–27). IntechOpen. doi:10.5772/intechopen.85919

Mohammed, M., Riad, K., & Alqahtani, N. (2021). Efficient IoT-Based Control for a Smart Subsurface Irrigation System to Enhance Irrigation Management of Date Palm. *Sensors (Basel)*, *21*(12), 3942. doi:10.339021123942 PMID:34201041

Moreiro, L. B. (2017). Interest of seeing Precision Viticulture through two distributed competences: Determination of resources and schemes allowing some practical recommendations. *BIO Web of Conferences*, *9*, 01023. 10.1051/bioconf/20170901023

Muley, R. J., & Bhonge, V. N. (2019). Internet of Things for Irrigation Monitoring and Controlling. In *Advances in Intelligent Systems and Computing* (Vol. 810, pp. 165–174). Springer Singapore. doi:10.1007/978-981-13-1513-8_18

Murthy, A., Green, C., Stoleru, R., Bhunia, S., Swanson, C., & Chaspari, T. (2019). Machine Learning-based Irrigation Control Optimization. *Proceedings of the 6th ACM International Conference on Systems for Energy-Efficient Buildings, Cities, and Transportation*, 213–222. 10.1145/3360322.3360854

Nawandar, N. K., & Satpute, V. R. (2019). IoT based low cost and intelligent module for smart irrigation system. *Computers and Electronics in Agriculture*, *162*, 979–990. doi:10.1016/j.compag.2019.05.027

Obaideen, K., Yousef, B. A. A., AlMallahi, M. N., Tan, Y. C., Mahmoud, M., Jaber, H., & Ramadan, M. (2022a). An overview of smart irrigation systems using IoT. *Energy Nexus*, *7*(July), 100124. doi:10.1016/j.nexus.2022.100124

Obaideen, K., Yousef, B. A. A., AlMallahi, M. N., Tan, Y. C., Mahmoud, M., Jaber, H., & Ramadan, M. (2022b). An overview of smart irrigation systems using IoT. *Energy Nexus*, *7*(January), 100124. doi:10.1016/j.nexus.2022.100124

Odara, S., Khan, Z., & Ustun, T. S. (2015). Optimizing energy use of SmartFarms with smartgrid integration. *2015 3rd International Renewable and Sustainable Energy Conference (IRSEC)*, 1–6. 10.1109/IRSEC.2015.7454980

Ortigara, A., Kay, M., & Uhlenbrook, S. (2018). A Review of the SDG 6 Synthesis Report 2018 from an Education, Training, and Research Perspective. *Water (Basel)*, *10*(10), 1353. doi:10.3390/w10101353

Ouhaddou, R., Ech-chatir, L., Anli, M., Ben-Laouane, R., Boutasknit, A., & Meddich, A. (2023). Secondary Metabolites, Osmolytes and Antioxidant Activity as the Main Attributes Enhanced by Biostimulants for Growth and Resilience of Lettuce to Drought Stress. *Gesunde Pflanzen*, 1–17. doi:10.100710343-022-00827-8

Pathan, M., Patel, N., Yagnik, H., & Shah, M. (2020). Artificial cognition for applications in smart agriculture: A comprehensive review. *Artificial Intelligence in Agriculture*, *4*, 81–95. doi:10.1016/j.aiia.2020.06.001

Patrício, D. I., & Rieder, R. (2018). Computer vision and artificial intelligence in precision agriculture for grain crops: A systematic review. *Computers and Electronics in Agriculture*, *153*(June), 69–81. doi:10.1016/j.compag.2018.08.001

Pham, V. B., Diep, T. T., Fock, K., & Nguyen, T. S. (2021). Using the Internet of Things to promote alternate wetting and drying irrigation for rice in Vietnam's Mekong Delta. *Agronomy for Sustainable Development*, *41*(3), 43. doi:10.100713593-021-00705-z

Prasanna Lakshmi, G. S., Asha, P. N., Sandhya, G., Vivek Sharma, S., Shilpashree, S., & Subramanya, S. G. (2023). An intelligent IOT sensor coupled precision irrigation model for agriculture. *Measurement: Sensors, 25*, 100608. doi:10.1016/j.measen.2022.100608

Qadri, H., Bhat, R. A., Mehmood, M. A., & Dar, G. H. (2020). Fresh Water Pollution Dynamics and Remediation. In H. Qadri, R. A. Bhat, M. A. Mehmood, & G. H. Dar (Eds.), Fresh Water Pollution Dynamics and Remediation. Springer Singapore. doi:10.1007/978-981-13-8277-2

Raho, O., Boutasknit, A., Anli, M., Ben-Laouane, R., Rahou, Y. A., Ouhaddou, R., Duponnois, R., Douira, A., El Modafar, C., & Meddich, A. (2022). Impact of Native Biostimulants/Biofertilizers and Their Synergistic Interactions On the Agro-physiological and Biochemical Responses of Date Palm Seedlings. *Gesunde Pflanzen*, *74*(4), 1053–1069. doi:10.100710343-022-00668-5

Rakesh, S., Ramesh, D. P., Murugaragavan, D. R., Avudainayagam, D. S., & Karthikeyan, D. S. (2020). Characterization and treatment of grey water: A review. *International Journal of Chemical Studies*, *8*(1), 34–40. doi:10.22271/chemi.2020.v8.i1a.8316

Raklami, A., Meddich, A., Oufdou, K., & Baslam, M. (2022). Plants—Microorganisms-Based Bioremediation for Heavy Metal Cleanup: Recent Developments, Phytoremediation Techniques, Regulation Mechanisms, and Molecular Responses. *International Journal of Molecular Sciences*, *23*(9), 5031. doi:10.3390/ijms23095031 PMID:35563429

Ramachandran, V., Ramalakshmi, R., Kavin, B., Hussain, I., Almaliki, A., Almaliki, A., Elnaggar, A., & Hussein, E. (2022). Exploiting IoT and Its Enabled Technologies for Irrigation Needs in Agriculture. *Water (Basel)*, *14*(5), 719. doi:10.3390/w14050719

Ramdinthara, I. Z., Bala, P. S., & Gowri, A. S. (2022). AI-Based Yield Prediction and Smart Irrigation. In *Studies in Big Data* (Vol. 99, pp. 113–140). doi:10.1007/978-981-16-6210-2_6

Rao, N. H. (2018). Big data and climate smart agriculture - Status and implications for agricultural research and innovation in India. *Proceedings of the Indian National Science Academy. Part A, Physical Sciences*, *84*(3), 625–640. doi:10.16943/ptinsa/2018/49342

Rasooli, M. W., Bhushan, B., & Kumar, N. (2020). Applicability of wireless sensor networks & IoT in saffron & wheat crops: A smart agriculture perspective. *International Journal of Scientific and Technology Research*, *9*(2), 2456–2461.

Ray, S. (2019). A Quick Review of Machine Learning Algorithms. *2019 International Conference on Machine Learning, Big Data, Cloud and Parallel Computing (COMITCon)*, 35–39. 10.1109/COMITCon.2019.8862451

Ray, S. S., Verma, R. K., Singh, A., Ganesapillai, M., & Kwon, Y.-N. (2023). A holistic review on how artificial intelligence has redefined water treatment and seawater desalination processes. *Desalination*, *546*, 116221. doi:10.1016/j.desal.2022.116221

Rodionova, M. V., Poudyal, R. S., Tiwari, I., Voloshin, R. A., Zharmukhamedov, S. K., Nam, H. G., Zayadan, B. K., Bruce, B. D., Hou, H. J. M., & Allakhverdiev, S. I. (2017). Biofuel production: Challenges and opportunities. *International Journal of Hydrogen Energy*, *42*(12), 8450–8461. doi:10.1016/j.ijhydene.2016.11.125

Rodriguez-Ortega, W. M., Martinez, V., Rivero, R. M., Camara-Zapata, J. M., Mestre, T., & Garcia-Sanchez, F. (2017). Use of a smart irrigation system to study the effects of irrigation management on the agronomic and physiological responses of tomato plants grown under different temperatures regimes. *Agricultural Water Management*, *183*, 158–168. doi:10.1016/j.agwat.2016.07.014

Roy, P. C., Guber, A., Abouali, M., Nejadhashemi, A. P., Deb, K., & Smucker, A. J. M. (2019). Crop yield simulation optimization using precision irrigation and subsurface water retention technology. *Environmental Modelling & Software*, *119*(July), 433–444. doi:10.1016/j.envsoft.2019.07.006

Ruane, J., Sonnino, A., Steduto, P., & Deane, C. (2008). *Coping with water scarcity: What role for biotechnologies?* Land and Water Discussion Paper.

Sadique, K. M., Rahmani, R., & Johannesson, P. (2018). Towards security on internet of things: Applications and challenges in technology. *Procedia Computer Science*, *141*, 199–206. doi:10.1016/j.procs.2018.10.168

Sanikhani, H., Kisi, O., Maroufpoor, E., & Yaseen, Z. M. (2019). Temperature-based modeling of reference evapotranspiration using several artificial intelligence models: Application of different modeling scenarios. *Theoretical and Applied Climatology*, *135*(1–2), 449–462. doi:10.100700704-018-2390-z

Satish, A., Nandhini, R., Poovizhi, S., Jose, P., Ranjitha, R., & Anila, S. (2017). Arduino based Smart Irrigation System using IoT. *3rd National Conference on Intelligent Information and Computing Technologies (IICT '17), December*, 1–5.

Shafi, U., Mumtaz, R., García-Nieto, J., Hassan, S. A., Zaidi, S. A. R., & Iqbal, N. (2019). Precision Agriculture Techniques and Practices: From Considerations to Applications. *Sensors (Basel)*, *19*(17), 3796. doi:10.339019173796 PMID:31480709

Sharma, A., Jain, A., Gupta, P., & Chowdary, V. (2021). Machine Learning Applications for Precision Agriculture: A Comprehensive Review. *IEEE Access : Practical Innovations, Open Solutions*, *9*, 4843–4873. doi:10.1109/ACCESS.2020.3048415

Shrestha, N. K., Du, X., & Wang, J. (2017). Assessing climate change impacts on fresh water resources of the Athabasca River Basin, Canada. *The Science of the Total Environment*, *601–602*, 425–440. doi:10.1016/j.scitotenv.2017.05.013 PMID:28570976

Singh, P., Pandey, P. C., Petropoulos, G. P., Pavlides, A., Srivastava, P. K., Koutsias, N., Deng, K. A. K., & Bao, Y. (2020). Hyperspectral remote sensing in precision agriculture: present status, challenges, and future trends. In *Hyperspectral Remote Sensing* (pp. 121–146). Elsevier. doi:10.1016/B978-0-08-102894-0.00009-7

Sinha, A., Shrivastava, G., & Kumar, P. (2019). Architecting user-centric internet of things for smart agriculture. *Sustainable Computing: Informatics and Systems, 23*, 88–102. doi:10.1016/j.suscom.2019.07.001

Sishodia, R. P., Ray, R. L., & Singh, S. K. (2020). Applications of Remote Sensing in Precision Agriculture: A Review. *Remote Sensing (Basel), 12*(19), 3136. doi:10.3390/rs12193136

Slimani, A. Z., A. F., Oufdou, K., & & Meddich, A. (2023). Impact of Climate Change on Water Status: Challenges and Emerging Solutions. In Water in Circular Economy (pp. 3–20). Academic Press.

Slimani, A., Raklami, A., Oufdou, K., & Meddich, A. (2022). Isolation and Characterization of PGPR and Their Potenzial for Drought Alleviation in Barley Plants. *Gesunde Pflanzen*, 1–15. doi:10.100710343-022-00709-z

Straits-Research. (2022). *Precision Agriculture Market Size is projected to reach USD 19.24 Billion by 2030, growing at a CAGR of 14.95%: Straits Research.* Https://Www.Globenewswire.Com/En/News-Release/2022/08/01/2489650/0/En/Precision-Agriculture-Market-Size-Is-Projected-to-Reach-USD-19-24-Billion-by-2030-Growing-at-a-CAGR-of-14-95-Straits-Research.Html

Sudarmaji, A., Sahirman, S., Saparso, & Ramadhani, Y. (2019). Time based automatic system of drip and sprinkler irrigation for horticulture cultivation on coastal area. *IOP Conference Series. Earth and Environmental Science, 250*(1), 012074. doi:10.1088/1755-1315/250/1/012074

Sun, J. X., Yin, Y. L., Sun, S. K., Wang, Y. B., Yu, X., & Yan, K. (2021). Review on research status of virtual water: The perspective of accounting methods, impact assessment and limitations. *Agricultural Water Management, 243*, 106407. doi:10.1016/j.agwat.2020.106407

Syrmos, E., Sidiropoulos, V., Bechtsis, D., Stergiopoulos, F., Aivazidou, E., Vrakas, D., Vezinias, P., & Vlahavas, I. (2023). An Intelligent Modular Water Monitoring IoT System for Real-Time Quantitative and Qualitative Measurements. *Sustainability (Basel), 15*(3), 2127. doi:10.3390u15032127

Tahiri, A. I., Meddich, A., Raklami, A., Alahmad, A., Bechtaoui, N., Anli, M., Göttfert, M., Heulin, T., Achouak, W., & Oufdou, K. (2022). Assessing the potential role of compost, PGPR, and AMF in improving tomato plant growth, yield, fruit quality, and water stress tolerance. *Journal of Soil Science and Plant Nutrition, 22*(1), 1–22. doi:10.100742729-021-00684-w

Talaviya, T., Shah, D., Patel, N., Yagnik, H., & Shah, M. (2020). Implementation of artificial intelligence in agriculture for optimisation of irrigation and application of pesticides and herbicides. *Artificial Intelligence in Agriculture*, *4*, 58–73. doi:10.1016/j.aiia.2020.04.002

Tantalaki, N., Souravlas, S., & Roumeliotis, M. (2019). Data-Driven Decision Making in Precision Agriculture: The Rise of Big Data in Agricultural Systems. *Journal of Agricultural & Food Information*, *20*(4), 344–380. doi:10.1080/10496505.2019.1638264

Thakur, D., Kumar, Y., Kumar, A., & Singh, P. K. (2019). Applicability of Wireless Sensor Networks in Precision Agriculture: A Review. *Wireless Personal Communications*, *107*(1), 471–512. doi:10.100711277-019-06285-2

Tien, J. M. (2017). Internet of Things, Real-Time Decision Making, and Artificial Intelligence. *Annals of Data Science*, *4*(2), 149–178. doi:10.100740745-017-0112-5

Togneri, R., Felipe dos Santos, D., Camponogara, G., Nagano, H., Custódio, G., Prati, R., Fernandes, S., & Kamienski, C. (2022). Soil moisture forecast for smart irrigation: The primetime for machine learning. *Expert Systems with Applications*, *207*(April), 1–23. doi:10.1016/j.eswa.2022.117653

Toubali, S., Ait-El-Mokhtar, M., Boutasknit, A., Anli, M., Ait-Rahou, Y., Benaffari, W., Ben-Ahmed, H., Mitsui, T., Baslam, M., & Meddich, A. (2022). Root Reinforcement Improved Performance, Productivity, and Grain Bioactive Quality of Field-Droughted Quinoa (*Chenopodium quinoa*). *Frontiers in Plant Science*, *13*(March), 1–20. doi:10.3389/fpls.2022.860484 PMID:35371170

Touil, S., Richa, A., Fizir, M., Argente García, J. E., & Skarmeta Gómez, A. F. (2022). A review on smart irrigation management strategies and their effect on water savings and crop yield. *Irrigation and Drainage*, *71*(5), 1396–1416. doi:10.1002/ird.2735

Tschand, A. (2023). Semi-supervised machine learning analysis of crop color for autonomous irrigation. *Smart Agricultural Technology, 3*, 100116. doi:10.1016/j.atech.2022.100116

Uddin, J., Smith, R. J., Gillies, M. H., Moller, P., & Robson, D. (2018). Smart Automated Furrow Irrigation of Cotton. *Journal of Irrigation and Drainage Engineering*, *144*(5), 1–10. doi:10.1061/(ASCE)IR.1943-4774.0001282

V, S. (2021). Internet of Things (IoT) based Smart Agriculture in India: An Overview. *Journal of ISMAC, 3*(1), 1–15. doi:10.36548/jismac.2021.1.001

van Vliet, M. T. H., Flörke, M., & Wada, Y. (2017). Quality matters for water scarcity. *Nature Geoscience*, *10*(11), 800–802. doi:10.1038/ngeo3047

van Vliet, M. T. H., Jones, E. R., Flörke, M., Franssen, W. H. P., Hanasaki, N., Wada, Y., & Yearsley, J. R. (2021). Global water scarcity including surface water quality and expansions of clean water technologies. *Environmental Research Letters*, *16*(2), 024020. doi:10.1088/1748-9326/abbfc3

Veerachamy, R., Ramar, R., Balaji, S., & Sharmila, L. (2022). Autonomous Application Controls on Smart Irrigation. *Computers & Electrical Engineering*, *100*(March), 107855. doi:10.1016/j.compeleceng.2022.107855

Veldkamp, T. I. E., Wada, Y., Aerts, J. C. J. H., Döll, P., Gosling, S. N., Liu, J., Masaki, Y., Oki, T., Ostberg, S., Pokhrel, Y., Satoh, Y., Kim, H., & Ward, P. J. (2017). Water scarcity hotspots travel downstream due to human interventions in the 20th and 21st century. *Nature Communications*, *8*(1), 15697. doi:10.1038/ncomms15697 PMID:28643784

Vij, A., Vijendra, S., Jain, A., Bajaj, S., Bassi, A., & Sharma, A. (2020). IoT and Machine Learning Approaches for Automation of Farm Irrigation System. *Procedia Computer Science, 167*(2019), 1250–1257. doi:10.1016/j.procs.2020.03.440

Vollmer, D., Shaad, K., Souter, N. J., Farrell, T., Dudgeon, D., Sullivan, C. A., Fauconnier, I., MacDonald, G. M., McCartney, M. P., Power, A. G., McNally, A., Andelman, S. J., Capon, T., Devineni, N., Apirumanekul, C., Ng, C. N., Rebecca Shaw, M., Wang, R. Y., Lai, C., ... Regan, H. M. (2018). Integrating the social, hydrological and ecological dimensions of freshwater health: The Freshwater Health Index. *The Science of the Total Environment*, *627*, 304–313. doi:10.1016/j.scitotenv.2018.01.040 PMID:29426153

Vrchota, J., Pech, M., & Švepešová, I. (2022). Precision Agriculture Technologies for Crop and Livestock Production in the Czech Republic. *Agriculture*, *12*(8), 1080. doi:10.3390/agriculture12081080

Wang, E., Attard, S., Linton, A., McGlinchey, M., Xiang, W., Philippa, B., & Everingham, Y. (2020). Development of a closed-loop irrigation system for sugarcane farms using the Internet of Things. *Computers and Electronics in Agriculture*, *172*(March), 105376. doi:10.1016/j.compag.2020.105376

Yin, J., Deng, Z., Ines, A. V. M., Wu, J., & Rasu, E. (2020). Forecast of short-term daily reference evapotranspiration under limited meteorological variables using a hybrid bi-directional long short-term memory model (Bi-LSTM). *Agricultural Water Management*, *242*(February), 106386. doi:10.1016/j.agwat.2020.106386

Zanfei, A., Menapace, A., & Righetti, M. (2023). An artificial intelligence approach for managing water demand in water supply systems. *IOP Conference Series. Earth and Environmental Science*, *1136*(1), 012004. doi:10.1088/1755-1315/1136/1/012004

Zhuo, L., & Hoekstra, A. Y. (2017). The effect of different agricultural management practices on irrigation efficiency, water use efficiency and green and blue water footprint. *Frontiers of Agricultural Science and Engineering*, *4*(2), 185. doi:10.15302/J-FASE-2017149

Chapter 7
A Study on Machine Learning-Based Water Quality Assessment and Wastewater Treatment

Satakshi Singh
https://orcid.org/0000-0002-5361-1898
Sam Higginbottom University of Agriculture, Technology, and Sciences, India

Suryanshi Mishra
Sam Higginbottom University of Agriculture, Technology, and Sciences, India

Tinku Singh
Indian Institute of Information Technology, Allahabad, India

Shobha Thakur
Sam Higginbottom University of Agriculture, Technology, and Sciences, India

ABSTRACT

The chapter will present state-of-the-art water assessment and treatment methods as well as the current issues and challenges of the domain. Modern techniques for water evaluation and treatment will be covered in this chapter, along with the current problems and difficulties facing the industry.

LEARNING OBJECTIVES

The chapter will present state-of-the-art water assessment and treatment methods

DOI: 10.4018/978-1-6684-6791-6.ch007

as well as the current issues and challenges of the domain. Modern techniques for water evaluation and treatment will be covered in this chapter, along with the current problems and difficulties facing the industry.

1. Usages of water.
2. Problems related to water: Water pollution, Water borne diseases, etc.
3. Survey of approaches to water-related problems.
4. Highlighting the application of machine learning to Water quality assessment and wastewater treatment

1. INTRODUCTION

Artificial intelligence (AI) and machine learning (ML)-based algorithms have recently assisted people in finding solutions to a variety of challenging and complicated real-world issues. Each day, enormous amounts of data are produced, making it practically impossible for humans to interpret them. These laborious processes can be accomplished by AI/ML approaches, which are also very beneficial for uncovering hidden knowledge, drawing conclusions from the data, and, in most situations, solving problems. Applications of AI and ML may be used to enhance environmental quality. Here, we address whether ML can effectively process and interpret a large amount of data. This study focuses on the ability of one of the most valuable environmental resources, namely, water, to be conserved. The possibilities of wastewater treatment for different reasons are discussed, as well as the uses of ML for water quality evaluation. This chapter will discuss current challenges and obstacles as well as state-of-the-art methods for assessing and treating water.

Data from the environment is continuously collected every day. As information increases exponentially, ML has become increasingly important for analyzing, classifying, and predicting. It can assist management and enhance water quality and reduce water pollution. Zhu et al. (Zhu et al., 2022), discuss occurrences where ML approaches have been employed to gauge water quality in several aquatic contexts, such as drinking water, sewage, salt water, and groundwater. Haghiabi et al. (2018) examine the effectiveness of AI approaches such as artificial neural network (ANN), group method of data handling (GMDH), and support vector machine (SVM) for estimating the quality of water of the Tireh River in Iran. Macrophytes are biological indicators (algae blooms, fish kills, and poor water clarity) used to gauge ecological status. They are called indicators and can even occur in river sectors where orthophosphate anion concentrations are high. The existence of macrophytes indicates the state of a river's hydro morphology and water quality. Due to the obvious non-linear connections between biological and chemical water quality measurements, some studies have employed AI models to predict the ecological state (Krtolica et al., 2022; Tarkowska-Kukuryk & Grzywna, 2022). AI techniques are analyzed as reliable resources for resolving non-linear issues, particularly in the fields of environmental and ecological engineering. Krtolica et al. (2022) compare eight cutting-edge ML classification models created in this study. The water quality index (WQI) is applied to detect threats and enable improved water resource oversight. The utilization of fractional derivatives approaches generates a model for predicting and evaluating the WQI by integrating an ML algorithm, the WQI, with remote sensing spectral indices (difference index, DI; ratio index, RI; and normalized difference index, NDI). The outcomes demonstrate that the computed WQI values vary from 56.61 to 2,886.51. They also investigate the connection between reflectance information and the WQI.

For the survival of business and agriculture, freshwater is a crucial resource. Poor water quality influences aquatic life and the ecosystem. Water quality evaluation is a fundamental aspect of freshwater management. According to the World Health Organization's latest study, several people are becoming sick or dying as a result of a dearth of safe drinking water. Kids and expectant mothers are particularly affected. Before utilizing water for any purpose, including drinking, chemical spraying (pesticides, etc.), or animal hydration, it is essential to assess its purity. Monitoring the quality of the water is one method of locating potable water. Khan et al. (2016) developed a water quality forecasting model with water quality variables through ANN and time-series analysis. It employs water quality preceding data from 2014, with a six minute interval. Data includes measurements of four parameters that affect and influence water quality. Mean-squared error (MSE), root mean square error (RMSE), and regression analysis are performance evaluation metrics used to assess models' efficacy. Hassan et al. (2021) identifies the performance of ML technologies to predict water quality parameters from satellite data. Popular ML models include ANN, random forest (RF), SVM, regression, cubists, genetic programming, and decision tree (DT). The most prevalent water quality parameters extracted were chlorophyll-a (Chl-a), temperature, salinity, colored dissolved organic matter (CDOM), suspended solids, and turbidity. To provide reliable efficiency for ecological monitoring and assessment, Qian et al. (2022) created an integrated innovative architecture with three modules: remote sensing technology (RST), cruise monitoring technology (CMT), and deep learning. It is conducted on the Qingcaosha Reservoir (QCSR) and demonstrated that deep neural network (DNN) outperformed more conventional ML techniques like multiple linear regression, SVM, and RF regression. Ragi et al. (2019) employ the Levenberg-Marquardt algorithm to give a concise methodology for estimating undetermined parameters, like Alkalinity, Chloride, and Sulfate values, according to known factors, such as pH, Electrical Conductivity, and total dissolved solids. This prediction method categorizes distinct bodies of water for use in a variety of applications. The accuracy of the predictions for chloride, total hardness, sulfate, and total alkalinity is 83.94%, 87.9%, 81.736%, and 79.48%, respectively.

According to a study by Tejoyadav et al. (2022), the Ganga, the greatest river in India, has been severely polluted by unsustainable industry and urbanization along its banks. To estimate Ganga pollution levels, they offer a multivariate hybrid model called vector auto regression-long short-term memory (VAR- LSTM). The VAR-LSTM model integrates the statistical model VAR, which uses multivariate time series analysis to describe the interdependency of different water contaminants, and the deep learning model LSTM, which employs temporal aspects of time-series data on water quality to make predictions. The LSTM model is then employed to predict the four water contaminants, total coliform (TC), dissolved oxygen (DO), fecal coliform (FC), and biological oxygen demand (BOD), as well as the corresponding WQI, from the VAR method-fitted values. Baek et al. (2020) employing a convolutional neural network (CNN) and LSTM network to precisely mimic the Nakdong River basin's total nitrogen, total phosphorus, and organic carbon levels. The CNN model integrates water level, while the LSTM model analyzes water quality. The simulation period was from January 1, 2016, to November 16, 2017. It was split into two stages: calibration (1 January 2016–1 March 2017) and testing (2 March 2017–16 November 2017). They demonstrate that CNN and LSTM models performed better than a Nash-Sutcliffe efficiency value of 0.75, adequately capturing pollutants' temporal variations in the Nakdong River basin. Consequently, the proposed approach is an efficient method for accurately simulating water levels and quality.

2. USAGE OF WATER

"Water is life," which is true as no living organism can survive without it. Water is highly essential to the earth's entire ecosystem. Water is necessary for humans not just for daily physical activities but also for residential needs. Washing veggies, other food items, cutlery, and clothing all require water. Additionally, it is utilized for cleanliness, food preparation, etc. Many marine species are also consumed as food. Water is used in many industries, including manufacturing and the power sector, for the production process or cooling machinery. Industrial water is utilized for processing, dilution, fabrication, washing, product transportation, and cooling, according to a USGS (United States Geological Survey) survey. Around 40% of all water abstraction is used by industry. Without water, many businesses would be in danger of failing. By products of residential, commercial, and industrial activity include wastewater and industrial water. One of the major industries that use water for industrial purposes is the paper and pulp sector. The utilization entails the production of pulp and paper as well as related tasks like bleaching, boiling, and washing. These activities use a lot of water, which produces a lot of polluting effluents. The textile industry is another of the most polluting sectors. Globally, some 20,000 chemicals are utilized in the production of textiles. Organ chlorine, lead phthalates, and other dangerous compounds can be found in the wastewater produced by the textile industry. Additionally, oil refineries have a significant influence on the contamination of surface and groundwater. Mercury, boron, and hazardous metals that resemble arsenic are released into streams by coal-fired power plants. Depending on its source, the wastewater may contain a variety of toxins. Chemical, physical, and biological pollution fall into these three categories. Complex organic molecules, phosphorus and nitrogen-rich compounds, bacteria, viruses, protozoa, organic and inorganic chemicals, micro plastics, radioactive materials, and other contaminants are frequently found in wastewater.

3. SOURCES OF WATER

Water is incredibly important to the earth's entire ecosystem. Water is a necessity for humans not just for daily physical processes but also for residential and agricultural uses, where it is widely utilized. Numerous other businesses, like manufacturing and the power sector, heavily rely on water. Nature is a free supply of this most priceless treasure. Although water covers around two-thirds of the surface of the globe, there is a limited supply of freshwater because only 2.5% of the total amount of water is fresh and suitable for human consumption. The availability of clean, fresh water is a big concern as a result of the extensive usage of water in several aspects of life. One of the biggest important natural resource concerns facing humanity is the provision of enough hygienic, fresh water. Due to a growing population, a booming economy, and environmental degradation, there is an expanding global problem with water shortages. Some of the available resources of water are:

- *Oceans:* Oceans are the most plentiful source of water. Ocean water should not be consumed directly because of its high salt content. The ocean water must undergo several processes to be fit for ingestion. Although there are ways to filter or process this salty water, they are quite expensive and consequently out of the ordinary person's price range. So, in comparison to the ocean's water supply, extremely little of this water is fit for human consumption.
- *Surface Water:* Surface water is the concept that includes water that is found above the earth, such as the water found in ponds, rivers, lakes, streams, etc. Surface water also includes ocean water. The most common source of water is surface water. It is the most readily available, freely given resource that is suitable for most types of usage. But because of the increasing rate

of industrialization and exploitation, such as in religious rituals, etc., this resource is quickly becoming contaminated.

- *Underground Water:* The source of water that is found under the ground or beneath the layer of soil is considered underground or groundwater, including wells, springs, etc. As the rainwater falls on the ground, it enters the interconnected openings between rock particles and fills them with water, which is then considered groundwater. Groundwater is also a highly usable resource of fresh water, generally being of better quality in comparison to surface water. The only problem that arises with the usage of groundwater is, when it becomes contaminated, as it is generally difficult and costly to treat the contaminated groundwater.
- *Glaciers and Icecaps:* Water can also be found in plenty in the ice caps and glaciers. Although it might be eaten, it would be too expensive to prepare it that way. Our lives would not be complete without water. For our lives to run smoothly, we need water. However, there isn't much drinkable water left on the globe. Water supplies are also becoming increasingly scarce. For a brighter future, we must be aware of these warning signs and understand how to revive these water sources.

4. WATER CONTAMINATION

Many life-threatening diseases, such as diarrhea, cholera, shigella, dysentery, etc., are water-borne diseases that are transmitted through contaminated water. Ingesting contaminated water also results in many other severe health conditions, like gastroenteritis and amoebiasis. Most of these diseases are caused by an increased level of nitrates and heavy metals. Industrial wastes or toxic materials from factories are the main reasons for contaminated water. For the serious reasons above, there is an urgent need to accurately track, analyze, and solve the causes of water pollutants. ML provides data-driven, efficient solutions by intelligently monitoring the origin and nature of water pollutants. But it's crucial to understand that not all toxins are visible, so clear water does not mean necessarily it's clean. Surface, ground, and gathered rainwater are our primary sources of water for drinking, washing, farming, and industry. These resources are all reliant on precipitation and snowfall to cover the ground. Surface water is commonly contaminated and is subject to significant seasonal turbidity fluctuations. Surface water is frequently the simplest source of water to reach, but it is challenging to treat properly due to high levels of suspended particles. Generally, because the water is purified to a certain extent as it percolates below through into rock, groundwater becomes less probable than surface water to be contaminated by germs or solid particles. But its dissolved chemical concentrations might be higher. This indicates that while there are fewer microbiological contaminants in groundwater, its quality may be significantly impacted by dissolved materials and minerals like fluoride. Water with fluoride is recognized as unhealthy. The World Health Organization recommends a concentration of 1.5 mg/l. The highest concentrations, which can exceed 10 mg/l, are found in the waters of the Rift Valley zone. Dental and skeletal fluorosis in the local population is characterized by brown patches on the teeth, joint pain, restricted joint flexibility, and eventually crippling.

5. WATER SOURCE CONTAMINATION

There are several ways in which surface water might be contaminated. This could be a sewage or pipe directly flowing into the river. The disposal of wastewater, including residential and commercial waste, is thought to be convenient in streams and rivers. Without proper treatment, industrial sources may release suspended particles, organic materials, and dangerous compounds. Drainage system, which can introduce contaminated items into the water, is another source of surface water contamination. The uncontrolled disposal of solid waste and open defecation are two possible sources of drainage contamination.

6. WATER SOURCE DEPLETION

Despite being a renewable resource, water will become depleted if it is extracted too much. Lowering of the water table occurs when groundwater is withdrawn from aquifers faster than it can be recharged. Rivers and streams, which often receive a portion of their water from underground sources, are also impacted by the decline in the water table. Groundwater overuse can also have an impact on springs, which could become irregular or even totally dry up. Reduced forest cover is also a direct result of water supply depletion. Rainwater rushes off the surface without penetrating the soil to replenish the groundwater as a result of the loss of trees and other vegetation. Climate change uncertainty and the consequences of droughts, which have a significant impact on the availability of surface and groundwater sources, are additional factors.

7. WATER QUALITY

Water quality is a critical concern as it is used for agriculture, fisheries, aquaculture production, and domestic purposes. Water quality for different purposes can be assured by understanding the controlling factors concerning it. Also, users should understand and take responsibility to be protective of water quality and wastewater management and usage. Poor water quality is a major issue that impacts ecosystems, plant/animal life, and human health worldwide. Surface water quality prediction is a major issue in water resources and environmental systems. Water quality prediction is a crucial and extensively researched topic as it significantly affects national/ regional water resources management.

Many water quality measures are available for the assessment of water quality. According to the guidelines, the standards for water quality have been assigned to be aware of and protect the water from pollutants. As safe drinking water is one of the major health concerns, it should be free of excessive concentrations of minerals at the same time, and must not contain toxins or disease organisms, although livestock does not need high purity in water. Similarly, irrigation water must not contain phytotoxic substances or excessive concentrations of minerals. Adequate quality of water is required in industries also. Some processes need very high-quality water. To save the metals from corrosion, the water should not be acidic. Water quality standards may be recommended for natural bodies of water, and effluents have to meet certain requirements to prevent pollution and adverse effects on the flora and fauna.

Quality monitoring of water is very important as the water is used for drinking and irrigation purposes in most parts of the world. Water Quality Index (WQI) is one of the tools primarily used to measure the quality of water. To classify the water quality, generally, the National Sanitation Founder Water Quality Index (NSFWQI) technique is used. WQI ranges from 0 to 100 on the index scale. Zero WQI signifies the best quality and an increase in WQI indicates the deterioration in the quality of water.

The data collected for the amount of COD (chemical oxygen demand), BOD (biological oxygen demand), heavy metals (like Fe, Zn, Mn, Cu, Co, Cr), PO, and E Coli are tested accordingly as per WQI, and the water quality analysis is done based on these parameters. This analysis can be utilized to check and control the pollutants. The data collected using such approaches can be further analyzed for pattern generation and solution provision. Additionally, a data repository may be used by researchers and students of machine learning, data sciences, and artificial intelligence.

It is crucial that water companies are able to detect changes in water quality promptly. Machine learning approaches are extremely popular today because they work incredibly well with time-series data sets. The technology of sensors is now widely used in various fields to monitor data in real-time. Muharemi et al. (2019) have described the use of machine learning techniques to identify water quality anomalies in a real-world dataset. They explained various methods for identifying water quality in time-series data. On the water quality data, they employ deep learning approaches such as logistic regression, linear discriminant analysis, support vector machines (SVMs), artificial neural networks (ANNs), deep neural networks (DNNs), recurrent neural networks (RNNs), and long short-term memory (LSTMs). Researchers used a severely unbalanced data set to illustrate the limitations of machine learning methods. Using the F1 metric to assess the performance of the models. SVM, ANN, and logistic regression are slightly less vulnerable when all of the algorithms for assessing the performance of the models are applied; however, DNN, RNN, and LSTM are quite sensitive. Previous findings of the water quality (Muharemi et al., 2018) have demonstrated that the SVM underperformed on the time series dataset as an outcome of the dataset not being scaled. We recognize that scaling has enhanced prediction for the current time when using the SVM classifier. It is highly desirable to find a more universal approach to the problem of imbalanced data sets since they occur in many different domains of application.

Xiang and Jiang (2009) studied the application of the least squares support vector machine (LS-SVM) in a water quality prediction model. They combined LS-SVM with particle swarm optimization (PSO) for time series prediction. The LS-SVM overcomes shortcomings in the Multilayer Perceptron (MLP) and the PSO automatically adjusts the LS-SVM parameters. It enhances the efficiency and the capability of prediction. The model came out to be quite effective at predicting water quality of the Liuxi River in Guangzhou through simulation testing.

According to Xuan et al. (2010), the widely used conventional techniques including regression analysis, ARIMA, and neural networks have a limited accuracy. They offered a hybrid strategy that combined a support vector machine with particle swarm optimization to improve prediction accuracy. The approach is applied to predict Heishui river water quality of the Bei-bei, Chongqing. The results show that the suggested strategy can produce more accurate predictions. Mohammad pour et al. (2015) used feed forward back propagation (FFBP) and radial basis function (RBF) to predict the water quality index (WQI) in a free constructed wetland. The results of this study demonstrate that the SVM and FFBP can be used to predict water quality in a free-surface built wetland environment.

Water is a crucial and indispensable component for the existence of life on earth. Water resources are becoming more contaminated because of the rapid growth of the population and increasing industrialization. In smart cities, the quality of the water becomes one of the key components for a high quality of life. Water quality has deteriorated due to several types of pollution, including the disposal of human, industrial and automobile wastes. Due to the quality of the water impacted by rising pollutants, it is essential to evaluate, monitor and accurately predict the water quality. A review on water quality evaluation and prediction is given by Kang et al. (2017), which classifies and compares big data analytics techniques and the big data-based prediction models for evaluating water quality.

Building water quality prediction models is aided by modern big data implementation utilizing sensor networks and machine learning with environmental data. M Kumar et al. (2023) presents water quality index based water quality assessment of Ganga River and Sangam at Prayagraj. The data samples were collected over 15 months using an IoT-based smart water sensors kit. The relevant features were extracted using principal component analysis (PCA). The seasonal pattern of water quality assessment were presented and well sup- ported by results. The average WQI values were used to recommend that the water was suitable for irrigation but not suitable for drinking, Nair and Vijaya (Nair & Vijaya, 2021) analyze various models for water prediction and evaluation that were created using machine learning and big data approaches. Hemdan et al. (2021) presented a study on water quality analysis using IoT and big data analytics.

Dogo et al. (2019) used conventional machine learning techniques such as support vector machines, logistic regression, and ANNs to study the anomaly detection of water quality. Water quality anomalies are described, providing an overview of their development and the progress made in identifying them.

Most likely, in terms of feature learning accuracy and a few false positive rates, DL techniques perform better than conventional ML approaches. However, due to the many datasets, models, and parameters used, it is challenging to fairly compare studies. Additionally, they examine the advantages of the extreme learning machine (ELM), whose use in this field has not been fully utilized. Further, they proposed the hybrid DL-ELM (Deep Learning - Extreme Learning Machine) methodology as a potential remedy that might be looked into more closely and used to spot anomalies. By leveraging ELM against SVM, the hybrid DL-ELM technique for water quality anomaly detection meets some of the desirable criteria that the algorithm be effective in terms of accuracy time, anomaly detection time, and memory computational intricacies while also being capable of dealing with high-dimensional data search areas occupied by sensors deployed within the water distribution system in the time series domain. In a hybrid DL-ELM model, the final DL layer is typically substituted by ELM while DL methods train numerous hidden layers with unsupervised initiation and extraction of features. These trained features are combined by ELM as its training input for categorization. The hybrid anomaly detection model, which may be used for testing and evaluating the model, is created by combining two algorithms. The goal is to take advantage of the complementary characteristics of DL and ELM in order to increase detection rates and minimize training costs, and accuracy all factors that are important for detecting water quality anomalies.

The assessment of river water quality using sensors and a deep neural network (DNN) has been covered by Chopade et al. (2021). Initially, they discussed the WQI for labeling the provided laboratory samples. Additionally, they stand for an automatic annotation technique that labels the instances of sensor data utilized to

Figure 1. Accuracy and F1 score of river dataset during training

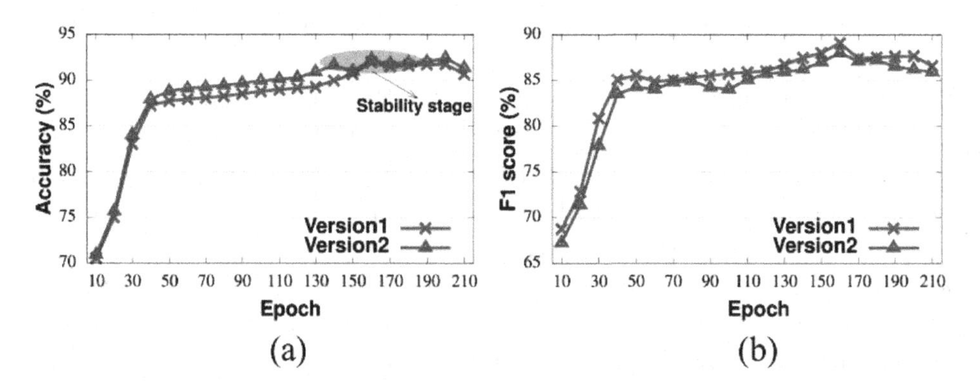

create a DNN classifier that forecasts water quality. Furthermore, a DNN classifier that forecasts water quality has been constructed using instances of labeled sensor data, and it also suggests a noise-handling loss function to contain noisy labels. A major part of the evaluation of performance is based on the Indian rivers. They are determining the training data by using two versions: Version 1 and Version 2. Version 1 selects six parameters, whereas Version 2 picks the remaining 17 parameters.

Figure 1 demonstrates that after 150 epochs, there is a minimal enhancement in accuracy during the training process. According to Figure 1(a), the system reaches its highest level of accuracy after around 160 epochs. As shown in Figure 1(b), the F1 score, which is comparable to accuracy, indicates impartial conduct on variations in epoch count. For DO, pH, BOD, FC, nitrates, and turbidity, they stood in for P1, P2, P3, P4, P5, and P6 parameters, respectively. The pH value outperformed all other parameters, and the F1 score of water quality is 75% and 76% respectively. They illustrate the importance of recall and precision in sensory data. Some classes' precision is greater than or almost matches the testing data set.

8. WASTEWATER TREATMENT

A process called wastewater treatment (WWaT) is applied to clean up the impurities from wastewater and turn it into an effluent that can be recycled into the hydrological (water) cycle. It is essential to secure both the health of individuals and creatures as well as our environment. When wastewater is not adequately handled, it can contaminate our water supplies, harm natural habitats, and lead to life-threatening infections. Several types of pollutants affect the water, including persistent inorganic pollutants, toxic materials, and chemical pollutants.

For several decades, various fields have benefited from artificial intelligence (AI) since it's become an indispensable tool for resolving real-world problems. This approach has been recently applied to water treatment and desalination to improve the process and provide feasible solutions to water pollution and scarcity of water. The use of AI has the potential to reduce operational costs and optimize chemical use in water treatment. MLAs and ANNs are more frequently utilized in wastewater treatment applications since they are easy to train and provide a solid grasp of the processes. Scientists (Pang et al., 2019; Torregrossa et al., 2018) in wastewater treatment has come to recognize and value the use of ML models. Icke et al. (2020) conducted research through the use of machine learning to enhance the efficiency of wastewater treatment operations. Due to stricter effluent quality regulations, the desire to minimize energy usage, and chemical dosing, upgrading wastewater treatment procedures is becoming more and more crucial. Two areas are addressed with machine learning: predictive control and advanced analytics. Currently, this method is being tested at the integrated validation plant of PUB, the National Water Agency (NWA) of Singapore. Essentially, predictive control (PC) is mostly used for effective nutrient (pollutant) removal. Load prediction with predictive control is 88% accurate according to operational outcomes, while approximately 15% less aeration is required than with conventional control approaches.

Operational support has been developed through advanced analytics (AA). Based on machine learning and self-learning algorithms, the ammonium effluent is fine-tuned to the self-learning feed-forward algorithm. A neural network model based on quantile regression is applied to identify anomalies. AA exhibits the capability to identify anomalies in processes and instruments automatically. An analysis of machine learning algorithms in the treatment of biological wastewater has been published by Sundui et al. (2021). A "paradigm shift" from the conventional wastewater treatment process is biological WWaT using algae-bacteria ensembles for nutrient uptake and resource recovery. This approach helps to reduce pollution and encourages a circular economy. Machine learning algorithms (MLAs) are beneficial in this regard for forecasting the ambiguous outcomes of therapeutic (treatment) processes. They observed that MLAs have been provided with good real-time monitoring, optimization, forecasting of uncertainties, and fault diagnosis of intricate environmental systems with favorable outcomes. The integration of transient operating conditions during this process, such as disruptions or failures caused by leaks in pipelines, damaged bioreactors, and unforeseen fluctuations in organic loading, rate of flow, and temperature, can be effectively predicted by using these algorithms in conjunction with virtual sensors. WWaT process monitoring and oversight are considered in (Yang et al., 2021). Intense coupling and significant nonlinearity make the process challenging for the conceptual model, and it is susceptible to unidentified disruptions. Implementing direct heuristic dynamic programming (dHDP)-based reinforcement learning management, Yang et al. (2021) tackle multivariable tracking management problems. By adjusting the oxygen transfer factor of the 5th aerobic zone (KLa5) and the inside recycle rate of flow (Qa), the objective is to maintain an ideal reference for the DO (dissolved oxygen) concentration of the 5th aerobic zone (SO_5) and the nitrate concentration of the 2nd anoxic zone (SNO_2). The dHDP attempts to manage the high coupling among SO_5 and SNO_2 and eliminate any unidentified process disruptions while obtaining a few aggregated WWaT process errors in tracking. They are analyzed through extensive and methodical computations that depend on

Table 1. Benefits and drawbacks of conventional approaches for treating contaminated industrial wastewater

Process	Benefits	Drawbacks
Chemical Precipitation	Accessible technology The physicochemical processes are integrated. Economically beneficial and effective for high-contamination loads-ready. Chemical oxygen consumption is reduced significantly. Incredibly effective at removing fluoride and metals.	Chemical consumption such as lime, oxidants, H2S, etc. Monitoring the effluent's physicochemical state. The necessity of oxidation while operating with complicated metals. Challenges with high sludge production, processing, and disposal: management, treatment, and cost. Neither metal ions nor concentrations are effectively removed
Flocculation Technique	Highly straightforward Capital-cost-cheap Ability to inactivate bacteria significant decreases in dissolved organic halogen and entire organic carbon such as the pulp and paper industry Perfect sludge balancing and dewatering properties	Low levels of arsenic elimination Capable of inactivating and removing insoluble pollutants e.g., pigments Arising in the volume of sludge production
Oxidation by chemicals Basic oxidation Ozone Treatment with hypochlorite Hydrogen peroxide	The capability of recycling water Not the production of sludge Effective odor and color removal Enhances the commodities bio-degradability Maximum throughput	Drop significantly in chemical oxygen consumption levels or perhaps a little impact (ozone) Produces sludge Chemicals are required Aromatic amines and unstable organic molecules released No change to salt concentration
Biological processes Bioreactors Biologically-activated sludge (BAS) Biological therapies Enzymatic breakdown Lagoon	The utilization of microorganisms to biodegrade organic pollutants is straight forward, economically advantageous, and popular with the general population In blended cultures or perfect cultures, numerous species are utilized Numerous extracellular enzymes having strong biodegradability capacities are produced by white-rot fungi	It is essential to establish a highly favorable environment This necessitates the physicochemical pre-treatment and/or control and preservation of the microorganisms The kinetic process challenge is exceedingly slow. BAS indicates low decolourization
Exchange the Ion Resins that corrode Specialized resins Microbial resins Polymer-based adsorbents (PBA) Hybrid adsorbents made of polymers	Various producers offer a wide variety of business commodities Procedures that have been tried and true ; simple control and management Simply combine with other methods (such as precipitation and filtration in combined WWaT) High regrowth with the potential for external developers to reuse Provide a treated effluent of the highest quality Efficacious and reasonably priced for removing metals	Financial restrictions Large columns are needed for huge volumes Saturation of the cationic converter prior to the addition of the anionic resin (metal precipitation and reactor blockage) The removal of impurities from beads requires a physicochemical pre-treatment (e.g., sand filtration). Beads are quickly clogged by particles and organic matter (organics and oils) Matrix deteriorates with time and certain scrap materials such as radioactive, powerful oxidants, etc. Not applicable to some target pollutants like disperse dyes, drugs, and so forth

the renowned BSM1 architecture of the WWaT process managed through the dHDP to assess their effectiveness with other models.

The positives and limitations of WWaT have been explored by Crini et al. (2019) employing technological approaches, specifically the conventional method, the well-established recovery procedure, and emerging removal techniques. It has evolved into a critical challenge for everyone, including citizens, academics, and decision-makers at the local, regional, national, and international levels. Some positives of WWaT are that it reduces impurities like nitrogen, phosphorus, and carbon nutrients (pollutants). The three approaches are employed in conjunction with some technology to remove the contaminants. Table 1 displays the benefits and deficiencies of industrial wastewater treatment.

Many life-threatening diseases, such as diarrhea, cholera, shigella, dysentery, etc., are water-borne diseases that are transmitted through contaminated water. Ingesting contaminated water also results in many other severe health conditions, like gastroenteritis and amoebiasis. Most of these diseases are caused by an increased level of nitrates and heavy metals. Industrial wastes or toxic materials from factories are the main reasons for contaminated water. For the serious reasons outlined above, there is an urgent need to accurately track, analyze, and solve the causes of water pollutants. ML provides data-driven, efficient solutions by intelligently monitoring the origin and nature of water pollutants.

9. CONCLUDING REMARKS

This chapter discusses the various water quality parameters. The chapter investigates the water quality assessment and wastewater treatment based research initiatives using Machine Learning and Data Analytics. Considering the existing works with applications of machine learning and data analytics, the related issues and challenges are discussed. Finally chapter highlights the wastewater management issues.

REFERENCES

Baek, S.-S., Pyo, J., & Chun, J. A. (2020). Prediction of water level and water quality using a cnn-lstm combined deep learning approach. *Water (Basel)*, *12*(12), 3399. doi:10.3390/w12123399

Chopade, S., Gupta, H. P., Mishra, R., Oswal, A., Kumari, P., & Dutta, T. (2021). A sensors based river water quality assessment system using deep neural network. *IEEE Internet of Things Journal*.

Crini, G., & Lichtfouse, E. (2019). Advantages and disadvantages of techniques used for wastewater treatment. *Environmental Chemistry Letters, 17*(1), 145–155. doi:10.100710311-018-0785-9

Dogo, E. M., Nwulu, N. I., Twala, B., & Aigbavboa, C. (2019). A survey of machine learning methods applied to anomaly detection on drinking-water quality data. *Urban Water Journal, 16*(3), 235–248. doi:10.1080/1573062X.2019.1637002

Haghiabi, A. H., Nasrolahi, A. H., & Parsaie, A. (2018). Water quality prediction using machine learning methods. *Water Quality Research Journal of Canada, 53*(1), 3–13. doi:10.2166/wqrj.2018.025

Hassan, N., & Woo, C. (2021). Machine learning application in water quality using satellite data. *IOP Conference Series. Earth and Environmental Science, 842*(1), 012018. doi:10.1088/1755-1315/842/1/012018

Hemdan, E. E.-D., Essa, Y. M., El-Sayed, A., Shouman, M., & Moustafa, A. N. (2021). Smart water quality analysis using iot and big data analytics: A review. In *2021 International Conference on Electronic Engineering (ICEEM)* (pp. 1–5). IEEE. 10.1109/ICEEM52022.2021.9480628

Icke, O., van Es, D., de Koning, M., Wuister, J., Ng, J., Phua, K., Koh, Y., Chan, W., & Tao, G. (2020). Performance improvement of wastewater treatment processes by application of machine learning. *Water Science and Technology, 82*(12), 2671–2680. doi:10.2166/wst.2020.382 PMID:33341761

Kang, G., Gao, J. Z., & Xie, G. (2017). Data-driven water quality analysis and pre diction: A survey. *2017 IEEE Third International Conference on Big Data Computing Service and Applications (BigDataService),* 224–232. 10.1109/BigDataService.2017.40

Khan, Y., & See, C. S. (2016). Predicting and analyzing water quality using machine learning: a comprehensive model. In *2016 IEEE Long Island Systems, Applications and Technology Conference (LISAT)* (pp. 1–6). IEEE. 10.1109/LISAT.2016.7494106

Krtolica, I., Savić, D., Bajić, B., & Radulović, S. (2022). Machine learning for water quality assessment based on macrophyte presence. *Sustainability (Basel), 15*(1), 522. doi:10.3390u15010522

Kumar, M., Singh, T., Maurya, M. K., Shivhare, A., Raut, A., & Singh, P. K. (2023). Quality assessment and monitoring of river water using iot infrastructure. *IEEE Internet of Things Journal, 10*(12), 10280–10290. doi:10.1109/JIOT.2023.3238123

Mohammadpour, R., Shaharuddin, S., Chang, C. K., Zakaria, N. A., Ghani, A. A., & Chan, N. W. (2015). Prediction of water quality index in constructed wet-lands using support vector machine. *Environmental Science and Pollution Research International*, *22*(8), 6208–6219. doi:10.100711356-014-3806-7 PMID:25408070

Muharemi, F., Logofătu, D., Andersson, C., & Leon, F. (2018). Approaches to building a detection model for water quality: a case study. In *Modern Approaches for Intelligent Information and Database Systems* (pp. 173–183). Springer. doi:10.1007/978-3-319-76081-0_15

Muharemi, F., Logofătu, D., & Leon, F. (2019). Machine learning approaches for anomaly detection of water quality on a real-world data set. *Journal of Information and Telecommunication*, *3*(3), 294–307. doi:10.1080/24751839.2019.1565653

Nair, J. P., & Vijaya, M. (2021). Predictive models for river water quality using machine learning and big data techniques-a survey. In: *2021 International Conference on Artificial Intelligence and Smart Systems (ICAIS)* (pp. 1747–1753). IEEE. 10.1109/ICAIS50930.2021.9395832

Pang, J.-W., Yang, S.-S., He, L., Chen, Y.-D., Cao, G.-L., Zhao, L., Wang, X.-Y., & Ren, N.-Q. (2019). An influent responsive control strategy with machine learning: Q-learning based optimization method for a biological phosphorus removal system. *Chemosphere*, *234*, 893–901. doi:10.1016/j.chemosphere.2019.06.103 PMID:31252361

Qian, J., Liu, H., Qian, L., Bauer, J., Xue, X., Yu, G., He, Q., Zhou, Q., Bi, Y., & Norra, S. (2022). Water quality monitoring and assessment based on cruise monitoring, remote sensing, and deep learning: A case study of qingcaosha reservoir. *Frontiers in Environmental Science*, *10*, 979133. doi:10.3389/fenvs.2022.979133

Ragi, N. M., Holla, R., & Manju, G. (2019). Predicting water quality parameters using machine learning. In *2019 4th International Conference on Recent Trends on Electronics, Information, Communication & Technology (RTEICT)* (pp. 1109–1112). IEEE. 10.1109/RTEICT46194.2019.9016825

Sundui, B., Ramirez Calderon, O. A., Abdeldayem, O. M., Lázaro-Gil, J., Rene, E. R., & Sambuu, U. (2021). Applications of machine learning algorithms for biological wastewater treatment: Updates and perspectives. *Clean Technologies and Environmental Policy*, *23*(1), 127–143. doi:10.100710098-020-01993-x

Tarkowska-Kukuryk, M., & Grzywna, A. (2022). Macrophyte communities as indicators of the ecological status of drainage canals and regulated rivers (eastern Poland). *Environmental Monitoring and Assessment*, *194*(3), 210. doi:10.100710661-022-09777-0 PMID:35194688

Tejoyadav, M., Nayak, R., & Pati, U. C. (2022). Multivariate water quality forecasting of river ganga using var-lstm based hybrid model. In *2022 IEEE 19th India Council International Conference (INDICON)* (pp. 1–6). IEEE. 10.1109/INDICON56171.2022.10040146

Torregrossa, D., Leopold, U., Hernández-Sancho, F., & Hansen, J. (2018). Machine learning for energy cost modelling in wastewater treatment plants. *Journal of Environmental Management*, *223*, 1061–1067. doi:10.1016/j.jenvman.2018.06.092 PMID:30096746

Xiang, Y., & Jiang, L. (2009). Water quality prediction using ls-svm and particle swarm optimization. In *2009 Second International Workshop on Knowledge Discovery and Data Mining* (pp. 900–904). IEEE. 10.1109/WKDD.2009.217

Xuan, W., Jiake, L., & Deti, X. (2010). A hybrid approach of support vector machine with particle swarm optimization for water quality prediction. In *2010 5th International Conference on Computer Science & Education* (pp. 1158–1163). IEEE. 10.1109/ICCSE.2010.5593697

Yang, Q., Cao, W., Meng, W., & Si, J. (2021). Reinforcement-learning-based tracking control of waste water treatment process under realistic system conditions and control performance requirements. *IEEE Transactions on Systems, Man, and Cybernetics. Systems*, *52*(8), 5284–5294. doi:10.1109/TSMC.2021.3122802

Zhu, M., Wang, J., Yang, X., Zhang, Y., Zhang, L., Ren, H., Wu, B., & Ye, L. (2022). *A review of the application of machine learning in water quality evaluation*. Eco-Environment & Health. doi:10.1016/j.eehl.2022.06.001

Chapter 8
Explaining the Importance of Water Quality Parameters for Prediction of the Quality of Water Using SHAP Value

Siddhartha Kumar Arjaria
Rajkiya Engineering College, Banda, India

Abhishek Singh Rathore
https://orcid.org/0000-0002-5513-2639
Shri Vaishnav Vidyapeeth Vishwavidyalaya, Indore, India

Shailendra Badal
Rajkiya Engineering College, Banda, India

ABSTRACT

Water is key to life on planet Earth, and hence, maintaining water quality is a critical issue in contemporary times. The water quality index decides the quality of drinking water. The presented work first explores different machine learning algorithms on the already collected water samples to decide the water quality and then applies the coalition game theory-based SHapley Additive exPlanations (SHAP) approach to decide the significance of each parameter in deciding the class of water sample based on quality. The potential of popular algorithms like K-NN, support vector machine, decision tree, etc. are being explored to find out the quality of water samples. All the machine learning algorithms used in the work give over 80% accuracy while the performance of neural network is 96% proving to be the best among all other algorithms. The presented work demonstrates the model agnostic, coalition game theoretic SHAP value-based method for explaining the importance and impact of each of the given parameter pH, HCO_3^-, Cl^-, NO_3^-, F^-, Ca, Mg, Na, Ec, etc. in deciding the quality of the water.

DOI: 10.4018/978-1-6684-6791-6.ch008

1. INTRODUCTION

Increase in economic growth results in the production of wastewater that contains a variety of contaminants, posing major risks to natural water ecosystems. As a result, several water-pollution mitigation strategies have been created. Analysis and water quality assessment have greatly increased effectiveness of water pollution control. The water quality depends upon various parameters like pH, Total Dissolved Solids (TDS), Hardness, alkalinity, Bicarbonate (HCO_{3},) Chloride (Cl), Sodium (Na,), Sulphate (SO_{4}), Fluoride (F), Iron (Fe), Magnesium (Mg) ions (Kothari et al., 2021). The Water Quality Index (WQI) and Trophic State Index (TSI) are commonly applied for the analysis of water quality (Rauen et al., 2018) and both are useful as thumbnail indicators and non-regulatory measurements for water quality assessment. TSI is used for lakes and estuaries, whereas WQI is used for streams, springs, and natural tea- and coffee-colored/black waters (Badal et al., n.d.).

Human health is in danger due to the presence of chemical contaminants in the source of potable water which may have instant health consequences. These pollutants are depending on man-made activities, and industrial, geological, and agricultural conditions. In the modern age, urbanization is one of the biggest problems which affect water quality.

Over-exploitation and improper use of water resources are making things more complex. The Bureau of Indian Standards (BIS, 2012) have been used for estimate the quality of groundwater for drinking purposes and the estimation of WQI. The following attributes are used in this paper to estimate water quality:

- pH – pH of water is finding the acidity and basicity of drinking water. As per the Bureau of Indian Standards (BIS, 2012) 6.5 to 8.5 is acceptable range of pH for drinking water. Higher and lower value of pH can cause inflammation in the eyes, skin problem, and mucous membrane in humans (Akter et al., 2016).
- Turbidity – Turbidity is the unclearness or cloudiness in water which is caused due to presence of sand, silt, mud, organic and inorganic matter algae, and microorganism and it can be seen with the naked eye. Turbidity reduces the scattering of light. Turbidity reduces the filtering capacity of a treatment plant and can choke the filtering apparatus. As per Bureau of Indian Standards permissible limit of turbidity is 1 NTU.
- Total dissolved solids (TDS) – TDS refers to amount of inorganic and organic matter dissolved in water. The main constituents of TDS are calcium, magnesium, sodium, and potassium, and the anions carbonate, bicarbonate, chloride, and sulphate (Priyanka et al., 2017). As per Bureau of Indian Standards (BIS, 1993) acceptable limit of TDS is 500 mg/l.

- Total Hardness – The Total Hardness (TH) is due to the presence of divalent ion (salt of Ca and Mg). This is most important parameters of water, it signifies that water can be used for domestic, industrial, or agricultural purposes. (Jameel & Sirajudeen, 2006). Hardness prevents lather formation with soap and increases the consumption of soap, increasing the boiling points of water and corrosion and incrustation of pipes. According to the Bureau of Indian Standards (BIS, 2012) permissible limit of Total Hardness is 200 mg/l.
- Fluoride – Higher value of fluoride in water causes fluorosis and molting of teeth. Over-consumption of fluoride for a long period causes crippling of bone. According to the Bureau of Indian Standards (BIS, 1993) acceptable limit of Fluoride is 1 mg/l.
- Arsenic – Excess arsenic in water causes very harmful health effects on human beings such as cancer, diabetes, and liver damage. It also increases the risk of hypertension and causes acute arsenic toxicity. Arsenic in water is due to both natural and anthropogenic activities of humans. According to the Bureau of Indian Standards (BIS, 1993) acceptable limit of arsenic in water is 0.01mg/l.

Laboratory experimentation on deciding water quality can prove expensive and time-consuming. Hence, machine learning based approaches which are increasingly being used in almost all domains to leverage existing data to predict meaningful decisions can be employed to study water quality of our natural water system. The learning algorithm learns from the samples and provide predictions on unseen data. In the proposed work, the machine learning algorithms are trained on the dataset to predict water quality. The learning model built from machine learning algorithms perform well but are unable to explain their predictions. They are not able to give the impact (role) of each input parameter in prediction/decision-making. The models are black boxes in nature and thus results/decisions are not easily interpretable. For the interpretation of the decisions, a coalition game theory-based SHAP value approach is being used. The SHAP method is a model independent methodology that can be applied to any machine-learning model to explain its predictions (Lundberg & Lee, 2017). The paper explores the explain ability and interpretability of black box machine learning models using various plots like feature summary plots, summary plots, force plots, waterfall plots, etc. Throughout the paper, five sections are discussed. Section two explores the related work and background study done by other researchers in the given domain. Proposed work is discussed in Section for deciding the quality of water and exploring the different machine learning algorithms with interpretability. Section four provides the interpretability of the results. Section five provides a result analysis, and the next section provides a conclusion based on the findings.

2. LITERATURE REVIEW

This section provides a quick overview of existing works by various authors to predict water quality. Many multivariate statistical tools with machine learning have been applied, among them, Principal Component Analysis (Kumar et al., 2020; Tripathi & Singal, 2019), Support Vector Machines, and Neural Networks are frequently used (Alnazer et al., 2021; Haghiabi et al., 2018; Liu & Lu, 2014; Ma et al., 2020; Park et al., 2015; G. Wang et al., 2022). Abba et.al. (2020) applied two hybrid approaches namely Extreme Gradient Boosting (XGB) with Extreme Machine Learning (EML) and Genetic Programming (GP) with EML for water quality assessment. Goz et.al. (2019) proposed kernel-based EML to predict the dissolved oxygen (DO) concentration from water temperature, pH, and conductivity. Gazza et.al. (2012) used a three-layered neural network to estimate Kinta River's (Malaysia) water quality index. Leong et.al. (2021) applied least squares support vector machine models for the prediction of WQI.

Wan et al. (2011) applied association rule mining to draw a relationship between water quality and water quantity of Wang Yu River at WangTing gate. Bouzid et. al (2019) proposed a drinking water distribution system to regulate water quality by controlling chlorine concentrations in water. The learning algorithm uses nonlinear fuzzy models with adaptable learning rates to keep chlorine levels within the permissible range. Gakii et al. (2019) applied five decision tree classifiers J48, LMT, Random forest, Hoeffding tree, and Decision stump for accessing the water quality for Kenya. Stump was used to build the model and then the performance is evaluated using the accuracy comparison. The result shows that J48 decision tree achieves the accuracy of 94% and the Decision Stump achieves the lowest accuracy of 83%.

Ambiga et al. (2015) collected Groundwater samples from 35 locations in Ranipet, Vellore district; Tamil Nadu, India. The pH, Total hardness, TDS, alkalinity, Electrical conductivity, Cl, Ca, Mg, Na, K, Nitrate, Sulphate, Phosphate, Fe, and Cr were used to assess the WQI using regression. During the monsoon season, the WQI values vary from 48.69 to 245.24, from 0 to 351.02 during the post-monsoon season, and from 0 to 344.62 during the pre-monsoon season. Bilali et al. (2020) applied machine learning models to find water suitability for irrigation purposes on 300 samples. The machine learning models were trained on 10 Irrigation Water Quality samples(El Bilali & Taleb, 2020). The adaptive neuro-fuzzy inference system algorithm using ANN and K-NN was designed by Al-Adhaileh (2021) to forecast the WQI.

Zhang et al. (2022) used SHAP values to explain the relationship between land use and river water quality. Wu et al. (2022) used a dynamic multi-mode optimization-based intelligent traceability technique for water pollution. The simple models explain the linear relations between parameters to predict WQI. However, complex models provide better predictions, but the nonlinearity remains unexplained. In addition, the impact of each feature is still an open issue with traditional machine-learning models. To pursue trade-off values between accuracy, efficiency, and resilience, various self-adjusting hyperparameter techniques are required (Gao et al., 2019; J. Wang & Kumbasar, 2019). Thus, applications of machine learning models in real-world contexts need evaluations of reliability, interpretability, and uncertainty of results. The paper presents alternative methods to decide the WQI quickly in an inexpensive way along interpretation of results.

Explainable AI (XAI) is used for water quality analysis. This is critical in the field of water quality analysis because the decisions made based on the AI model's output can have significant impacts on public health and the environment. XAI is used to build trust and confidence in the AI model's results, as scientists can see the factors and variables that were considered in the analysis. Furthermore, XAI can help to improve the interpretability and transparency of water quality data, making it easier for stakeholders to understand and act on the information. This can lead to more informed decision-making and better management of water resources.

3. METHODS AND MATERIALS USED

Assessing the water quality quickly and accurately at the least cost is a big issue. To date, various techniques have been developed as discussed in previous sections to assess water quality using WQI. It is impractical to consider all water quality criteria since doing so is not only costly and technically challenging but also ignores the fluctuation in water quality. The presented work (refer to Figure 1) explores the different methods namely k-nearest neighbors (KNN), Decision Tree, Artificial Neural Network, Random Forest, Logistic Regression, and Support Vector Machines (SVM) to estimate the water quality. In the next phase, the best-performing model is chosen and the SHAP method is applied to explain the result of the selected model using the different plots.

The WQI data for the present work is taken from data collected by Verma et al. (Verma et al., 2013). The datasets contain 9 features and are divided into three classes good, poor, and very poor as shown in Figure 2. Feature space of the dataset (refer to Figure 3) shows that data is not linearly separable and thus, linear models are unable to create an efficient separating hyperplane.

The feature statistics of the dataset are shown in Table 1. The table shows the mean, median, dispersion, minimum, and maximum of each feature. Data preprocessing is applied to make the data suitable for processing purposes. Missing and inconsistent values are dealt with before processing. The data is split into sets: training and testing. The presented work uses 5 cross-fold cross-validation post which data was analyzed using machine learning models. Machine Learning techniques are widely

Figure 1. Proposed methodology

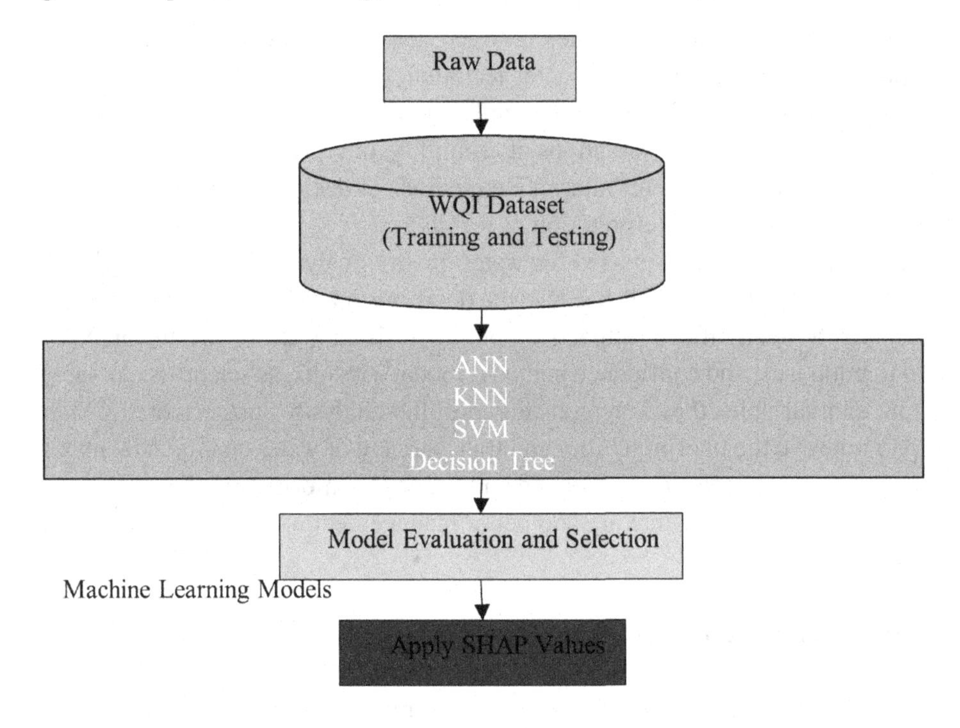

use for the decision making. The present work demonstrates the performance of the different machine learning algorithm namely K-Nearest Neighbor, Decision tree, Random Forest, Neural Network, Logistic Regression, SVM for the finding the quality of water sample.

3.1 Artificial Neural Network (ANN)

The neural network is the popular supervised learning algorithm used for classification. The neural network simulates the behavior of brain neurons. A neural recognizes the relationships in a set of training data by adjusting its weights. A neural network has input, hidden and output layers of interconnected nodes. Each node acts as multiple linear regression in functionality. An activation function is applied to the weighted sum produced by the neuron. The training phase involves searching for the best set of weights where the Root mean square is minimized. In the paper, the single hidden layer configuration 9neuron(input)-15neuron(hidden)-3 neuron(output) with relu as activation function, and the maximum epochs are 200. Adam solver is used with global minima for error set at 0.0001.

Figure 2. Class-wise distribution of data

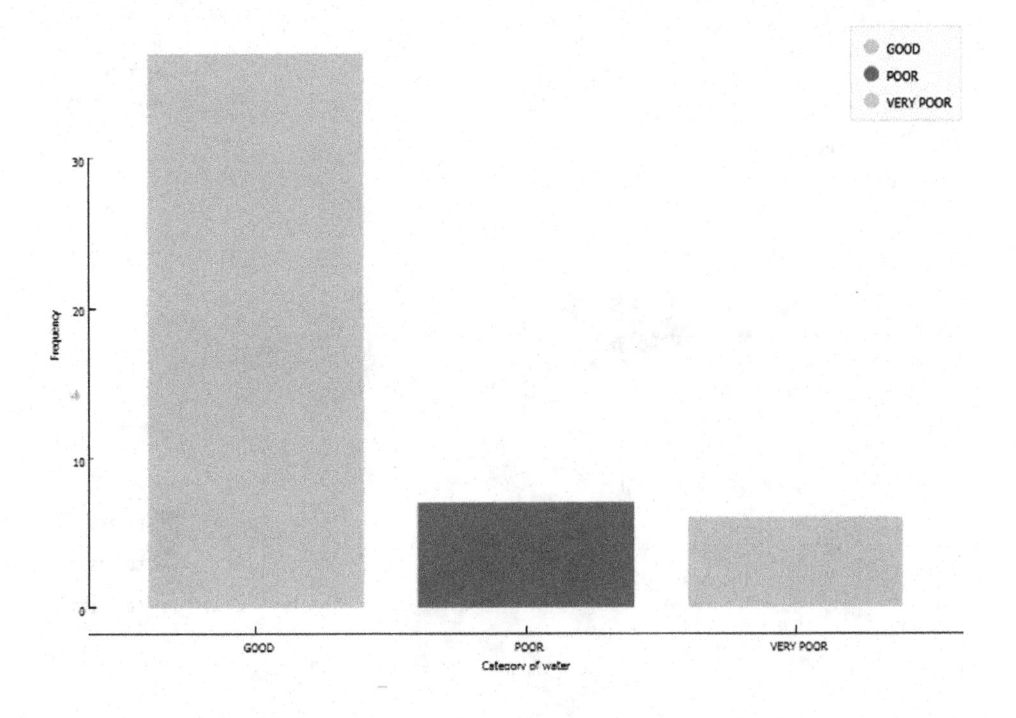

3.2 Support Vector Machine (SVM

It is a reliable supervised machine learning that is regularly used for data classification. It works very well with a small dataset. A support vector machine takes the training data and creates a hyperplane or decision boundary to classify the data. For nonlinear data classification, SVM maps our space to a higher dimension to create the nonlinear decision boundary. In our case a polynomial kernel (g.x.y+c)d with the value of g=0.14,c=0.51,d=1 with cost (C)=0.70, regression loss epsilon=0.10 and numerical tolerance=0.0001 was used.

3.3 Logistic Regression

It is one of the most popular classifiers that gives an output value that reflects the probability of data belonging to a given class. This model calculates a weighted sum of the input features along with a bias term and it outputs the *logistic* of this result. The logit is also known as the *log-odds* since it is the log of the ratio between the estimated probability for the positive case and the estimated probability for the negative case. In this work, the Ridge regularization(L2) with C=1 is being used.

Figure 3. Feature space of data set

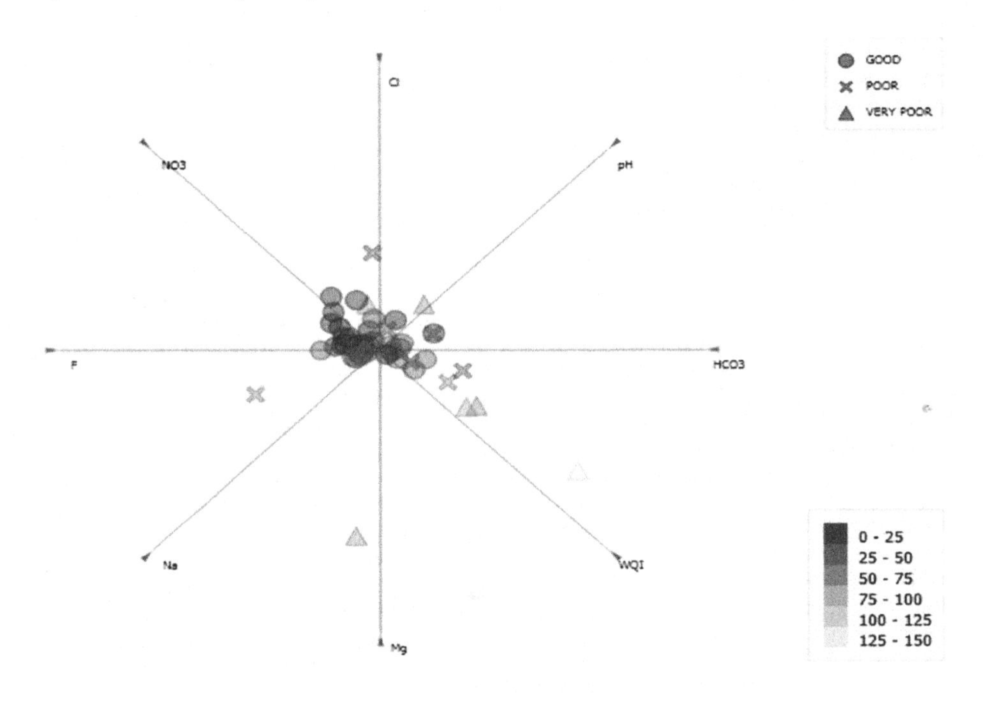

3.4 Random Forest (RF)

It is a flexible, simple supervised machine learning algorithm. It builds the forest which is an ensemble of decision trees that use the bagging method to increase the overall classification performance. This algorithm does not use hyper-parameter

Table 1. Feature statistics

Feature	Mean	Median	Dispersion	Min.	Max.
pH	7.964	8.0	0.024	7.500	8.2
HCO_3	277.4	247.5	0.387	51.000	659
Cl	32.126	14.0	1.050	7.000	142
NO3	10.470	2.70	1.977	0.700	118
F	0.495	0.445	0.502	0.050	1.2
Ca	30.480	27.0	0.433	8.000	80
Mg	34.180	25.50	0.656	15.000	141
Na	41.450	27.0	1.377	3.500	400
Ec	581.680	480.0	0.552	230.000	1440

tuning and can effectively perform for both classification and regression tasks. In this paper number of trees =10 and we do not split the subset smaller than 5.

3.5 K-Nearest Neighbor (K-NN)

It is also known as a lazy learner. It is one of the simplest and most popularly used algorithms for classifying data. It stores all the existing samples and categorizes the unseen sample on the basis of similarity measure. The similarity will be measured on the basis distance metric between two data points. Euclidean distance, Manhattan, and Mahalanobis are some popular similarity measures. K-NN provides the class to new sample based on the majority of votes among its K nearest neighbors. The K is only a hyperparameter needed to be tuned for good performance. In our case, nearest neighbors K=3 and Euclidean distance similarity measure were used.

3.6 Decision Tree

Decision Trees are nonparametric models and are applicable in both classification and regression problems. They have two kinds of elements: nodes and branches. At every internal node, one selected feature of the dataset is evaluated for splitting the samples of the dataset, thus decision trees algorithm recursively split the training samples using the features selected using feature selection measures like the *Gini index*, Information gain, Chi-Square, etc. Figure 4 shows the decision tree generated for our WQI dataset.

4. INTERPRETATION OF PREDICTION

Determining the water quality using conventional lab methods can prove to be time-consuming and expensive. The presented work applies various popular machine learning algorithms to decide the water quality. Although machine learning algorithms predict the quality of water with a good degree of accuracy, these methods are unable to answer the question of how the classifier achieves these results. The explanation of the prediction increases the trust in the model. To make a transparent prediction system, the presented work uses the SHAP method. SHAP explains predictions by finding the contribution of each feature in the prediction. Each feature value will be treated as players in a coalition. Shapley values talk about the distribution of prediction among the features and gives the average contribution of each feature. Shapley values give the signed feature importance (magnitude of the contribution). As an example, value, +2.0 of Feature x indicates that feature x contributes(encourages)

Figure 4. Decision tree generated on WQI dataset

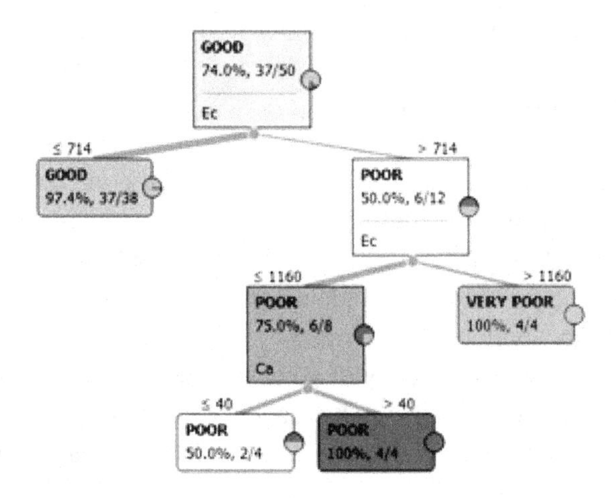

the prediction of class I with the value of 2, in the same way, if the value of feature y is -2.0 means that feature y discourages the prediction of class I with the value of 2

The importance of a feature *i* is defined by the Shapley[1] value in Eq. 1

$$\varnothing i = \frac{1}{N!} \sum_{S \subseteq N \setminus \{i\}|} S!(N! - S - 1)!\left[f \cup \{i\}) - f(S)\right] \quad (1)$$

Here

S: feature set

f(S) indicate the output of the Machine learning model required to be described using feature set S,

N is a set of all features.

Φi is final influence or Shapley value of feature *I*,

Φi is the average of feature *i* contributions across all probable combinations of a feature set. The features are added one by one to the set and the corresponding alteration in the model output shows the significance of that feature.

SHAP method can be used for both local (instance-wise) explanation and global (considering all the instances of the dataset) explanation of prediction

Table 2. Confusion matrix

		Class Predicted by Classifier	
		Class I	**Class II**
Real class of sample	**Class I**	No. of samples Predicted Positive by classifier and it's Positive originally (TP)(a)	No. of samples Predicted Negative by classifier, and it's Positive originally (FN)(b)
	Class II	No. of samples Predicted Positive by classifier, and it's Negative originally. (FP)(c)	No. of samples Predicted Negative by classifier and it's truly Negative originally (TN) (d)

5. RESULTS ANALYSIS

This section elaborates on the result part of the analysis. The work applies and studies the performance of various popular machine learning algorithms for prediction of the quality of water. Performance of the machine learning algorithms is judged using Classification accuracy, Precision, recall, F1 measure, and Specificity.

To find out the values of these metrics, the program starts with the construction of the confusion matrix. For the binary class classification problem, the classification model's prediction gives the following four possible outcomes:

- **True Positive**: Categorized in Positive class by classifier and it's Positive originally.
- **True Negative**: Categorized in Negative by classifier and it's truly Negative originally.
- **False Positive**: Categorized in Positive by classifier, and it's Negative originally.
- **False Negative**: Categorized in Negative by classifier, and it's Positive originally.

Table 2 is the confusion matrix for the binary classification. Each cell in the matrix should record the number of observations that fall into one of the four entry of confusion matrix

On the basis of Confusion matrix entries, the following values can be calculated as

Accuracy = (a + d) / (a+b+c+d).

Precision = a / (a + c)

Table 3. Computed confusion matrix on WQI dataset

Logistic Regression	Predicted				
Actual		GOOD	POOR	VERY POOR	Σ
	GOOD	37	0	0	37
	POOR	1	6	0	7
	VERY POOR	0	2	4	6
	Σ	38	8	4	50

Recall = Sensitivity = TPR = a/ (a + b)

F1- measure = 2* (Precision * Recall) / (Precision + Recall)

The confusion matrix of logistic regression classifier on WQI dataset is shown in Table 3.

Table 4 presents the performance card of all the applied methods. All the models give an accuracy above 86%. The logistic regression, neural network, and SVM achieve an accuracy of 94%. To demonstrate the next part of SHAP based explanation we choose the logistic regression model.

5.1 Shap-Based Explanation

The first part of the paper analyzed the performance of different machine learning algorithms to decide the water quality. It is observed that almost all machine learning model achieve over 85% accuracy. But the performance of these models has no solid

Table 4. Performance of different machine learning algorithms on WQI dataset

	All Features				
Model	Classification Accuracy	F1-Measure	Precision	Recall	Specificity
kNN	0.92	0.92	0.92	0.92	0.88
Logistic Regression	0.94	0.94	0.95	0.94	0.94
Neural Network	0.94	0.94	0.94	0.94	0.88
Random Forest	0.9	0.89	0.89	0.9	0.88
SVM	0.94	0.94	0.94	0.94	0.88
Decision Tree	0.86	0.86	0.86	0.86	0.92

Figure 5. Feature importance

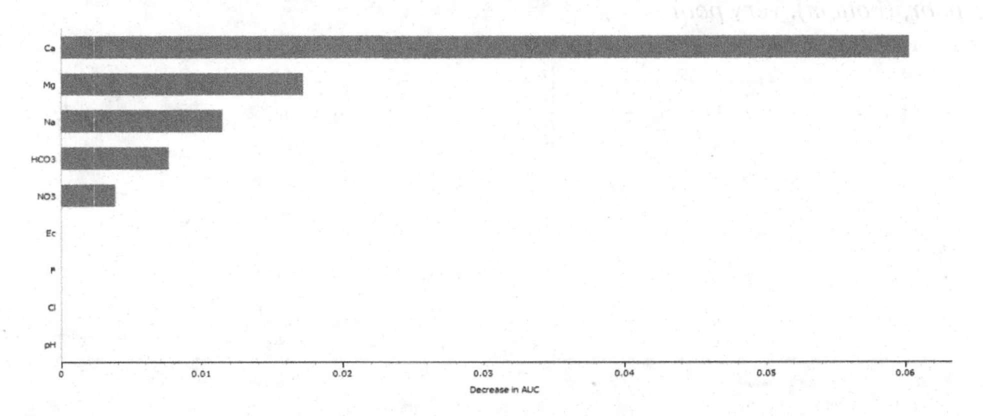

explanation. Which features play important role in deciding the class? To further explore the model a Coalition game theory-based approach is used to interpret the model performance, sample-wise (Local) and overall performance-wise (Global). In Figure 5 feature importance of the logistic regression is shown. All the features are sorted based on their importance. The quantity of Ca, Mg, NA, HC03, and NO_3 play a prominent role in deciding the water quality as compared to the remaining features.

Figure 6 is showing the summary of features for each class. Each point on the summary plot is a Shapley value for a feature and an instance. The Y axis of the plot indicates all the features arranged in descending order of importance. The X-axis represents the Shapley value. Jittered overlapping points per feature give the glimpses of distribution of the Shapley values. As clear from the figure -- the low value of calcium (ca), Magnesium (Mg), and EC caused low predictions (class good). The high value of calcium (ca), Magnesium (Mg), and EC cause higher predictions (class good). The low value of NO_3 caused higher predictions (class good). In the same way, a lower value of the pH causes low predictions (class good) and he higher value of the pH cause high predictions (class good)

Figure 7 presents the correlation between features. The figure suggests that the features are independent of each other.

6. CONCLUSION

This article presents a comparative study of machine learning-based water quality decision-making algorithms. Different feature selection methods are used to identify the best features, and the performance of the model is examined using different feature sets. This reduces training time for machine learning algorithms.

Figure 6. Feature summary of each class: (top left) class: good; (top right) class: poor; (bottom): very poor

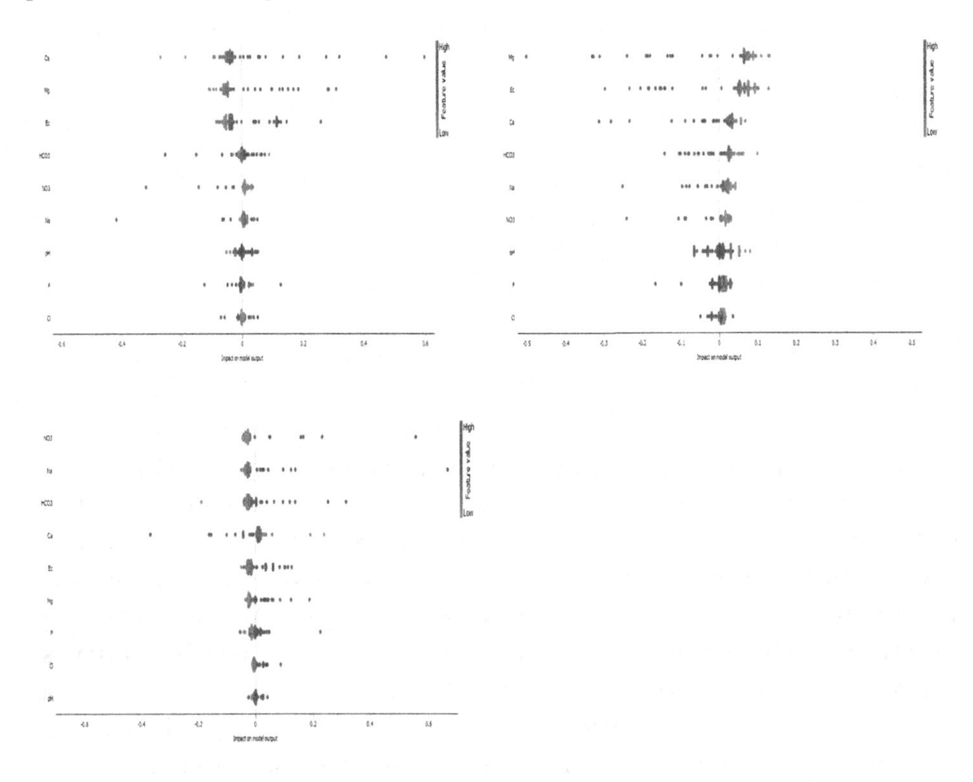

This research contribution will help explore different classification algorithms in the field of applied chemistry. This study also proposes an optimal function for

Figure 7. Summary plots with the impact of features on model output: class: good (left); class: poor (center); class: very poor (right)

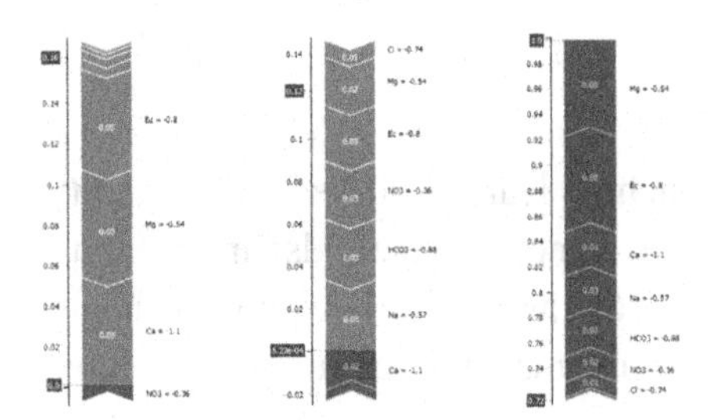

Table 5. Conclusion of the study

Feature	Good	Poor	Very Poor
pH	L	L+DI	L+II
HCO_3	M+DI	M+DI	H+II
Cl	L	L	L+II
NO_3	M	M+DI	H+II
F	L	L	L+II
Ca	H+II	H+DI	M+DI
Mg	H+II	H+DI	L+II
Na	L	M+DI	H+II
Ec	H +II	H+DI	L+II

Where Table 5 can be read by using the following nomenclature:
II Chances Rises with Rise in Value
DI Chances decline with a Rise in Value
I No matter value is high or low, chances rise
R No matter value is high or low, chances decrease
L Low effects on prediction
H High effect on prediction
M Medium effect on prediction
D Affecting prediction when interacting with another feature

determining water quality. We can conclude that the predictive effectiveness of the model depends on the number of features and the feature selection techniques. Here, key features for water quality determination are identified, and classification accuracies of over 90% and up to about 98% are observed in all cases. It turns out to be a better alternative compared to traditional formula-based methods. This work can be further extended by applying other ensemble classification techniques to selected features, increasing accuracy up to 100%.

REFERENCES

Abba, S. I., Hadi, S. J., Sammen, S. S., Salih, S. Q., Abdulkadir, R. A., Pham, Q. B., & Yaseen, Z. M. (2020). Evolutionary computational intelligence algorithm coupled with self-tuning predictive model for water quality index determination. *Journal of Hydrology (Amsterdam)*, *587*, 124974. doi:10.1016/j.jhydrol.2020.124974

Akter, T., Jhohura, F. T., Akter, F., Chowdhury, T. R., Mistry, S. K., Dey, D., Barua, M. K., Islam, M. A., & Rahman, M. (2016). Water Quality Index for measuring drinking water quality in rural Bangladesh: A cross-sectional study. *Journal of Health, Population and Nutrition*, *35*(1), 4. doi:10.118641043-016-0041-5 PMID:26860541

Alnazer, I., Bourdon, P., Urruty, T., Falou, O., Khalil, M., Shahin, A., & Fernandez-Maloigne, C. (2021). Recent advances in medical image processing for the evaluation of chronic kidney disease. *Medical Image Analysis*, *69*, 101960. doi:10.1016/j.media.2021.101960 PMID:33517241

Ambiga, K., & Annadurai, R. (2015). Development of Water Quality Index and Regression Model for Assessment of Groundwater Quality. *International Journal of Advanced Remote Sensing and GIS*, *4*(1), 931–943. doi:10.23953/cloud.ijarsg.88

BIS. (2012). Indian Standard Drinking Water Specification IS 10500. *Bureau of Indian Standards*, *25*(May), 1–3. http://cgwb.gov.in/Documents/WQ-standards.pdf

Bouzid, S., Ramdani, M., & Chenikher, S. (2019). Quality Fuzzy Predictive Control of Water in Drinking Water Systems. *Automatic Control and Computer Sciences*, *53*(6), 492–501. doi:10.3103/S0146411619060026

El Bilali, A., & Taleb, A. (2020). Prediction of irrigation water quality parameters using machine learning models in a semi-arid environment. *Journal of the Saudi Society of Agricultural Sciences*, *19*(7), 439–451. doi:10.1016/j.jssas.2020.08.001

Gakii, C., & Jepkoech, J. (2019). A Classification Model for Water Quality Analysis using Decision Tree. *European Journal of Computer Science and Information Technology*, *7*(3), 1–2.

Gao, S., Zhou, M., Wang, Y., Cheng, J., Yachi, H., & Wang, J. (2019). Dendritic Neuron Model With Effective Learning Algorithms for Classification, Approximation, and Prediction. *IEEE Transactions on Neural Networks and Learning Systems*, *30*(2), 601–614. doi:10.1109/TNNLS.2018.2846646 PMID:30004892

Gazzaz, N. M., Yusoff, M. K., Aris, A. Z., Juahir, H., & Ramli, M. F. (2012). Artificial neural network modeling of the water quality index for Kinta River (Malaysia) using water quality variables as predictors. *Marine Pollution Bulletin*, *64*(11), 2409–2420. doi:10.1016/j.marpolbul.2012.08.005 PMID:22925610

Goz, E., Yuceer, M. & Karadurmus, E. (2019, April). *Machine Learning Application of Dissolved Oxygen Prediction in River Water Quality*. doi:10.11159/iceptp19.119

Haghiabi, A. H., Nasrolahi, A. H., & Parsaie, A. (2018). Water quality prediction using machine learning methods. *Water Quality Research Journal of Canada*, *53*(1), 3–13. doi:10.2166/wqrj.2018.025

Hmoud Al-Adhaileh, M., & Waselallah Alsaade, F. (2021). Modelling and Prediction of Water Quality by Using Artificial Intelligence. *Sustainability (Basel)*, *13*(8), 4259. doi:10.3390u13084259

Jameel, A. A., & Sirajudeen, J. (2006). Risk Assessment of Physico-Chemical Contaminants in Groundwater of Pettavaithalai Area, Tiruchirappalli, Tamilnadu – India. *Environmental Monitoring and Assessment*, *123*(1–3), 299–312. doi:10.100710661-006-9198-5 PMID:17054009

Kothari, V., Vij, S., Sharma, S., & Gupta, N. (2021). Correlation of various water quality parameters and water quality index of districts of Uttarakhand. *Environmental and Sustainability Indicators, 9*, 100093. https://doi.org/https://doi.org/10.1016/j.indic.2020.100093

Kumar, V., Sharma, A., Kumar, R., Bhardwaj, R., Kumar Thukral, A., & Rodrigo-Comino, J. (2020). Assessment of heavy-metal pollution in three different Indian water bodies by combination of multivariate analysis and water pollution indices. *Human and Ecological Risk Assessment*, *26*(1), 1–16. doi:10.1080/10807039.2018.1497946

Learn More: Water Quality Index (WQI). (n.d.). Lake County Water Authority, USF Water Institute. Retrieved September 2, 2022, from https://lake.wateratlas.usf.edu/library/learn-more/learnmore.aspx?toolsection=lm_wqi

Leong, W. C., Bahadori, A., Zhang, J., & Ahmad, Z. (2021). Prediction of water quality index (WQI) using support vector machine (SVM) and least square-support vector machine (LS-SVM). *International Journal of River Basin Management*, *19*(2), 149–156. doi:10.1080/15715124.2019.1628030

Liu, M., & Lu, J. (2014). Support vector machine—An alternative to artificial neuron network for water quality forecasting in an agricultural nonpoint source polluted river? *Environmental Science and Pollution Research International*, *21*(18), 11036–11053. doi:10.100711356-014-3046-x PMID:24894753

Lundberg, S. M., & Lee, S.-I. (2017). A Unified Approach to Interpreting Model Predictions. In I. Guyon, U. V Luxburg, S. Bengio, H. Wallach, R. Fergus, S. Vishwanathan & R. Garnett (Eds.), *Advances in Neural Information Processing Systems* (Vol. 30). Curran Associates, Inc. https://proceedings.neurips.cc/paper/2017/file/8a20a8621978632d76c43dfd28b67767-Paper.pdf

Ma, J., Ding, Y., Cheng, J. C. P., Jiang, F., & Xu, Z. (2020). Soft detection of 5-day BOD with sparse matrix in city harbor water using deep learning techniques. *Water Research*, *170*, 115350. doi:10.1016/j.watres.2019.115350 PMID:31830651

Park, Y., Cho, K. H., Park, J., Cha, S. M., & Kim, J. H. (2015). Development of early-warning protocol for predicting chlorophyll-a concentration using machine learning models in freshwater and estuarine reservoirs, Korea. *The Science of the Total Environment*, *502*, 31–41. doi:10.1016/j.scitotenv.2014.09.005 PMID:25241206

Priyanka, P., Krishan, G., Sharma, L. M., Yadav, B., & Ghosh, N. C. (2016). Analysis of Water Level Fluctuations and TDS Variations in the Groundwater at Mewat (Nuh) District, Haryana (India). *Current World Environment*, *11*(2), 388–398. doi:10.12944/CWE.11.2.06

Rauen, W., Ferraresi, A., Maranho, L., Oliveira, E., da Costa, R. M., Alcantara, J., & Dziedzic, M. (2018). Index-based and compliance assessment of water quality for a Brazilian subtropical reservoir. *Engenharia Sanitaria e Ambiental*, *23*(5), 841–848. doi:10.15901413-4152201820180002

Tripathi, M., & Singal, S. K. (2019). Use of Principal Component Analysis for parameter selection for development of a novel Water Quality Index: A case study of river Ganga India. *Ecological Indicators*, *96*, 430–436. doi:10.1016/j.ecolind.2018.09.025

Verma, A., Thakur, B., Kartiyar, S., Singh, D., & Rai, M. (2013). Evaluation of ground water quality in Lucknow, Uttar Pradesh using remote sensing and geographic information systems (GIS). *Int. J. Water Res. Environ. Eng.*, *5*(2), 67–76. https://doi.org/https://doi.org/10.5897/IJWREE11.142

Wan, D., Wu, H., Li, S., & Cheng, F. (2011). Application of Data Mining in Relationship between Water Quantity and Water Quality. In H. Deng, D. Miao, J. Lei, & F. L. Wang (Eds.), *Artificial Intelligence and Computational Intelligence* (pp. 404–412). Springer Berlin Heidelberg. doi:10.1007/978-3-642-23881-9_53

Wang, G., Jia, Q.-S., Zhou, M., Bi, J., Qiao, J., & Abusorrah, A. (2022). Artificial neural networks for water quality soft-sensing in wastewater treatment: A review. *Artificial Intelligence Review*, *55*(1), 565–587. doi:10.100710462-021-10038-8

Wang, J. & Kumbasar, T. (2019). Parameter optimization of interval Type-2 fuzzy neural networks based on PSO and BBBC methods. *IEEE/CAA Journal of Automatica Sinica, 6*(1), 247–257. doi:10.1109/JAS.2019.1911348

Wu, Q., Wu, B., & Yan, X. (2022). An intelligent traceability method of water pollution based on dynamic multi-mode optimization. *Neural Computing & Applications*. Advance online publication. doi:10.100700521-022-07002-0 PMID:35221540

Zhang, Z., Huang, J., Duan, S., Huang, Y., Cai, J., & Bian, J. (2022). Use of interpretable machine learning to identify the factors influencing the nonlinear linkage between land use and river water quality in the Chesapeake Bay watershed. *Ecological Indicators*, *140*, 108977. doi:10.1016/j.ecolind.2022.108977

ENDNOTE

1 https://christophm.github.io/interpretable-ml-book/shapley.html

Chapter 9
Classification of Quality of Water Using Machine Learning

P. Umamaheswari
https://orcid.org/0000-0003-2007-697X
SASTRA University (Deemed), India

M. Kamaladevi
SASTRA University (Deemed), India

ABSTRACT

Keeping a check on the quality of water is necessary for protecting both the health of humans and of the environment. AI can be used to classify and predict water quality. The proposed system uses several machine learning algorithms to manage water quality data gathered over a protracted period. Water quality index (WQI) is used to categorize the given samples by using machine learning and ensemble approaches. The studied classifiers included random forest classifier, CatBoost classifier, k nearest neighbors, logistic regression, etc. The authors used precision-recall curves, ROC curves and confusion matrices as performance metrics for the ML classifiers used. With an accuracy of 95.43%, the random forest model was shown to be the most accurate classifier. Furthermore, CatBoost classifier and k nearest neighbors provided satisfactory results with 94.86% and 94.08% accuracy, respectively. Therefore, CatBoost algorithm is considered to be a more reliable approach for the quality of water classification.

DOI: 10.4018/978-1-6684-6791-6.ch009

1. INTRODUCTION

In water studies, machine learning presents excellent prospects for rating, categorizing, and predicting indicators of water quality. For instance, given access to sufficient quantities of data, machine learning models may successfully simulate hydrological processes and the transfer of pollutants. Photo sensors help in detecting the individual characteristics of collected water samples. The detection of phosphorus can be found using colorimetric techniques, using the color obtained from chemical reaction occurring between phosphorus and other reagents. Various other sensors are also capable of finding the dissolved pollutants in the samples.

The extensively used arithmetic algorithm, the water quality index (WQI), was used to classify the groundwater quality as per World Health Organization (WHO) recommendations and Bangladesh (BD) standards (Rahman et al., 2023). The outputs of these procedures provide the required information which are precisely and easily accessed by artificial intelligence (AI). Mechanism-oriented models are versatile and can mimic water quality using typically complex data structures found in sophisticated systems. For various aquatic systems, other models have been introduced. Researchers now show that artificial intelligence is practical and useful for estimating water quality (Uddin, 2023). About 71% of Earth's surface is water covered, that is, the Earth is about two-thirds water-filled. Less than 3% of total water is only fresh water while the rest is completely salty. Water Evaluation and Planning (WEAP), a numerical simulation tool, was used to model river water quality using two scenarios, namely business as usual (BAU) and scenario with measures (Kumar, 2019). Freshwater is present in glaciers and icecaps. The water which can be used for drinking is less than 1%. Also, rivers make up about 0.49% of surface freshwater. The majority of this water is replenished by precipitation. Precipitation is a phase of water cycle which affects the life on Earth. There are numerous water borne diseases hence, keeping a check on water quality is necessary. Acharya et al. (2020) revealed that Arcobacter and Aeromonas were prime faecal pollution threats. High Arcobacter butzleri loads detected in Ethiopia were endorsed by qPCR. Some more researchers found that introducing disclosure mandates led to significant improvements in water quality by examining salts that are considered signatures for hydraulic fracturing (HF) impact (Bonetti, Leuz, & Michelon, 2020).

Monitoring water quality is essential for the people and organizations. It paves way to analyze the changes undergone by the samples and to provide us with reliable strategies to decrease the unsustainable exploitation of water. The traditional approaches comprised of the manual procedures of sample collection using the sensors that were placed at the locations (Wei et al., 2020). In addition, machine learning provides ample occasions to improve the results prediction and classification

(Mazhar et al., 2023). Hence, we use machine learning (ML) algorithms to classify water based on its quality.

2. RELATED WORK

The major reason behind the topic is that nowadays, there are many issues arose due to the water pollution. So water quality prediction is playing an important role in many areas related to agriculture, fishery and people investing more money in real estate business, etc. Mainly due to the increase in the content of nitrates, magnesium (Xu & Su, 2019) and lead percentages in ground water (drinking). In order to predict quality of water, many researchers had to predict the safe water using redox conditions and pH level present (Larocque et al., 2019) in ground water. As far as ground water is concerned, when pH condition increases then it leads to the increase in calcium and magnesium carbonates in the pipes. While this is higher, then pH does not pose any health risk, it can cause skin problems, such as becoming dry, itchy and irritated. When pH level decreases then acidity in water increases and it becomes toxic. It has been observed that aquatic pesticide pollution trends are mainly driven (Chow et al., 2020) by pesticide use and hydrology. Moreover they suggested baseline monitoring is essential to infer long-term water quality trends.

It is understood that this development in water quality is 'short-lived' would fail once the normal industrial activities are resumed, indicating a strong impact of crude commercial industrial waste water. The river can be rejuvenated if issues of wastewater and adequate flow releases are addressed (Dutta et al., 2020). By knowing the redox state of ground water, we can say that whether the water is toxic or anoxic and also helps us to determine whether it contains elevated levels of many contaminants including arsenic, nitrate and even some man made contaminants. The concentrations of arsenic and manganese are more likely to be present at levels that exceed human health bench marks in anoxic ground water, and concentrations of uranium, selenium and nitrate are more likely to exceed this benchmarks in toxic groundwater. As stated in (M'nassri & Majdoub, 2022) the redox conditions of the ground water is an important factor in predicting the contaminants and constituents might be present in groundwater at levels of concern for human health. In this proposed work, gradient boosted regression tree has been applied as the main machine learning model for the application. The analysis of the north eastern U.S.A aquifers are collected for the data.

The application of many of machine learning (Sakata & Ohama, 2018) have been applied to compare the efficiency of all models. Hence, it was finalized to visualize the coastal mapping of north-eastern U.S.A has been shown in results. The existing system is difficult in order to find the exact location in the map as it is worked on the volumetric and areal views. This shows that it is difficult to find the small areas in map and to find the location approximately. The approach in the existing system is time taking and not optimizing to the expected level to find the needy. The Supervised Learning Algorithm of Machine Learning (Xu et al., 2020) decision tree has been checked for the classification of the data into testing one and training one. Then the classified data has been used for the diagram representation of the tree in the code. Decision tree has been used to classify data into safe and unsafe drinking water. This allows the data classification proper and efficient in execution. Existing work had been done by using boosted regression directly as this project is a bit different in choosing the machine learning models and its classification techniques.

The predictors have been chosen in accordance with the dataset. Water Quality research work (Ferhati et al., 2023) identified the mineralization origin and discriminate between the different classes of groundwater quality in several areas of the semiarid basin of Hodna in central Algeria. Several multivariate statistical techniques are applied to a dataset composed of 64 georeferenced individuals with 19 chemical variables. The combined heat and water (CHW) system has been studied (Yang et al., 2023) by many scholars since it was proposed, and a project has been implemented in Shandong, China in 2021.

Groundwater from wells is heavily used for water offer and ecological concern in wells. Another approach (Kedia, 2015) uses sensors and actuator to develop a method and it seems to be involving technical challenges. IOT based prediction system used for small scale application (Rahman et al., 2019) and also it was not able to find large scale application because involving hidden cost of investing more on equipment related to hardware. Water quality soft sensing model implemented based on ANN. (Wang et al., 2022) hyper parameter tuning and initial weight parameter must have been considered.

3. PROPOSED METHODOLOGY

First we collect the required water samples from different locations and then we send them to laboratories. Sample collection and lab analysis are expensive and they are time consuming and can provide irrelevant results in some cases. Furthermore, it is becoming more common to monitor water quality using intelligent systems, particularly when real-time information is required. One of the reasons we use artificial intelligence is that the algorithms are efficient enough in providing better solutions within a given time frame without knowledge in programming. Machine learning approaches are capable of taking insights and spot trends in given historical information.

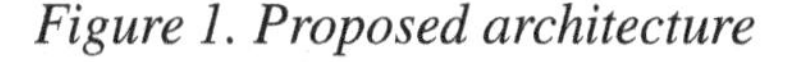

Figure 1. Proposed architecture

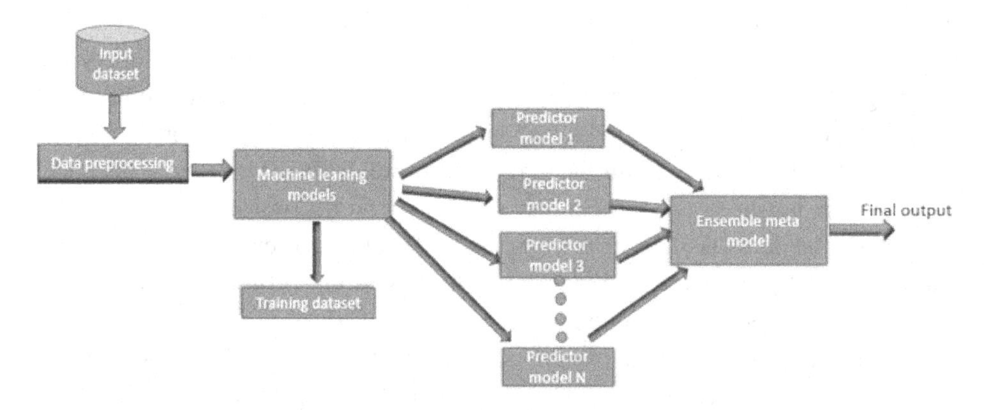

The aim of this study is to evaluate the water quality in real time basis. Monitoring water quality is beneficial in assessing water quality at regular intervals and providing strategies that help minimize damage caused by water. This proves to be advantageous to humans as well as to the environment. As manual evaluation and testing is time consuming and expensive, we use machine learning classifiers for water quality classification. Minimal usage of hardware and resources can be achieved using this study. Hence this study is a comparative approach between ten machine learning classifiers. We then conclude the best classifier with the highest accuracy in this aspect. In predictive modeling classification, we equate a mapping function from the input to output. The process included splitting the data into training and test sets for model training and performance evaluation. Water quality classification can be done by using machine learning model which includes various algorithm as shown in fig.1

3.1. Dataset

The dataset consisted of 1991 samples collected from over different states over India between the years 2005 and 2014 and fig.2 depicts the sample dataset of our proposed approach. The features that were taken into consideration were Biological oxygen demand, dissolved oxygen, pH of sample, conductivity, fecal coliform and total coliform. Furthermore, the collected data was subjected to preprocessing to achieve desired results. Data preprocessing is cleaning the data by removing the unnecessary details and labeling the data.

Figure 2. Sample dataset

	STATION CODE	LOCATIONS	STATE	Temp	D.O. (mg/l)	PH	CONDUCTIVITY (µmhos/cm)	B.O.D. (mg/l)	NITRATENAN N+ NITRITENANN (mg/l)	FECAL COLIFORM (MPN/100ml)	TOTAL COLIFORM (MPN/100ml)Mean	year
0	1393	DAMANGANGA AT D/S OF MADHUBAN, DAMAN	DAMAN & DIU	30.6	6.7	7.5	203	NAN	0.1	11	27	2014
1	1399	ZUARI AT D/S OF PT. WHERE KUMBARJRIA CANAL JOI...	GOA	29.8	5.7	7.2	189	2	0.2	4953	8391	2014
2	1475	ZUARI AT PANCHAWADI	GOA	29.5	6.3	6.9	179	1.7	0.1	3243	5330	2014
3	3181	RIVER ZUARI AT BORIM BRIDGE	GOA	29.7	5.8	6.9	64	3.8	0.5	5382	8443	2014
4	3182	RIVER ZUARI AT MARCAIM JETTY	GOA	29.5	5.8	7.3	83	1.9	0.4	3428	5500	2014
...	...	...	...	...	...	...	...	...	...	...	...	...
1986	1330	TAMBIRAPARANI AT ARUMUGANERI, TAMILNADU	NAN	NAN	7.9	738	7.2	2.7	0.518	0.518	202	2003
1987	1450	PALAR AT VANIYAMBADI WATER SUPPLY HEAD WORK, T...	NAN	29	7.5	585	6.3	2.6	0.155	0.155	315	2003
1988	1403	GUMTI AT U/S SOUTH TRIPURA,TRIPURA	NAN	28	7.6	98	6.2	1.2	NAN	NAN	570	2003
1989	1404	GUMTI AT D/S SOUTH TRIPURA, TRIPURA	NAN	28	7.7	91	6.5	1.3	NAN	NAN	562	2003
1990	1726	CHANDRAPUR, AGARTALA D/S OF HAORA RIVER, TRIPURA	NAN	29	7.6	110	5.7	1.1	NAN	NAN	546	2003

1991 rows × 12 columns

3.2. Water Quality Index Calculation

The water quality index (WQI) is the average of different index values of all the parameters used.

$$WQI = \frac{\sum_{i=1}^{N} q_i * w_i}{\sum_{i=1}^{N} w_i} \tag{1}$$

$$q_i = 100 * \frac{v_i - v_{ideal}}{s_i - v_{ideal}} \tag{2}$$

$$w_i = \frac{k}{S_i} \tag{3}$$

Figure 3. SMOTE visualization

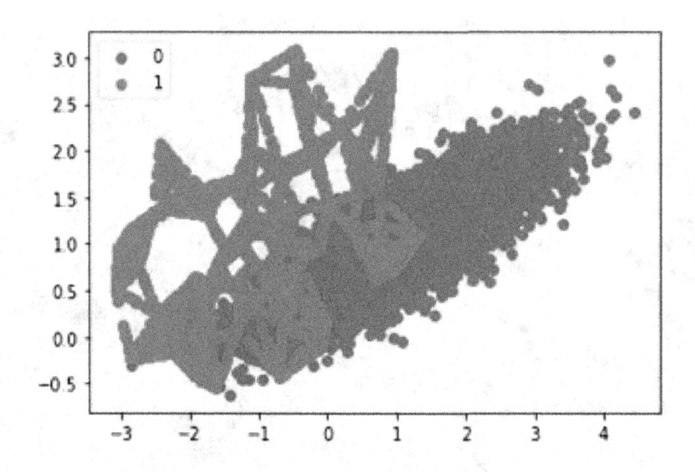

$$K = \frac{1}{\sum_{i=1}^{N} s_i} \tag{4}$$

Where,

N = total number of parameters
q_i = Quality estimate of parameter i
w_i = Each parameter's unit weight
V_i = Computed value of tested water samples
V_{ideal} = Pure water reference value
S_i = Desired value of parameter i
K = Proportionality constant

3.3. Data Analysis With SMOTE

The Synthetic Minority Oversampling Technique is abbreviated as SMOTE. We use this approach to oversample the minority class over unbalanced datasets. Imbalanced datasets may not yield desired results and may even perform poor. Hence, we use SMOTE to boost accuracy. It takes a subset from the minority class and generates new samples. Then the generated samples are combined with the dataset. In this way, the imbalance between different classes is reduced. The new dataset is used for training the model. This method can be used to overcome the problem of overfitting. The visualization of synthetic instances obtained due to SMOTE process is given in

Figure 4. Heat map

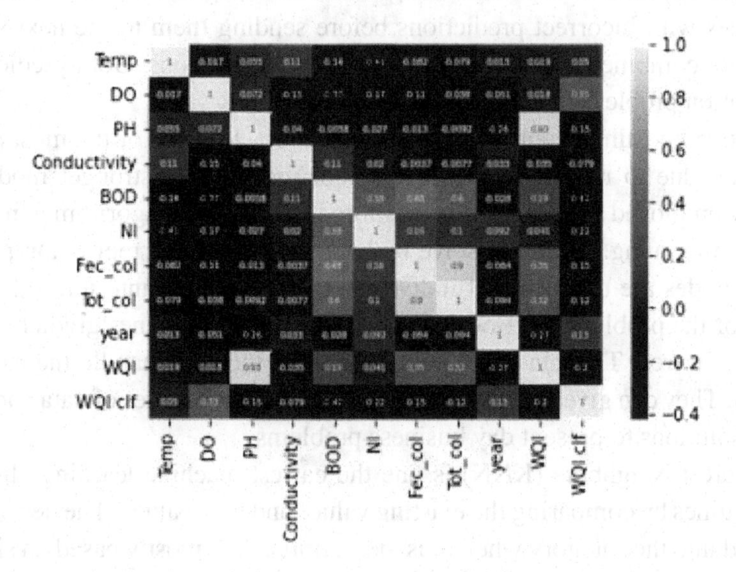

fig 3. The correlation between the attributes is depicted using heat maps. The heat map visualization is given in fig 4.

3.4 Description of Classifiers

Support Vector Machine (SVM) is an approach which operates by creating a hyper plane that separates the classes which can be used for classification as well as regression. The dimension of the hyper plane is based on the number of features. It is a supervised learning technique. This approach works well for handling high dimensional data. Random forest is simply a combination of many decision trees. It predicts output by comparing the outputs of the decision trees. The prediction of the class is based on the most count of votes of each decision tree. It is an ensemble classifier.

Multilayer perception (MLP) is a type of neural network which works under feed forward mechanism. In simple words, it is networks comprising of multiple layers of perceptron as its name implies. The input nodes are connected in the form of directed graphs between the input and output layers.

Logistic regression comes under supervised learning. It predicts the output based on given independent variable. It is a statistical model; hence the output values lie between 0 and 1. Unlike linear regression, this approach can be used in classification too.

XGBoost is elaborated as extreme gradient boosting. More weights are applied to the trees with incorrect predictions before sending them to the next step. The outputs are combined into a model for accurate predictions. Both prediction and classification problems can be solved using XGBoost.

Adaptive boosting is abbreviated as ADABoost. It is one of the most preferred approaches due to its ability to convert poor models into stronger models. It is primarily employed in binary classification. Decision tree algorithm comes under supervised learning. They can solve both classification and regression problems. The leaf nodes are the labels of individual classes and the interior nodes are the features of the problem. CatBoost is an approach that combines gradient boosting and decision tree. This classifier is pre-built and they can handle the categorical data well. They can give good performance for diverse varieties of data and provide reliable solutions to present day business problems.

K Nearest Neighbors (KNN) is one the easiest machine learning algorithms. KNN assumes by comparing the existing values and new values. The new sample is combined into the category where it is most similar. It is mostly based on similarity. Using KNN, new data can be classified in less time. The Naïve Bayes algorithm is a type of supervised learning technique that used Bayes theorem. It is quick and is based on probability. Hence, theory based assumptions can be made using naïve Bayes algorithm. It gives the likelihood of occurrence.

4. RESULTS AND DISCUSSION

The dataset is split into 75% and 25%, respectively. According to the estimates, Random Forest is the best classifier for classifying water based on its quality. Normal distribution plots are used to detect deviations in the dataset, hence the normal distribution plots were plotted. The plots were like Gaussian (bell shaped curve). Further, we use z score normalization to normalize the values.

The z score formula is given by,

$$z = \frac{x - \mu}{\sigma} \tag{5}$$

The negative values were removed as they were considered as outliers. The outcomes are shown in the plots. Confusion matrices are tables to determine the model errors. The actual classes are the rows and columns are the predicted classes. It helps in identifying incorrect predictions. Given below are the confusion matrices obtained for Random Forest and K Nearest Neighbors Classifiers respectively. The

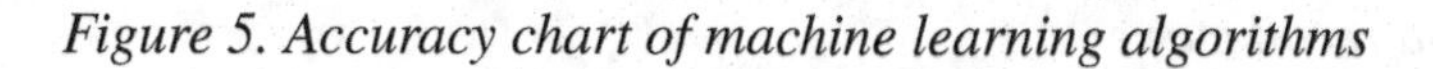

Figure 5. Accuracy chart of machine learning algorithms

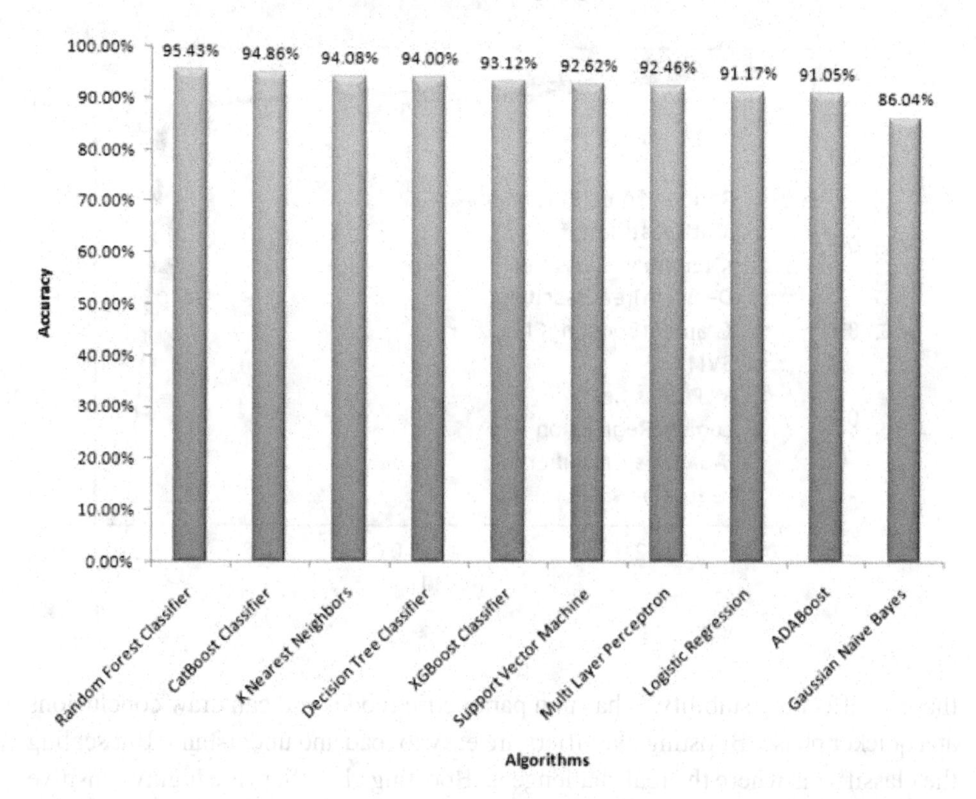

accuracy of each algorithm is given in fig 5. According to the study, random forest produced which is the ensemble model gives the best accuracy with 95.43% followed by CatBoost (94.86%) and KNN (94.08%). In terms of accuracy, the performances of the ML classifiers are:

Random Forest (95.43%), CatBoost (94.86%), K Nearest Neighbors (94.08%), Decision Tree Classifier (94.00%), XGBoost (93.12%), Support Vector Machine (92.62%), Multi Layer Perceptron (92.46%), Logistic Regression (91.17%), ADABoost (91.05%), and Gaussian Naïve Bayes (86.04%). Hence, Random Forest is the best classifier for improving accuracy which is given in Figure 5.

The outcome is influenced by the caliber of the dataset used while learning. Apart from regression, random forests can be used for classification as well. Though random forest is difficult to implement, it performs well for data with different properties. Support vector machine (SVM), on the other hand, works well with high dimensionality. But as the training period is slow for SVM, the performance for larger dataset is not up to mark. Comparatively, decision tree requires lesser data preprocessing hence small changes can show up significant changes in the output

Figure 6. Precision and recall curve

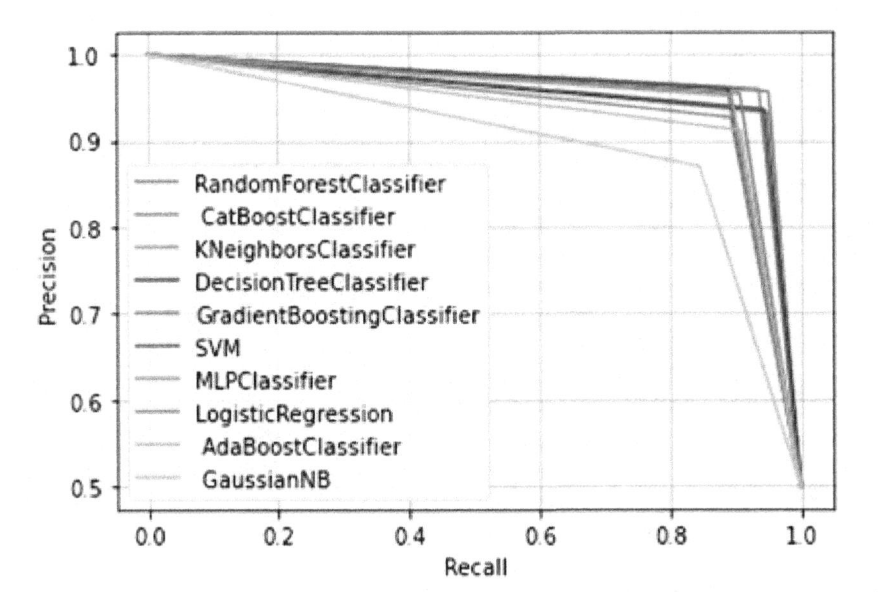

that can affect the stability. It has non parametric model and can draw conclusions at a quicker phase. Boosting classifiers are easy to read and understand. But scaling the classifier is where the real challenge is. Boosting classifiers are highly sensitive to outliers. MLP can provide faster results as they can be applied on nonlinear datasets as well.

The advantage of using ensemble algorithms is that they combine the benefits of different algorithms into one meta classifier. Hence they can assist the researchers worldwide in choosing the best classifier to employ in their search for the best algorithm for their study. The precision recall curves of the ML classifiers are presented in Figure 6 and the Receiver Operating Characteristics curve (ROC) curve for the algorithms are depicted in Figure 7.

The plots of MSE, RMSE and MAE are created for the all machine leaning models and proposed approach is shown in Figures 8-10. It is able to identify that our proposed model has the lowest error value.

CONCLUSION AND FUTURE WORK

This proposed approach is based on classifying water quality with respect to its water quality index and it has proposed a real time monitoring of water quality. The dataset included water samples collected from various states over a period of 2005 and 2014.

Figure 7. ROC curve of ML models

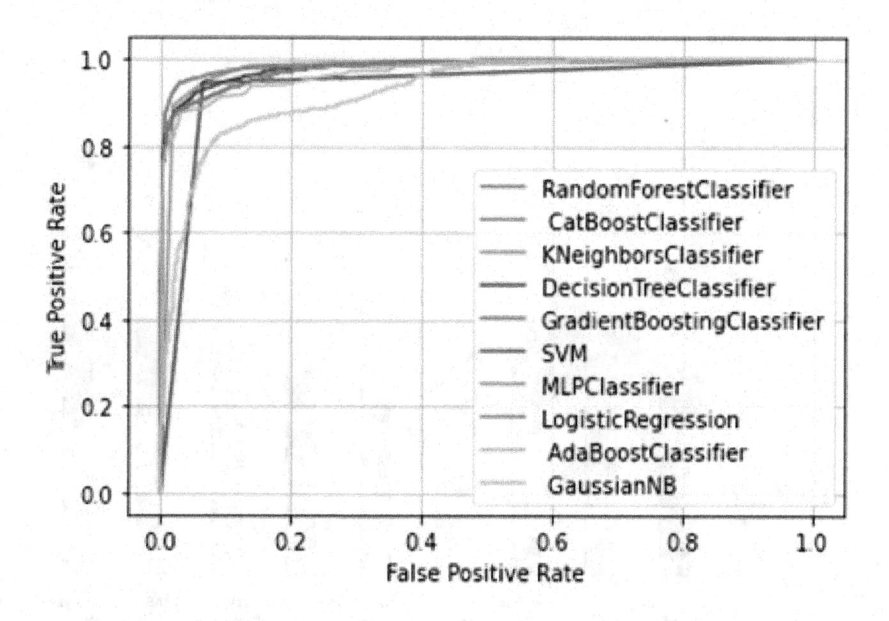

Figure 8. RMSE chart of ML models

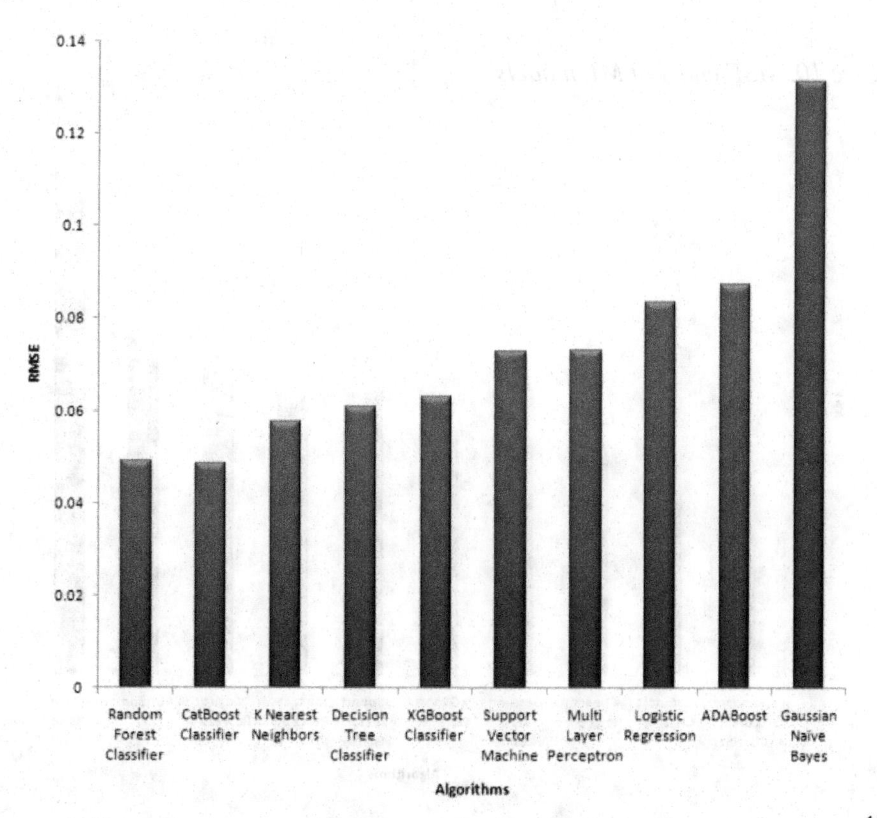

Figure 9. MAE chart of ML models

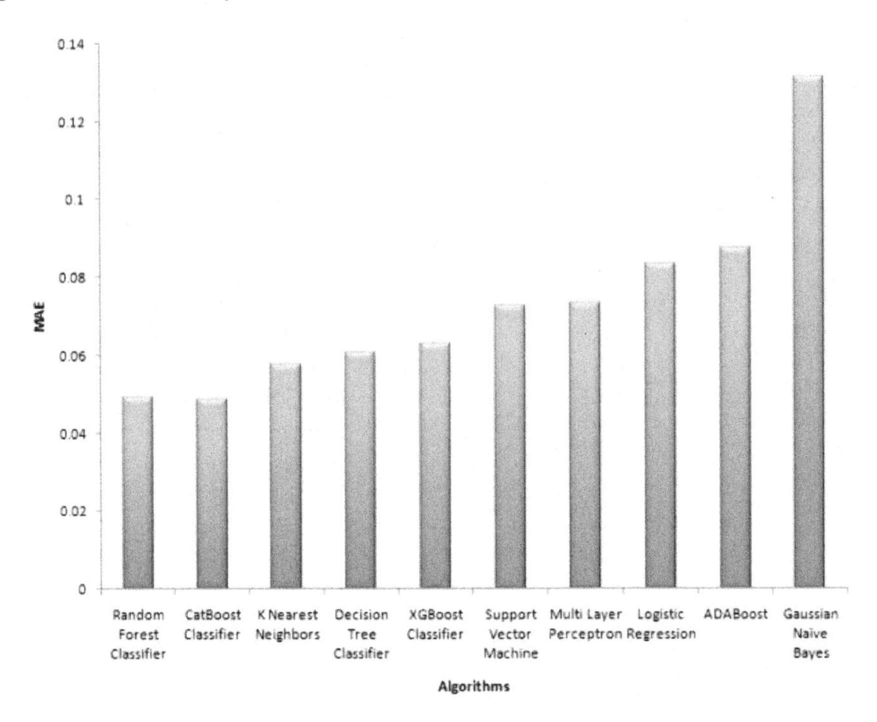

Figure 10. MSE chart of ML models

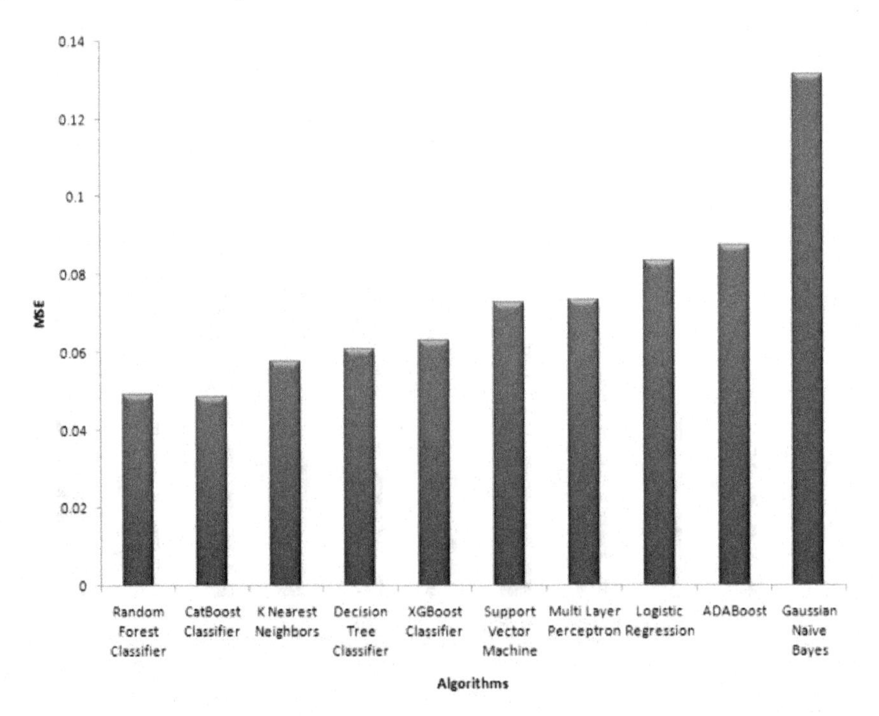

The attributes that were used for evaluation were biological oxygen demand, dissolved oxygen, presence of nitrates, total and fecal coliform, pH and conductivity. The ML classifiers used were Random forest, decision tree, catboost, XGboost, ADAboost, KNN, SVM, MLP, logistic regression and naïve bayes. Performance metrics used were precision recall and ROC curves. Also confusion matrices were constructed. Future applications of this research cannot be restricted to hardware. It may include deep learning techniques and AI techniques to draw conclusions. This study can be expanded to Internet of Things (IoT) and biomedical studies.

REFERENCES

Acharya, K., Blackburn, A., Mohammed, J., Haile, A. T., Hiruy, A. M., & Werner, D. (2020). Metagenomic water quality monitoring with a portable laboratory. *Water Research, 184*, 116112.

Bonetti, P., Leuz, C., & Michelon, G. (2023). *Internalizing externalities: Disclosure regulation for hydraulic fracturing, drilling activity and water quality (No. w30842).* National Bureau of Economic Research. doi:10.3386/w30842

Chow, R., Scheidegger, R., Doppler, T., Dietzel, A., Fenicia, F., & Stamm, C. (2020). A review of long-term pesticide monitoring studies to assess surface water quality trends. *Water Research X, 9*, 100064. doi:10.1016/j.wroa.2020.100064 PMID:32995734

Dutta, V., Dubey, D., & Kumar, S. (2020). Cleaning the River Ganga: Impact of lockdown on water quality and future implications on river rejuvenation strategies. *The Science of the Total Environment, 743*, 140756. doi:10.1016/j.scitotenv.2020.140756 PMID:32758842

Ferhati, A., Mitiche-Kettab, R., Belazreg, N. E. H., Khodja, H. D., Djerbouai, S., & Hasbaia, M. (2023). Hydrochemical analysis of groundwater quality in central Hodna Basin, Algeria: A case study. *International Journal of Hydrology Science and Technology, 15*(1), 22–39. doi:10.1504/IJHST.2023.127889

Goi, C. L. (2020). The river water quality before and during the Movement Control Order (MCO) in Malaysia. *Case Studies in Chemical and Environmental Engineering*, 2, 100027. doi:10.1016/j.cscee.2020.100027

Ke, G., Menu, Q., Finley, T., Wang, T., Chen, W., Ma, W., ... Liu, T. Y. (2017). Lightgbm: A highly efficient gradient boosting decision tree. *Advances in Neural Information Processing Systems*, 30.

Kedia, N. (2015, September). Water quality monitoring for rural areas-a sensor cloud based economical project. In *2015 1st International Conference on Next Generation Computing Technologies (NGCT)* (pp. 50-54). IEEE. 10.1109/NGCT.2015.7375081

Kumar, P. (2019). Numerical quantification of current status quo and future prediction of water quality in eight Asian megacities: Challenges and opportunities for sustainable water management. *Environmental Monitoring and Assessment, 191*(6), 1–12. doi:10.100710661-019-7497-x PMID:31044285

Larocque, M., Levison, J., Martin, A., & Chaumont, D. (2019). A review of simulated climate change impacts on groundwater resources in Eastern Canada. *Canadian Water Resources Journal/Revue canadienne des ressources hydriques, 44*(1), 22-41.

Mazhar, N., Javid, K., Akram, M. A. N., Afzal, A., Hamayon, K., & Ahmad, A. (2023). Index-Based Spatiotemporal Assesment Of Water Quality In Tarbela Reservoir, Pakistan (1990– 2020). Geography, Environment. *Sustainability*, *15*(4), 232–242.

Min, F., Hu, Q., & Zhu, W. (2014). Feature selection with test cost constraint. *International Journal of Approximate Reasoning*, *55*(1), 167–179. doi:10.1016/j.ijar.2013.04.003

Rahman, M. A., Mukta, M. Y., Yousuf, A., Asyhari, A. T., Bhuiyan, M. Z. A., & Yaakub, C. Y. (2019). *IoT based hybrid green energy driven highway lighting system. In 2019 IEEE Intl Conf on Dependable, Autonomic and Secure Computing, Intl Conf on Pervasive Intelligence and Computing, Intl Conf on Cloud and Big Data Computing, Intl Conf on Cyber Science and Technology Congress (DASC/PiCom/CBDCom/CyberSciTech)*. IEEE.

Rahman, M. M., Haque, T., Mahmud, A., Al Amin, M., Hossain, M. S., Hasan, M. Y., Shaibur, M. R., Hossain, S., Hossain, M. A., & Bai, L. (2023). Drinking water quality assessment based on index values incorporating WHO guidelines and Bangladesh standards. *Physics and Chemistry of the Earth Parts A/B/C, 129*, 103353. doi:10.1016/j.pce.2022.103353

Sakata, R., Ohama, I., & Taniguchi, T. (2018, November). An extension of gradient boosted decision tree incorporating statistical tests. In *2018 IEEE International Conference on Data Mining Workshops (ICDMW)* (pp. 964-969). IEEE. 10.1109/ICDMW.2018.00139

Uddin, M. G., Nash, S., Rahman, A., & Olbert, A. I. (2023). A novel approach for estimating and predicting uncertainty in water quality index model using machine learning approaches. *Water Research, 229*, 119422. doi:10.1016/j.watres.2022.119422 PMID:36459893

Wang, G., Jia, Q. S., Zhou, M., Bi, J., Qiao, J., & Abusorrah, A. (2022). Artificial neural networks for water quality soft-sensing in wastewater treatment: A review. *Artificial Intelligence Review*, *55*(1), 565–587. doi:10.100710462-021-10038-8

Wei, W., Gao, Y., Huang, J., & Gao, J. (2020). Exploring the effect of basin land degradation on lake and reservoir water quality in China. *Journal of Cleaner Production*, *268*, 122249. doi:10.1016/j.jclepro.2020.122249

Wen, Z., He, B., Kotagiri, R., Lu, S., & Shi, J. (2018, May). Efficient gradient boosted decision tree training on GPUs. In *2018 IEEE International Parallel and Distributed Processing Symposium (IPDPS)* (pp. 234-243). IEEE. 10.1109/IPDPS.2018.00033

Xu, T., Coco, G., & Neale, M. (2020). A predictive model of recreational water quality based on adaptive synthetic sampling algorithms and machine learning. *Water Research*, *177*, 115788. doi:10.1016/j.watres.2020.115788 PMID:32330740

Xu, W., & Su, X. (2019). Challenges and impacts of climate change and human activities on groundwater-dependent ecosystems in arid areas–A case study of the Nalenggele alluvial fan in NW China. *Journal of Hydrology (Amsterdam)*, *573*, 376–385. doi:10.1016/j.jhydrol.2019.03.082

Xu, Z., Shen, J., Qu, Y., Chen, H., Zhou, X., Hong, H., Sun, H., Lin, H., Deng, W., & Wu, F. (2022). Using simple and easy water quality parameters to predict trihalomethane occurrence in tap water. *Chemosphere*, *286*, 131586. doi:10.1016/j.chemosphere.2021.131586 PMID:34303907

Yang, X., Liu, Z., & Xia, J. (2023). Optimization and analysis of combined heat and water production system based on a coal-fired power plant. *Energy*, *262*, 125611. doi:10.1016/j.energy.2022.125611

Compilation of References

Abba, S. I., Hadi, S. J., Sammen, S. S., Salih, S. Q., Abdulkadir, R. A., Pham, Q. B., & Yaseen, Z. M. (2020). Evolutionary computational intelligence algorithm coupled with self-tuning predictive model for water quality index determination. *Journal of Hydrology (Amsterdam)*, *587*, 124974. doi:10.1016/j.jhydrol.2020.124974

Abba, S. I., Pham, Q. B., Usman, A. G., Linh, N. T. T., Aliyu, D. S., Nguyen, Q., & Bach, Q. V. (2020). Emerging evolutionary algorithm integrated with kernel principal component analysis for modeling the performance of a water treatment plant. *Journal of Water Process Engineering*, *33*, 101081. doi:10.1016/j.jwpe.2019.101081

Abbasnia, A., Yousefi, N., Mahvi, A. H., Nabizadeh, R., Radfard, M., Yousefi, M., & Alimohammadi, M. (2019). Evaluation of groundwater quality using water quality index and its suitability for assessing water for drinking and irrigation purposes: Case study of Sistan and Baluchistan province (Iran). *Human and Ecological Risk Assessment*, *25*(4), 988–1005. doi:10.1080/10807039.2018.1458596

Abdi, J., & Mazloom, G. (2022). Machine learning approaches for predicting arsenic adsorption from water using porous metal–organic frameworks. *Scientific Reports*, *12*(1), 1–13. doi:10.103841598-022-20762-y PMID:36180503

Abdurahman, R. (2022). *IoT-Based Smart Farming Using Machine Learning For Red Spinach*. Academic Press.

Abioye, E. A., Abidin, M. S. Z., Aman, M. N., Mahmud, M. S. A., & Buyamin, S. (2021). A model predictive controller for precision irrigation using discrete Lagurre networks. *Computers and Electronics in Agriculture, 181*, 105953. doi:10.1016/j.compag.2020.105953

Acharya, K., Blackburn, A., Mohammed, J., Haile, A. T., Hiruy, A. M., & Werner, D. (2020). Metagenomic water quality monitoring with a portable laboratory. *Water Research, 184*, 116112.

Adhiguru, P., & Mruthyunjaya. (2004). *Institutional innovations for using information and communication technology in agriculture.* Policy Brief 18. New Delhi: National Centre for Agricultural Economics and Policy Research.

Aftab, S., Shah, A., Nisar, J., Ashiq, M. N., Akhter, M. S., & Shah, A. H. (2020). Marketability Prospects of Microbial Fuel Cells for Sustainable Energy Generation. *Energy & Fuels*, *34*(8), 9108–9136. doi:10.1021/acs.energyfuels.0c01766

Afzaal, H., Farooque, A. A., Esau, T. J., Schumann, A. W., Zaman, Q. U., Abbas, F., & Bos, M. (2023). Artificial neural modeling for precision agricultural water management practices. In *Precision Agriculture* (pp. 169–186). Elsevier. doi:10.1016/B978-0-443-18953-1.00005-2

Agarwal, A., Upadhyay, U., Sreedhar, I., Singh, S. A., & Patel, C. M. (2020). A review on valorization of biomass in heavy metal removal from wastewater. *Journal of Water Process Engineering*, *38*, 101602. doi:10.1016/j.jwpe.2020.101602

Aghababaie, M., Farhadian, M., Jeihanipour, A., & Biria, D. (2015). Effective factors on the performance of microbial fuel cells in wastewater treatment–a review. *Environmental Technology Reviews*, *4*(1), 71–89. doi:10.1080/09593330.2015.1077896

Ahring, B. K. (2003). Perspectives for anaerobic digestion. *Advances in Biochemical Engineering/ Biotechnology*, *81*, 1–30. doi:10.1007/3-540-45839-5_1 PMID:12747559

Aina, I. O., Kaniki, A. M., & Ojiambo, J. B. (1995). *Agricultural information in Africa*. Third World Information Services.

Ait Rahou, Y., Ait-El-Mokhtar, M., Anli, M., Boutasknit, A., Ben-Laouane, R., Douira, A., Benkirane, R., El Modafar, C., & Meddich, A. (2021). Use of mycorrhizal fungi and compost for improving the growth and yield of tomato and its resistance to *Verticillium dahliae*. *Archiv für Phytopathologie und Pflanzenschutz*, *54*(13–14), 665–690. doi:10.1080/03235408.2020.1854938

Ait-El-Mokhtar, M. (2022). Cereals and Phytohormones Under Drought Stress. *Sustainable Remedies for Abiotic Stress in Cereals*. https://doi.org/https://doi.org/10.1007/978-981-19-5121-3_13

Akensous, F., Anli, M., Boutasknit, A., Ben-Laouane, R., Ait-Rahou, Y., Ahmed, H., Nasri, N., Hafidi, M., & Meddich, A. (2022). Boosting Date Palm (*Phoenix dactylifera* L.) Growth under Drought Stress: Effects of Innovative Biostimulants. *Gesunde Pflanzen*, *74*(4), 961–982. doi:10.100710343-022-00651-0

Akensous, F.-Z., Anli, M., & Meddich, A. (2022). Biostimulants as Innovative Tools to Boost Date Palm (*Phoenix dactylifera* L.) Performance under Drought, Salinity, and Heavy Metal(Oid) s′ Stresses: A Concise Review. *Sustainability (Basel)*, *14*(23), 15984. doi:10.3390u142315984

Akinremi, O. O., McGinn, S. M., & Barr, A. G. (1996). Simulation of soil moisture and other components of the hydrological cycle using a water budget approach. *Canadian Journal of Soil Science*, *76*(2), 133–142. doi:10.4141/cjss96-020

Akter, T., Jhohura, F. T., Akter, F., Chowdhury, T. R., Mistry, S. K., Dey, D., Barua, M. K., Islam, M. A., & Rahman, M. (2016). Water Quality Index for measuring drinking water quality in rural Bangladesh: A cross-sectional study. *Journal of Health, Population and Nutrition*, *35*(1), 4. doi:10.118641043-016-0041-5 PMID:26860541

Akullo, W. N., & Mulumba, O. (2016). *Making ICTs relevant to rural farmers in Uganda: a case of Kamuli district.* Available at: https://library.ifla.org/1488/1/110-akullo-en.pdf

Alam, G., Ihsanullah, I., Naushad, M., & Sillanpää, M. (2022). Applications of artificial intelligence in water treatment for optimization and automation of adsorption processes: Recent advances and prospects. *Chemical Engineering Journal, 427*, 130011. doi:10.1016/j.cej.2021.130011

Alitane, A., Essahlaoui, A., Van Griensven, A., Yimer, E. A., Essahlaoui, N., Mohajane, M., Chawanda, C. J., & Van Rompaey, A. (2022). Towards a Decision-Making Approach of Sustainable Water Resources Management Based on Hydrological Modeling: A Case Study in Central Morocco. *Sustainability (Basel), 14*(17), 10848. doi:10.3390u141710848

Alkaisi, A., Mossad, R., & Sharifian-Barforoush, A. (2017). A Review of the Water Desalination Systems Integrated with Renewable Energy. *Energy Procedia, 110*, 268–274. doi:10.1016/j.egypro.2017.03.138

Alnazer, I., Bourdon, P., Urruty, T., Falou, O., Khalil, M., Shahin, A., & Fernandez-Maloigne, C. (2021). Recent advances in medical image processing for the evaluation of chronic kidney disease. *Medical Image Analysis, 69*, 101960. doi:10.1016/j.media.2021.101960 PMID:33517241

Alves, R. G., Maia, R. F., & Lima, F. (2023). Development of a Digital Twin for smart farming: Irrigation management system for water saving. *Journal of Cleaner Production, 388*, 135920. doi:10.1016/j.jclepro.2023.135920

Ambiga, K., & Annadurai, R. (2015). Development of Water Quality Index and Regression Model for Assessment of Groundwater Quality. *International Journal of Advanced Remote Sensing and GIS, 4*(1), 931–943. doi:10.23953/cloud.ijarsg.88

Angelidaki, I., Ellegaard, L., & Ahring, B. K. (2003). Applications of the anaerobic digestion process. *Advances in Biochemical Engineering/Biotechnology, 82*, 1–33. doi:10.1007/3-540-45838-7_1 PMID:12747564

Anitha, C., Komala, C. R., Vivekanand, C. V., Lalitha, S. D., Boopathi, S., & Revathi, R. (2023, February). Artificial Intelligence driven security model for Internet of Medical Things (IoMT). *Proceedings of 2023 3rd International Conference on Innovative Practices in Technology and Management, ICIPTM 2023.* 10.1109/ICIPTM57143.2023.10117713

Anli, M., Ait-El-Mokhtar, M., Akensous, F.-Z., Boutasknit, A., Ben-Laouane, R., Fakhech, A., Ouhaddou, R., Raho, O., & Meddich, A. (2023). Biofertilizers in Date Palm Cultivation. In *Date Palm* (pp. 266–296). CABI. doi:10.1079/9781800620209.0009

Anli, M., Baslam, M., Tahiri, A., Raklami, A., Symanczik, S., Boutasknit, A., Ait-El-Mokhtar, M., Ben-Laouane, R., Toubali, S., Ait Rahou, Y., Ait Chitt, M., Oufdou, K., Mitsui, T., Hafidi, M., & Meddich, A. (2020). Biofertilizers as Strategies to Improve Photosynthetic Apparatus, Growth, and Drought Stress Tolerance in the Date Palm. *Frontiers in Plant Science, 11*, 1–22. doi:10.3389/fpls.2020.516818 PMID:33193464

Antony, A. P., Leith, K., Jolley, C., Lu, J., & Sweeney, D. J. (2020). A review of practice and implementation of the internet of things (IoT) for smallholder agriculture. *Sustainability (Basel), 12*(9), 1–19. doi:10.3390u12093750

Arabloo, M., Mirabi, A., Ghaedi, M., & Asfaram, A. (2019). Machine learning-based prediction of adsorption capacity of modified carbon nanotubes for heavy metal ions removal from water. *Journal of Environmental Chemical Engineering*, *7*(6), 103409. doi:10.1016/j.jece.2019.103409

Ardabili, S., Mosavi, A., Dehghani, M., & Várkonyi-Kóczy, A. R. (2020). Deep Learning and Machine Learning in Hydrological Processes Climate Change and Earth Systems a Systematic Review. *Lecture Notes in Networks and Systems*, *101*, 52–62. doi:10.1007/978-3-030-36841-8_5

Asfaram, Ghaedi, Goudarzi, & Hajati. (2015). *Ternary dyes adsorption onto MnO2 nanoparticle-loaded activated carbon: derivative spectrophotometry and modeling*. doi:10.1039/C5RA10815B

Aziz, M., Khan, M., Anjum, N., Sultan, M., Shamshiri, R. R., Ibrahim, S. M., Balasundram, S. K., & Aleem, M. (2022). Scientific Irrigation Scheduling for Sustainable Production in Olive Groves. *Agriculture*, *12*(4), 564. doi:10.3390/agriculture12040564

Babaeian, E., Sadeghi, M., Jones, S. B., Montzka, C., Vereecken, H., & Tuller, M. (2019). Ground, Proximal, and Satellite Remote Sensing of Soil Moisture. *Reviews of Geophysics*, *57*(2), 530–616. doi:10.1029/2018RG000618

Babu, B. S., Kamalakannan, J., Meenatchi, N., M, S. K. S., S, K., & Boopathi, S. (2023). Economic impacts and reliability evaluation of battery by adopting Electric Vehicle. *IEEE Explore*, 1–6. doi:10.1109/ICPECTS56089.2022.10046786

Babu, S. C., Joshi, P. K., Glendenning, C. J., Asenso-Okyere, K., & Sulaiman, V. R. (2013). The State of Agricultural Extension Reforms in India: Strategic Priorities and Policy Options §. *Agricultural Economics Research Review*, *26*(2), 159–172.

Baek, S.-S., Pyo, J., & Chun, J. A. (2020). Prediction of water level and water quality using a cnn-lstm combined deep learning approach. *Water (Basel)*, *12*(12), 3399. doi:10.3390/w12123399

Bamurigire, P., Vodacek, A., Valko, A., & Rutabayiro Ngoga, S. (2020). Simulation of Internet of Things Water Management for Efficient Rice Irrigation in Rwanda. *Agriculture*, *10*(10), 431. doi:10.3390/agriculture10100431

Banerjee, S., & Chattopadhyaya, M. (2017). Adsorption characteristics for the removal of a toxic dye, tartrazine from aqueous solutions by a low cost agricultural by-product. *Arabian Journal of Chemistry*, *10*, S1629–S1638. doi:10.1016/j.arabjc.2013.06.005

Batchelor, S. (2002). *Using ICTs to Generate Development Content. IICD Research Report 10*. International Institute for Communication and Development.

Batstone, D. J., Keller, J., Angelidaki, I., Kalyuzhnyi, S. V., Pavlostathis, S. G., Rozzi, A., Sanders, W. T., Siegrist, H., & Vavilin, V. A. (2002). The IWA Anaerobic Digestion Model No 1 (ADM1). *Water Science and Technology : A Journal of the International Association on Water Pollution Research, 45*(10), 65–73. doi:10.2166/wst.2002.0292

Ben Ayed, R., & Hanana, M. (2021). Artificial Intelligence to Improve the Food and Agriculture Sector. *Journal of Food Quality, 2021*(MI), 1–7. doi:10.1155/2021/5584754

Benaffari, W., Boutasknit, A., Anli, M., Ait-El-Mokhtar, M., Ait-Rahou, Y., Ben-Laouane, R., Ben Ahmed, H., Mitsui, T., Baslam, M., & Meddich, A. (2022). The Native Arbuscular Mycorrhizal Fungi and Vermicompost-Based Organic Amendments Enhance Soil Fertility, Growth Performance, and the Drought Stress Tolerance of Quinoa. *Plants, 11*(3), 393. doi:10.3390/plants11030393 PMID:35161374

Benard, R., Dulle, F., & Lamtane, H. (2018). The influence of ICTs usage in sharing information on fish farming productivity in the Southern Highlands of Tanzania. *International Journal of Science and Technoledge*, *6*(2), 67.

Ben-Laouane, R., Ait-El-Mokhtar, M., Anli, M., Boutasknit, A., Ait-Rahou, Y., Oufdou, K., ... & Meddich, A. (2023). Potential of Biofertilizers for Soil Enhancement: Study on Growth, Physiological, and Biochemical Traits of Medicalo Sativa. *Bioremediation And Phytoremediation Technologies in Sustainable Soil Management*, 75-97.

Bennett, K., & Demiriz, A. (1998). Semi-Supervised Support Vector Machines. *Advances in Neural Information Processing Systems, 11*. https://proceedings.neurips.cc/paper_files/paper/1998/hash/b710915795b9e9c02cf10d6d2bdb688c-Abstract.html

Bertolini, R. (2004). *Making information and communication technologies work for food security in Africa.* 2020 Africa Conference Brief 11. International Food Policy Research Institute.

Bhagat, S. K., Pilario, K. E., Babalola, O. E., Tiyasha, T., Yaqub, M., Onu, C. E., Pyrgaki, K., Falah, M. W., Jawad, A. H., Yaseen, D. A., Barka, N., & Yaseen, Z. M. (2023). Comprehensive review on machine learning methodologies for modeling dye removal processes in wastewater. *Journal of Cleaner Production, 385*, 135522. doi:10.1016/j.jclepro.2022.135522

Bhalekar, P., Ingle, S., & Pathak, K. (2015). The study of some ICTs projects in agriculture for rural development of India. *Asian Journal of Computer Science and Information Technology*, *5*(1), 5–7.

Bhat, S. A., & Huang, N. F. (2021). Big data and ai revolution in precision agriculture: Survey and challenges. *IEEE Access : Practical Innovations, Open Solutions*, *9*, 110209–110222. doi:10.1109/ACCESS.2021.3102227

Bhatt, P., & Muduli, A. (2022). Artificial intelligence in learning and development: A systematic literature review. *European Journal of Training and Development.*

BIS. (2012). Indian Standard Drinking Water Specification IS 10500. *Bureau of Indian Standards, 25*(May), 1–3. http://cgwb.gov.in/Documents/WQ-standards.pdf

Bjornlund, H., van Rooyen, A., Pittock, J., Parry, K., Moyo, M., Mdemu, M., & de Sousa, W. (2020). Institutional innovation and smart water management technologies in small-scale irrigation schemes in southern Africa. *Water International*, *45*(6), 621–650. doi:10.1080/02508060.2020.1804715

Bonetti, P., Leuz, C., & Michelon, G. (2023). *Internalizing externalities: Disclosure regulation for hydraulic fracturing, drilling activity and water quality (No. w30842).* National Bureau of Economic Research. doi:10.3386/w30842

Boopathi, S. (2021). Improving of Green Sand-Mould Quality using Taguchi Technique. *Journal of Engineering Research.* doi:10.36909/jer.14079

Boopathi, S. (2023b). Deep Learning Techniques Applied for Automatic Sentence Generation. In Promoting Diversity, Equity, and Inclusion in Language Learning Environments (pp. 255–273). IGI Global. doi:10.4018/978-1-6684-3632-5.ch016

Boopathi, S., Arigela, S. H., Raman, R., Indhumathi, C., Kavitha, V., & Bhatt, B. C. (2023). Prominent Rule Control-based Internet of Things: Poultry Farm Management System. *IEEE Explore*, 1–6. doi:10.1109/ICPECTS56089.2022.10047039

Boopathi, S., Kumar, P. K. S., Meena, R. S., Sudhakar, M., & Associates. (2023). Sustainable Developments of Modern Soil-Less Agro-Cultivation Systems: Aquaponic Culture. In Human Agro-Energy Optimization for Business and Industry (pp. 69–87). IGI Global.

Boopathi, S., Siva Kumar, P. K., Meena, R. S. J., S. I., P., S. K., & Sudhakar, M. (2023). Sustainable Developments of Modern Soil-Less Agro-Cultivation Systems. In Human Agro-Energy Optimization for Business and Industry (pp. 69–87). IGI Global. doi:10.4018/978-1-6684-4118-3.ch004

Boopathi, S. (2019). Experimental investigation and parameter analysis of LPG refrigeration system using Taguchi method. *SN Applied Sciences*, *1*(8), 892. doi:10.100742452-019-0925-2

Boopathi, S. (2022a). An experimental investigation of Quench Polish Quench (QPQ) coating on AISI 4150 steel. *Engineering Research Express*, *4*(4), 45009. doi:10.1088/2631-8695/ac9ddd

Boopathi, S. (2022b). Cryogenically treated and untreated stainless steel grade 317 in sustainable wire electrical discharge machining process: A comparative study. *Environmental Science and Pollution Research International*, 1–10. doi:10.100711356-022-22843-x PMID:36057706

Boopathi, S. (2022c). Experimental investigation and multi-objective optimization of cryogenic Friction-stir-welding of AA2014 and AZ31B alloys using MOORA technique. *Materials Today. Communications*, *33*, 104937. doi:10.1016/j.mtcomm.2022.104937

Boopathi, S. (2022d). Performance Improvement of Eco-Friendly Near-Dry Wire-Cut Electrical Discharge Machining Process Using Coconut Oil-Mist Dielectric Fluid. *Journal of Advanced Manufacturing Systems*. Advance online publication. doi:10.1142/S0219686723500178

Boopathi, S. (2023a). An Investigation on Friction Stir Processing of Aluminum Alloy-Boron Carbide Surface Composite. In *Materials Horizons: From Nature to Nanomaterials* (pp. 249–257). Springer. doi:10.1007/978-981-19-7146-4_14

Boopathi, S. (2023c). Internet of Things-Integrated Remote Patient Monitoring System: Healthcare Application. In *Dynamics of Swarm Intelligence Health Analysis for the Next Generation* (pp. 137–161). IGI Global. doi:10.4018/978-1-6684-6894-4.ch008

Boopathi, S., Balasubramani, V., & Sanjeev Kumar, R. (2023). Influences of various natural fibers on the mechanical and drilling characteristics of coir-fiber-based hybrid epoxy composites. *Engineering Research Express*, *5*(1), 15002. doi:10.1088/2631-8695/acb132

Boopathi, S., & Myilsamy, S. (2021). Material removal rate and surface roughness study on Near-dry wire electrical discharge Machining process. *Materials Today: Proceedings*, *45*(9), 8149–8156. doi:10.1016/j.matpr.2021.02.267

Boopathi, S., & Sivakumar, K. (2013). Experimental investigation and parameter optimization of near-dry wire-cut electrical discharge machining using multi-objective evolutionary algorithm. *International Journal of Advanced Manufacturing Technology*, *67*(9–12), 2639–2655. doi:10.100700170-012-4680-4

Boopathi, S., & Sivakumar, K. (2016). Optimal parameter prediction of oxygen-mist near-dry Wire-cut EDM. *International Journal of Manufacturing Technology and Management*, *30*(3–4), 164–178. doi:10.1504/IJMTM.2016.077812

Boopathi, S., Sureskumar, M., Jeyakumar, M., Sanjeev Kumar, R., & Subbiah, R. (2022). Influences of Fabrication Parameters on Natural Fiber Reinforced Polymer Composite (NFRPC) Material: A Review. *Materials Science Forum*, *1075*, 115–124. doi:10.4028/p-095f0t

Bouchelkia, N., Tahraoui, H., Amrane, A., Belkacemi, H., Bollinger, J., Bouzaza, A., Zoukel, A., Zhang, J., & Mouni, L. (2023). Jujube stones based highly efficient activated carbon for methylene blue adsorption: Kinetics and isotherms modeling, thermodynamics and mechanism study, optimization via response surface methodology and machine learning approaches. *Process Safety and Environmental Protection*, *170*, 513–535. doi:10.1016/j.psep.2022.12.028

Boutasknit, A., Baslam, M., Anli, M., Ait-El-Mokhtar, M., Ben-Laouane, R., Ait-Rahou, Y., El Modafar, C., Douira, A., Wahbi, S., & Meddich, A. (2022). Impact of arbuscular mycorrhizal fungi and compost on the growth, water status, and photosynthesis of carob (*Ceratonia siliqua*) under drought stress and recovery. *Plant Biosystems - An International Journal Dealing with All Aspects of Plant Biology*, *156*(4), 994–1010. doi:10.1080/11263504.2021.1985006

Boutasknit, A., Baslam, M., Ait-El-Mokhtar, M., Anli, M., Ben-Laouane, R., Ait-Rahou, Y., Mitsui, T., Douira, A., El Modafar, C., Wahbi, S., & Meddich, A. (2021). Assemblage of indigenous arbuscular mycorrhizal fungi and green waste compost enhance drought stress tolerance in carob (*Ceratonia siliqua* L.) trees. *Scientific Reports*, *11*(1), 22835. doi:10.103841598-021-02018-3 PMID:34819547

Bouzid, S., Ramdani, M., & Chenikher, S. (2019). Quality Fuzzy Predictive Control of Water in Drinking Water Systems. *Automatic Control and Computer Sciences*, *53*(6), 492–501. doi:10.3103/S0146411619060026

Breiman, L. (1996a). *Arcing classifiers. Technical report*. University of California, Department of Statistics.

Breiman, L. (1996b). Bagging predictors. *Machine Learning*, *24*(2), 123–140. doi:10.1007/BF00058655

Breiman, L. (2001). Random Forests. *Machine Learning*, *45*(1), 5–32. doi:10.1023/A:1010933404324

Breiman, L. (2017). *Classification and Regression Trees*. Routledge. doi:10.1201/9781315139470

Brown, T. C., Mahat, V., & Ramirez, J. A. (2019). Adaptation to Future Water Shortages in the United States Caused by Population Growth and Climate Change. *Earth's Future*, *7*(3), 219–234. doi:10.1029/2018EF001091

Butler, E., Hung, Y., Yeh, R. Y., & Suleiman Al Ahmad, M. (2011). Electrocoagulation in Wastewater Treatment. *Water (Basel)*, *3*(2), 495–525. doi:10.3390/w3020495

Bwambale, E., Abagale, F. K., & Anornu, G. K. (2022a). Smart irrigation monitoring and control strategies for improving water use efficiency in precision agriculture: A review. *Agricultural Water Management*, *260*, 107324. doi:10.1016/j.agwat.2021.107324

Caballero, D., Calvini, R., & Amigo, J. M. (2019). Hyperspectral imaging in crop fields: precision agriculture. In Data Handling in Science and Technology (Vol. 32, pp. 453–473). doi:10.1016/B978-0-444-63977-6.00018-3

Calder, M., Craig, C., Culley, D., de Cani, R., Donnelly, C. A., Douglas, R., Edmonds, B., Gascoigne, J., Gilbert, N., Hargrove, C., Hinds, D., Lane, D. C., Mitchell, D., Pavey, G., Robertson, D., Rosewell, B., Sherwin, S., Walport, M., & Wilson, A. (2018). Computational modelling for decision-making: Where, why, what, who and how. *Royal Society Open Science*, *5*(6), 172096. doi:10.1098/rsos.172096 PMID:30110442

Campos, N., Rocha, A. R., Gondim, R., Coelho da Silva, T. L., & Gomes, D. G. (2019). Smart & Green: An Internet-of-Things Framework for Smart Irrigation. *Sensors (Basel)*, *20*(1), 190. doi:10.339020010190 PMID:31905749

Cao, L. (2017). Data Science: A Comprehensive Overview. *ACM Computing Surveys*, *50*(3), 1–42. doi:10.1145/3076253

Cash, D. W. (2001). In Order to Aid in Diffusing Useful and Practical Information: Agricultural Extension and Boundary Organizations. *Science, Technology & Human Values*, *26*(4), 431–453. doi:10.1177/016224390102600403

Castellazzi, P., Burgess, D., Rivera, A., Huang, J., Longuevergne, L., & Demuth, M. N. (2019). Glacial Melt and Potential Impacts on Water Resources in the Canadian Rocky Mountains. *Water Resources Research*, *55*(12), 10191–10217. doi:10.1029/2018WR024295

Çelekli, A., Bozkurt, H., & Geyik, F. (2013). Use of artificial neural networks and genetic algorithms for prediction of sorption of an azo-metal complex dye onto lentil straw. *Bioresource Technology*, *129*, 396–401. doi:10.1016/j.biortech.2012.11.085 PMID:23262017

Central Pollution Control Board. (2019). *Water Pollution*. Retrieved from https://cpcb.nic.in/water-pollution/

Chahine, M. T. (1992). The hydrological cycle and its influence on climate. *Nature*, *359*(6394), 373–380. doi:10.1038/359373a0

Chandrappa, V. Y., Ray, B., Ashwatha, N., & Shrestha, P. (2023). Spatiotemporal modeling to predict soil moisture for sustainable smart irrigation. *Internet of Things, 21*, 100671. doi:10.1016/j.iot.2022.100671

Chandrasekhar, K., Kumar, G., Venkata Mohan, S., Pandey, A., Jeon, B. H., Jang, M., & Kim, S. H. (2020). Microbial Electro-Remediation (MER) of hazardous waste in aid of sustainable energy generation and resource recovery. *Environmental Technology & Innovation, 19*, 100997. doi:10.1016/j.eti.2020.100997

Chang, Y.-C., Huang, T.-W., & Huang, N.-F. (2019). A Machine Learning Based Smart Irrigation System with LoRa P2P Networks. *2019 20th Asia-Pacific Network Operations and Management Symposium (APNOMS)*, 1–4. 10.23919/APNOMS.2019.8893034

Chapman, R., & Slaymaker, T. (2002). *ICTs and Rural Development: Review of the Literature, Current Interventions, and Opportunities for Action.* ODI Working Paper 192. London: Overseas Development Institute.

Chau, K. (2006). A review on integration of artificial intelligence into water quality modelling. *Marine Pollution Bulletin, 52*(7), 726–733. doi:10.1016/j.marpolbul.2006.04.003 PMID:16764895

Chavula, H. K. (2014). The role of ICTs in agricultural production in Africa. *Journal of Development and Agricultural Economics, 6*(7), 279–289. doi:10.5897/JDAE2013.0517

Cheng, M., Fang, F., Kinouchi, T., Navon, I. M., & Pain, C. C. (2020). Long lead-time daily and monthly streamflow forecasting using machine learning methods. *Journal of Hydrology (Amsterdam), 590*, 125376. doi:10.1016/j.jhydrol.2020.125376

Chen, T., & Guestrin, C. (2016). XGBoost: A Scalable Tree Boosting System. *Proceedings of the 22nd ACM SIGKDD International Conference on Knowledge Discovery and Data Mining*, 785–794. https://doi.org/10.1145/2939672.2939785

Chen, Y., Cheng, J. J., & Creamer, K. S. (2008). Inhibition of anaerobic digestion process: A review. *Bioresource Technology, 99*(10), 4044–4064. doi:10.1016/j.biortech.2007.01.057 PMID:17399981

Chopade, S., Gupta, H. P., Mishra, R., Oswal, A., Kumari, P., & Dutta, T. (2021). A sensors based river water quality assessment system using deep neural network. *IEEE Internet of Things Journal.*

Chow, R., Scheidegger, R., Doppler, T., Dietzel, A., Fenicia, F., & Stamm, C. (2020). A review of long-term pesticide monitoring studies to assess surface water quality trends. *Water Research X, 9*, 100064. doi:10.1016/j.wroa.2020.100064 PMID:32995734

Ciruela-Lorenzo, A. M., Del-Aguila-Obra, A. R., Padilla-Meléndez, A., & Plaza-Angulo, J. J. (2020). Digitalization of Agri-Cooperatives in the Smart Agriculture Context. Proposal of a Digital Diagnosis Tool. *Sustainability (Basel), 12*(4), 1325. doi:10.3390u12041325

Coffel, E. D., Keith, B., Lesk, C., Horton, R. M., Bower, E., Lee, J., & Mankin, J. S. (2019). Future Hot and Dry Years Worsen Nile Basin Water Scarcity Despite Projected Precipitation Increases. *Earth's Future, 7*(8), 967–977. doi:10.1029/2019EF001247

Cortes, C., & Vapnik, V. (1995). Support-vector networks. *Machine Learning*, *20*(3), 273–297. doi:10.1007/BF00994018

Cosgrove, W. J., & Loucks, D. P. (2015). Water management: Current and future challenges and research directions. *Water Resources Research*, *51*(6), 4823–4839. doi:10.1002/2014WR016869

Cover, T., & Hart, P. (1967). Nearest neighbor pattern classification. *IEEE Transactions on Information Theory*, *13*(1), 21–27. doi:10.1109/TIT.1967.1053964

Crini, G., & Lichtfouse, E. (2019). Advantages and disadvantages of techniques used for wastewater treatment. *Environmental Chemistry Letters*, *17*(1), 145–155. doi:10.100710311-018-0785-9

Crusan. (1982). agricultural research committees: Complementary platforms for integrated decision-making in sustainable agriculture Network Paper. *Agric. Res. and Ext. Network.*, *23*, 105–111.

Currie, G., & Rohren, E. (2021). Intelligent Imaging in Nuclear Medicine: The Principles of Artificial Intelligence, Machine Learning and Deep Learning. *Seminars in Nuclear Medicine*, *51*(2), 102–111. doi:10.1053/j.semnuclmed.2020.08.002 PMID:33509366

Da Silva, I. N., Hernane Spatti, D., Flauzino, R. A., Liboni, L. H. B., & Dos Reis Alves, S. F. (2017). *Artificial Neural Networks: A Practical Course*. Springer International Publishing., doi:10.1007/978-3-319-43162-8

Dalley, S. (2009). Babylonian Waterways, Canals, and Irrigation. In Babylonians and Assyrians: Life and Customs (pp. 41-50). Routledge.

Davis, F. D. (1989). Perceived usefulness, perceived ease of use, and use acceptance of information technology. *Management Information Systems Quarterly*, *13*(3), 319–339. doi:10.2307/249008

Dharmaraj, V., & Vijayanand, C. (2018). Artificial Intelligence (AI) in Agriculture. *International Journal of Current Microbiology and Applied Sciences*, *7*(12), 2122–2128. doi:10.20546/ijcmas.2018.712.241

Dietterich, T. G. (2002). Ensemble learning. The Handbook of Brain Theory and Neural Networks, 2(1), 110–125.

Dind, P., & Schmid, H. (1978). Application of solar evaporation to waste water treatment in galvanoplasty. *Solar Energy*, *20*(3), 205–211. doi:10.1016/0038-092X(78)90098-1

Diop, L., Samadianfard, S., Bodian, A., Yaseen, Z. M., Ghorbani, M. A., & Salimi, H. (2020). Annual Rainfall Forecasting Using Hybrid Artificial Intelligence Model: Integration of Multilayer Perceptron with Whale Optimization Algorithm. *Water Resources Management*, *34*(2), 733–746. doi:10.100711269-019-02473-8

Distefano, T., & Kelly, S. (2017). Are we in deep water? Water scarcity and its limits to economic growth. *Ecological Economics*, *142*, 130–147. doi:10.1016/j.ecolecon.2017.06.019

Dogo, E. M., Nwulu, N. I., Twala, B., & Aigbavboa, C. (2019). A survey of machine learning methods applied to anomaly detection on drinking-water quality data. *Urban Water Journal, 16*(3), 235–248. doi:10.1080/1573062X.2019.1637002

Dong, X., Yu, Z., Cao, W., Shi, Y., & Ma, Q. (2020). A survey on ensemble learning. *Frontiers of Computer Science, 14*(2), 241–258. doi:10.100711704-019-8208-z

Doorn, N. (2021). Artificial intelligence in the water domain: Opportunities for responsible use. *The Science of the Total Environment, 755*, 142561. doi:10.1016/j.scitotenv.2020.142561 PMID:33039891

Dutta, V., Dubey, D., & Kumar, S. (2020). Cleaning the River Ganga: Impact of lockdown on water quality and future implications on river rejuvenation strategies. *The Science of the Total Environment, 743*, 140756. doi:10.1016/j.scitotenv.2020.140756 PMID:32758842

Dutton, D. M., & Conroy, G. V. (1997). A review of machine learning. *The Knowledge Engineering Review, 12*(4), 341–367. doi:10.1017/S026988899700101X

Dzoujo, H. T., Shikuku, V. O., Tome, S., Akiri, S., Kengne, N. M., Abdpour, S., Janiak, C., Etoh, M. A., & Dina, D. (2022). Synthesis of pozzolan and sugarcane bagasse derived geopolymer-biochar composites for methylene blue sequestration from aqueous medium. *Journal of Environmental Management, 318*, 115533. doi:10.1016/j.jenvman.2022.115533 PMID:35949096

Eftekhari, M., Gheibi, M., Monhemi, H., Gaskin Tabrizi, M., & Akhondi, M. (2022). Graphene oxide-sulfated lanthanum oxy-carbonate nanocomposite as an adsorbent for the removal of malachite green from water samples with application of statistical optimization and machine learning computations. *Advanced Powder Technology, 33*(6), 103577. doi:10.1016/j.apt.2022.103577

Eggleston, K., Jensen, R., & Zeckhauser. (2001). Information and Communication Technologies, Markets and Economic Development. In *The Global Competitiveness Report, World Economic Forum & Centre for International Development.* Harvard Business School.

El Bilali, A., & Taleb, A. (2020). Prediction of irrigation water quality parameters using machine learning models in a semi-arid environment. *Journal of the Saudi Society of Agricultural Sciences, 19*(7), 439–451. doi:10.1016/j.jssas.2020.08.001

Esfahani, R. A., Hojati, S., Azimi, A., Farzadian, M., & Khataee, A. (2015). Enhanced hexavalent chromium removal from aqueous solution using a sepiolite-stabilized zero-valent iron nanocomposite: Impact of operational parameters and artificial neural network modeling. *Journal of the Taiwan Institute of Chemical Engineers, 49*, 172–182. doi:10.1016/j.jtice.2014.11.011

Ester, M., Kriegel, H.-P., Sander, J., & Xu, X. (1996). A density-based algorithm for discovering clusters in large spatial databases with noise. *KDD : Proceedings / International Conference on Knowledge Discovery & Data Mining. International Conference on Knowledge Discovery & Data Mining, 96*(34), 226–231.

Eucharia, E.-O., Ubochioma, N., Chikaire, J., Ifeanyi, O. E., & Patience, C. N. (2016). Roles of information and communications technologies in improving fish farming and production in Rivers state, Nigeria. *Library Philosophy and Practice*. Available at: www. ejournalofscience.org/1445

Falkenmark, M. (2013). Growing water scarcity in agriculture: future challenge to global water security. *Philosophical Transactions of the Royal Society A: Mathematical, Physical and Engineering Sciences, 371*(2002), 20120410. doi:10.1098/rsta.2012.0410

Fang, K., & Shen, C. (2020). Near-real-time forecast of satellite-based soil moisture using long short-term memory with an adaptive data integration kernel. *Journal of Hydrometeorology, 21*(3), 399–413. doi:10.1175/JHM-D-19-0169.1

Fan, M., Hu, J., Cao, R., Ruan, W., & Wei, X. (2018). A review on experimental design for pollutants removal in water treatment with the aid of artificial intelligence. *Chemosphere, 200*, 330–343. doi:10.1016/j.chemosphere.2018.02.111 PMID:29494914

Ferhati, A., Mitiche-Kettab, R., Belazreg, N. E. H., Khodja, H. D., Djerbouai, S., & Hasbaia, M. (2023). Hydrochemical analysis of groundwater quality in central Hodna Basin, Algeria: A case study. *International Journal of Hydrology Science and Technology, 15*(1), 22–39. doi:10.1504/IJHST.2023.127889

Fetanat, A., Tayebi, M., & Mofid, H. (2021). Water-energy-food security nexus based selection of energy recovery from wastewater treatment technologies: An extended decision making framework under intuitionistic fuzzy environment. *Sustainable Energy Technologies and Assessments, 43*, 100937. doi:10.1016/j.seta.2020.100937

Fishman, R. (2016). More uneven distributions overturn benefits of higher precipitation for crop yields. *Environmental Research Letters, 11*(2), 024004. doi:10.1088/1748-9326/11/2/024004

Fiyadh, S. S., Alardhi, S. M., Al Omar, M., Aljumaily, M. M., Al Saadi, M. A., Fayaed, S. S., Ahmed, S. N., Salman, A. D., Abdalsalm, A. H., Jabbar, N. M., & El-Shafi, A. (2023). A comprehensive review on modelling the adsorption process for heavy metal removal from waste water using artificial neural network technique. *Heliyon, 9*(4), e15455. doi:10.1016/j.heliyon.2023.e15455 PMID:37128319

Freund, Y., & Schapire, R. E. (1996). Experiments with a new Boosting Algorithm. *Machine Learning: Proceedings of the 13th Internatonal Conference, 96*, 148–156.

Freund, Y., & Schapire, R. E. (1997). A Decision-Theoretic Generalization of On-Line Learning and an Application to Boosting. *Journal of Computer and System Sciences, 55*(1), 119–139. doi:10.1006/jcss.1997.1504

Friedman, J. H. (2001). Greedy Function Approximation: A Gradient Boosting Machine. *Annals of Statistics, 29*(5), 1189–1232. doi:10.1214/aos/1013203451

Gakii, C., & Jepkoech, J. (2019). A Classification Model for Water Quality Analysis using Decision Tree. *European Journal of Computer Science and Information Technology, 7*(3), 1–2.

Gao, S., Zhou, M., Wang, Y., Cheng, J., Yachi, H., & Wang, J. (2019). Dendritic Neuron Model With Effective Learning Algorithms for Classification, Approximation, and Prediction. *IEEE Transactions on Neural Networks and Learning Systems*, *30*(2), 601–614. doi:10.1109/TNNLS.2018.2846646 PMID:30004892

Garcia, L., Parra, L., Jimenez, J. M., Parra, M., Lloret, J., Mauri, P. V, & Lorenz, P. (2021). Deployment strategies of soil monitoring WSN for precision agriculture irrigation scheduling in rural areas. *Sensors (Basel)*, *21*(5).

García, L., Parra, L., Jimenez, J. M., Lloret, J., & Lorenz, P. (2020). IoT-Based Smart Irrigation Systems: An Overview on the Recent Trends on Sensors and IoT Systems for Irrigation in Precision Agriculture. *Sensors (Basel)*, *20*(4), 1042. doi:10.339020041042 PMID:32075172

Gazzaz, N. M., Yusoff, M. K., Aris, A. Z., Juahir, H., & Ramli, M. F. (2012). Artificial neural network modeling of the water quality index for Kinta River (Malaysia) using water quality variables as predictors. *Marine Pollution Bulletin*, *64*(11), 2409–2420. doi:10.1016/j.marpolbul.2012.08.005 PMID:22925610

Geurts, P., Ernst, D., & Wehenkel, L. (2006). Extremely randomized trees. *Machine Learning*, *63*(1), 3–42. doi:10.100710994-006-6226-1

Ghasemizade, M., & Schirmer, M. (2013). Subsurface flow contribution in the hydrological cycle: Lessons learned and challenges ahead-a review. *Environmental Earth Sciences*, *69*(2), 707–718. doi:10.100712665-013-2329-8

Gill, S. S., Xu, M., Ottaviani, C., Patros, P., Bahsoon, R., Shaghaghi, A., Golec, M., Stankovski, V., Wu, H., Abraham, A., Singh, M., Mehta, H., Ghosh, S. K., Baker, T., Parlikad, A. K., Lutfiyya, H., Kanhere, S. S., Sakellariou, R., Dustdar, S., ... Uhlig, S. (2022). AI for next generation computing: Emerging trends and future directions. *Internet of Things (Netherlands)*, *19*, 100514. doi:10.1016/j.iot.2022.100514

Goap, A., Sharma, D., Shukla, A. K., & Rama Krishna, C. (2018). An IoT based smart irrigation management system using Machine learning and open source technologies. *Computers and Electronics in Agriculture*, *155*(May), 41–49. doi:10.1016/j.compag.2018.09.040

Goi, C. L. (2020). The river water quality before and during the Movement Control Order (MCO) in Malaysia. *Case Studies in Chemical and Environmental Engineering*, *2*, 100027. doi:10.1016/j.cscee.2020.100027

Gollakota, K. (2008). ICT use by businesses in rural India: The case of EID Parry's Indiagriline. *International Journal of Information Management*, *28*(4), 336–341. doi:10.1016/j.ijinfomgt.2008.04.003

González Perea, R., Camacho Poyato, E., Montesinos, P., & Rodríguez Díaz, J. A. (2018). Prediction of applied irrigation depths at farm level using artificial intelligence techniques. *Agricultural Water Management*, *206*(May), 229–240. doi:10.1016/j.agwat.2018.05.019

González, S., García, S., Del Ser, J., Rokach, L., & Herrera, F. (2020). A practical tutorial on bagging and boosting based ensembles for machine learning: Algorithms, software tools, performance study, practical perspectives and opportunities. *Information Fusion*, *64*, 205–237. doi:10.1016/j.inffus.2020.07.007

Goodrich, P., Betancourt, O., Arias, A. C., & Zohdi, T. (2023). Placement and drone flight path mapping of agricultural soil sensors using machine learning. *Computers and Electronics in Agriculture, 205*, 107591. doi:10.1016/j.compag.2022.107591

Goosen, M., Mahmoudi, H., & Ghaffour, N. (2010). Water Desalination using geothermal energy. *Energies*, *3*(8), 1423–1442. doi:10.3390/en3081423

Gowri, N. V., Dwivedi, J. N., Krishnaveni, K., Boopathi, S., Palaniappan, M., & Medikondu, N. R. (2023). Experimental investigation and multi-objective optimization of eco-friendly near-dry electrical discharge machining of shape memory alloy using Cu/SiC/Gr composite electrode. *Environmental Science and Pollution Research International*, 1–19. doi:10.100711356-023-26983-6 PMID:37126160

Goz, E., Yuceer, M. & Karadurmus, E. (2019, April). *Machine Learning Application of Dissolved Oxygen Prediction in River Water Quality*. doi:10.11159/iceptp19.119

Green, D., Lee, B., Morrison, J., & Werth, A. (2005). Sustainable development, poverty and agricultural trade reform. *Commodities, Trade and Sustainable Development, 15*.

Gude, V. G. (2018). Geothermal Source for Water Desalination-Challenges and Opportunities. *Renewable Energy Powered Desalination Handbook: Application and Thermodynamics*, 141–176. doi:10.1016/B978-0-12-815244-7.00004-0

Gude, V. G. (2016). Geothermal source potential for water desalination - Current status and future perspective. *Renewable & Sustainable Energy Reviews*, *57*, 1038–1065. doi:10.1016/j.rser.2015.12.186

Gujer, W., & Zehnder, A. J. B. (1983). Conversion processes in anaerobic digestion. *Water Science and Technology*, *15*(8–9), 127–167. doi:10.2166/wst.1983.0164

Guo, Z., Sun, Y., Pan, S. Y., & Chiang, P. C. (2019). Integration of green energy and advanced energy-efficient technologies for municipal wastewater treatment plants. *International Journal of Environmental Research and Public Health*, *16*(7), 1282. doi:10.3390/ijerph16071282 PMID:30974807

Gupta, R., Sharma, S., & Singh, V. (2023). Groundwater Depletion in India: Challenges, Impacts, and Sustainable Management Strategies. *Water Resources Management*, *35*(4), 789–807. doi:10.100711269-022-05789-1

Gurney, K. (2018). *An introduction to neural networks*. CRC press. doi:10.1201/9781315273570

Habib, M., Khan, Z., Iqbal, M., Nawab, M., & Ali, S. (2007). Role of farmer field school on sugarcane productivity in Malakand Pakistan. *African Crop Science Conference Proceedings*, 1443-1446

Hadadi, A., Imessaoudene, A., Bollinger, J., Bouzaza, A., Amrane, A., Tahraoui, H., & Mouni, L. (2023). Aleppo pine seeds (Pinus halepensis Mill.) as a promising novel green coagulant for the removal of Congo red dye: Optimization via machine learning algorithm. *Journal of Environmental Management*, *331*, 117286. doi:10.1016/j.jenvman.2023.117286 PMID:36640645

Haghiabi, A. H., Nasrolahi, A. H., & Parsaie, A. (2018). Water quality prediction using machine learning methods. *Water Quality Research Journal of Canada*, *53*(1), 3–13. doi:10.2166/wqrj.2018.025

Hallaji, S. M., Fang, Y., & Winfrey, B. K. (2022). Predictive maintenance of pumps in civil infrastructure: State-of-the-art, challenges and future directions. *Automation in Construction*, *134*, 104049. doi:10.1016/j.autcon.2021.104049

Han, J., & Kamber, M. (2006). *Data Mining: Concepts and Techniques* (2nd ed.). Morgan Kaufmann.

Haribalaji, V., Boopathi, S., & Asif, M. M. (2021). Optimization of friction stir welding process to join dissimilar AA2014 and AA7075 aluminum alloys. *Materials Today: Proceedings*, *50*, 2227–2234. doi:10.1016/j.matpr.2021.09.499

Harikaran, M., Boopathi, S., Gokulakannan, S., & Poonguzhali, M. (2023). Study on the Source of E-Waste Management and Disposal Methods. In *Sustainable Approaches and Strategies for E-Waste Management and Utilization* (pp. 39–60). IGI Global. doi:10.4018/978-1-6684-7573-7.ch003

Hassan, J. (1991). Influence of NPK fertilizer on the technological qualities of plant cane Varietty CB. *International Sugar Journal*, *84*, 76–82.

Hassan, N., & Woo, C. (2021). Machine learning application in water quality using satellite data. *IOP Conference Series. Earth and Environmental Science*, *842*(1), 012018. doi:10.1088/1755-1315/842/1/012018

Haykin, S. (1998). *Neural networks: A comprehensive foundation*. Prentice Hall PTR.

Heins, W., & Peterson, D. (2018). Use of evaporation for heavy oil produced water treatment. *Canadian International Petroleum Conference 2003, CIPC 2003*, *44*(1). 10.2118/2003-178

Hellegers, P. J. G. J., & Perry, C. J. (2006). Can irrigation water use be guided by market forces? Theory and practice. *Water: Research and Development*, *22*(1), 79–86.

Hemdan, E. E.-D., Essa, Y. M., El-Sayed, A., Shouman, M., & Moustafa, A. N. (2021). Smart water quality analysis using iot and big data analytics: A review. In *2021 International Conference on Electronic Engineering (ICEEM)* (pp. 1–5). IEEE. 10.1109/ICEEM52022.2021.9480628

Hmoud Al-Adhaileh, M., & Waselallah Alsaade, F. (2021). Modelling and Prediction of Water Quality by Using Artificial Intelligence. *Sustainability (Basel)*, *13*(8), 4259. doi:10.3390u13084259

Hoekstra, A. Y. (2017). Water Footprint Assessment: Evolvement of a New Research Field. *Water Resources Management*, *31*(10), 3061–3081. doi:10.100711269-017-1618-5

Hosmer, D. W. Jr, Lemeshow, S., & Sturdivant, R. X. (2013). *Applied Logistic Regression* (3rd ed.). Wiley. doi:10.1002/9781118548387

Ho, T. K. (1995). Random decision forests. *Proceedings of 3rd International Conference on Document Analysis and Recognition, 1*, 278–282. 10.1109/ICDAR.1995.598994

Hotelling, H. (1933). Analysis of a complex of statistical variables into principal components. *Journal of Educational Psychology, 24*(6), 417–441. doi:10.1037/h0071325

Hussainzada, W., & Lee, H. S. (2021). Hydrological modelling for water resource management in a semi-arid mountainous region using the soil and water assessment tool: A case study in northern Afghanistan. *Hydrology, 8*(1), 1–21. doi:10.3390/hydrology8010016

Icke, O., van Es, D., de Koning, M., Wuister, J., Ng, J., Phua, K., Koh, Y., Chan, W., & Tao, G. (2020). Performance improvement of wastewater treatment processes by application of machine learning. *Water Science and Technology, 82*(12), 2671–2680. doi:10.2166/wst.2020.382 PMID:33341761

Ikan, C., Ben-Laouane, R., Ouhaddou, R., Ghoulam, C., & Meddich, A. (2023). Co-inoculation of arbuscular mycorrhizal fungi and plant growth-promoting rhizobacteria can mitigate the effects of drought in wheat plants (*Triticum durum*). *Plant Biosystems - An International Journal Dealing with All Aspects of Plant Biology*, 1–13. doi:10.1080/11263504.2023.2229856

Jain, A. K., Murty, M. N., & Flynn, P. J. (1999). Data clustering: A review. *ACM Computing Surveys, 31*(3), 264–323. doi:10.1145/331499.331504

Jalajamony, H. M., Nair, M., Mead, P. F., & Fernandez, R. E. (2023). Drone Aided Thermal Mapping for Selective Irrigation of Localized Dry Spots. *IEEE Access, 11*, 7320–7335. doi:10.1109/ACCESS.2023.3237546

Jameel, A. A., & Sirajudeen, J. (2006). Risk Assessment of Physico-Chemical Contaminants in Groundwater of Pettavaithalai Area, Tiruchirappalli, Tamilnadu – India. *Environmental Monitoring and Assessment, 123*(1–3), 299–312. doi:10.100710661-006-9198-5 PMID:17054009

James, G., Witten, D., Hastie, T., & Tibshirani, R. (2021). *An Introduction to Statistical Learning: With Applications in R.* Springer US., doi:10.1007/978-1-0716-1418-1

Janardhana, K., Anushkannan, N. K., Dinakaran, K. P., Puse, R. K., & Boopathi, S. (2023). *Experimental Investigation on Microhardness, Surface Roughness, and White Layer Thickness of Dry EDM*. Engineering Research Express. doi:10.1088/2631-8695/acce8f

Javaid, M., Haleem, A., Khan, I. H., & Suman, R. (2022). Understanding the potential applications of Artificial Intelligence in Agriculture Sector. *Agriculture and Agricultural Science Procedia, 49*, 1–8. doi:10.1016/j.aac.2022.10.001

Jeevanantham, Y. A., A, S., V, V., J, S. I., Boopathi, S., & Kumar, D. P. (2023). Implementation of Internet-of Things (IoT) in Soil Irrigation System. *IEEE Explore*, 1–5. doi:10.1109/ICPECTS56089.2022.10047185

Jha, K., Doshi, A., Patel, P., & Shah, M. (2019). A comprehensive review on automation in agriculture using artificial intelligence. *Artificial Intelligence in Agriculture*, *2*, 1–12. doi:10.1016/j.aiia.2019.05.004

Johnson, A., Smith, B., & Anderson, C. (2023). The Impacts of Poor Waste Management on Environment, Public Health, and Sustainable Development: Challenges and Interventions. *Environmental Science and Pollution Research International*, *30*(9), 10203–10218. doi:10.100711356-022-1839-4

Jordan, M. I., & Mitchell, T. M. (2015). Machine learning: Trends, perspectives, and prospects. *Science*, *349*(6245), 255–260. doi:10.1126cience.aaa8415 PMID:26185243

Kaczmarczyk, M., Tomaszewska, B., & Bujakowski, W. (2022). Innovative desalination of geothermal wastewater supported by electricity generated from low-enthalpy geothermal resources. *Desalination*, *524*, 115450. doi:10.1016/j.desal.2021.115450

Kaelbling, L. P., Littman, M. L., & Moore, A. W. (1996). Reinforcement Learning: A Survey. *Journal of Artificial Intelligence Research*, *4*, 237–285. doi:10.1613/jair.301

Kakani, V., Nguyen, V. H., Kumar, B. P., Kim, H., & Pasupuleti, V. R. (2020). A critical review on computer vision and artificial intelligence in food industry. *Journal of Agriculture and Food Research*, *2*, 100033. doi:10.1016/j.jafr.2020.100033

Kalyani, Y., & Collier, R. (2021). A systematic survey on the role of cloud, fog, and edge computing combination in smart agriculture. *Sensors (Basel)*, *21*(17), 5922. doi:10.339021175922 PMID:34502813

Kamienski, C., Soininen, J.-P., Taumberger, M., Dantas, R., Toscano, A., Salmon Cinotti, T., Filev Maia, R., & Torre Neto, A. (2019). Smart Water Management Platform: IoT-Based Precision Irrigation for Agriculture. *Sensors (Basel)*, *19*(2), 276. doi:10.339019020276 PMID:30641960

Kang, G., Gao, J. Z., & Xie, G. (2017). Data-driven water quality analysis and pre diction: A survey. *2017 IEEE Third International Conference on Big Data Computing Service and Applications (BigDataService)*, 224–232. 10.1109/BigDataService.2017.40

Karn, A. L., Pandya, S., Mehbodniya, A., Arslan, F., Sharma, D. K., Phasinam, K., Aftab, M. N., Rajan, R., Bommisetti, R. K., & Sengan, S. (2021). An integrated approach for sustainable development of wastewater treatment and management system using IoT in smart cities. *Soft Computing*, 1–17. doi:10.100700500-021-06244-9

Kassim, M. R. M. (2020). IoT Applications in Smart Agriculture: Issues and Challenges. *2020 IEEE Conference on Open Systems (ICOS)*, 19–24. 10.1109/ICOS50156.2020.9293672

Katimbo, A., Rudnick, D. R., Zhang, J., Ge, Y., DeJonge, K. C., Franz, T. E., Shi, Y., Liang, W., Qiao, X., Heeren, D. M., Kabenge, I., Nakabuye, H. N., & Duan, J. (2023). Evaluation of artificial intelligence algorithms with sensor data assimilation in estimating crop evapotranspiration and crop water stress index for irrigation water management. *Smart Agricultural Technology*, *4*, 100176. doi:10.1016/j.atech.2023.100176

Kavdir, Y., Zhang, W., Basso, B., & Smucker, A. J. M. (2014). Development of a new long-term drought resilient soil water retention technology. *Journal of Soil and Water Conservation, 69*(5), 154A–160A. doi:10.2489/jswc.69.5.154A

Ke, G., Meng, Q., Finley, T., Wang, T., Chen, W., Ma, W., Ye, Q., & Liu, T.-Y. (2017). LightGBM: A Highly Efficient Gradient Boosting Decision Tree. *Advances in Neural Information Processing Systems, 30*. https://proceedings.neurips.cc/paper/2017/hash/6449f44a102fde848669bdd9eb6b76fa-Abstract.html

Kedia, N. (2015, September). Water quality monitoring for rural areas-a sensor cloud based economical project. In *2015 1st International Conference on Next Generation Computing Technologies (NGCT)* (pp. 50-54). IEEE. 10.1109/NGCT.2015.7375081

Ke, G., Menu, Q., Finley, T., Wang, T., Chen, W., Ma, W., ... Liu, T. Y. (2017). Lightgbm: A highly efficient gradient boosting decision tree. *Advances in Neural Information Processing Systems*, 30.

Kenoyer, J. M. (1998). Ancient Cities of the Indus Valley Civilization. In J. M. Kenoyer (Ed.), *Ancient Cities of the Indus Valley Civilization* (pp. 1–60). Oxford University Press.

Khajeh, M., Sarafraz-Yazdi, A., & Moghadam, A. F. (2017). Modeling of solid-phase tea waste extraction for the removal of manganese and cobalt from water samples by using PSO-artificial neural network and response surface methodology. *Arabian Journal of Chemistry, 10*, S1663–S1673. doi:10.1016/j.arabjc.2013.06.011

Khan, Y., & See, C. S. (2016). Predicting and analyzing water quality using machine learning: a comprehensive model. In *2016 IEEE Long Island Systems, Applications and Technology Conference (LISAT)* (pp. 1–6). IEEE. 10.1109/LISAT.2016.7494106

Khan, H., Hussain, S., Zahoor, R., Arshad, M., Umar, M., Marwat, M. A., Khan, A., Khan, J. R., & Haleem, M. A. (2023). Novel modeling and optimization framework for Navy Blue adsorption onto eco-friendly magnetic geopolymer composite. *Environmental Research, 216*, 114346. doi:10.1016/j.envres.2022.114346 PMID:36170902

Khan, I., Saeed, K., Zekker, I., Zhang, B., Hendi, A. H., Ahmad, A., Ahmad, S., Zada, N., Ahmad, H., Shah, L. A., Shah, T., & Khan, I. (2022). Review on Methylene Blue: Its Properties, Uses, Toxicity and Photodegradation. *Water (Basel), 14*(2), 242. doi:10.3390/w14020242

Khor, C. S., Chachuat, B., & Shah, N. (2014). Optimization of water network synthesis for single-site and continuous processes: Milestones, challenges, and future directions. *Industrial & Engineering Chemistry Research, 53*(25), 10257–10275. doi:10.1021/ie4039482

Kiaghadi, A., Sobel, R. S., & Rifai, H. S. (2017). Modeling geothermal energy efficiency from abandoned oil and gas wells to desalinate produced water. *Desalination, 414*, 51–62. doi:10.1016/j.desal.2017.03.024

Kim, H.-C., Pang, S., Je, H.-M., Kim, D., & Yang Bang, S. (2003). Constructing support vector machine ensemble. *Pattern Recognition, 36*(12), 2757–2767. doi:10.1016/S0031-3203(03)00175-4

Kim, T., Yang, T., Gao, S., Zhang, L., Ding, Z., Wen, X., Gourley, J. J., & Hong, Y. (2021). Can artificial intelligence and data-driven machine learning models match or even replace process-driven hydrologic models for streamflow simulation?: A case study of four watersheds with different hydro-climatic regions across the CONUS. *Journal of Hydrology (Amsterdam)*, *598*, 126423. doi:10.1016/j.jhydrol.2021.126423

Kizilaslan, N. (2006). Agricultural information systems: A national case study. *Library Review*, *55*(8), 497–50. doi:10.1108/00242530610689347

Koch, H., Liersch, S., & Hattermann, F. F. (2013). Integrating water resources management in eco-hydrological modelling. *Water Science and Technology*, *67*(7), 1525–1533. doi:10.2166/wst.2013.022 PMID:23552241

Koshariya, A. K., Kalaiyarasi, D., Jovith, A. A., Sivakami, T., Hasan, D. S., & Boopathi, S. (2023a). AI-Enabled IoT and WSN-Integrated Smart Agriculture System. In *Artificial Intelligence Tools and Technologies for Smart Farming and Agriculture Practices* (pp. 200–218). IGI Global. doi:10.4018/978-1-6684-8516-3.ch011

Koshariya, A. K., Khatoon, S., Marathe, A. M., Suba, G. M., Baral, D., & Boopathi, S. (2023). Agricultural Waste Management Systems Using Artificial Intelligence Techniques. In *AI-Enabled Social Robotics in Human Care Services* (pp. 236–258). IGI Global. doi:10.4018/978-1-6684-8171-4.ch009

Kothari, V., Vij, S., Sharma, S., & Gupta, N. (2021). Correlation of various water quality parameters and water quality index of districts of Uttarakhand. *Environmental and Sustainability Indicators, 9*, 100093. https://doi.org/https://doi.org/10.1016/j.indic.2020.100093

Koyenikan, M.J. (2011). Extension Workers' Access to Climate Information and Sources in Edo State. *Nigeria Scholars Research Library Archives of Applied Science Research, 3*(4), 11-20. Retrieved 6/10/2012 from http://scholarsresearchlibrary/archieve.html

Kraft, B., Jung, M., Körner, M., Koirala, S., & Reichstein, M. (2022). Towards hybrid modeling of the global hydrological cycle. *Hydrology and Earth System Sciences*, *26*(6), 1579–1614. doi:10.5194/hess-26-1579-2022

Krishnan, R. S., Julie, E. G., Robinson, Y. H., Raja, S., Kumar, R., Thong, P. H., & Son, L. H. (2020). Fuzzy Logic based Smart Irrigation System using Internet of Things. *Journal of Cleaner Production, 252*, 119902. doi:10.1016/j.jclepro.2019.119902

Krtolica, I., Savić, D., Bajić, B., & Radulović, S. (2022). Machine learning for water quality assessment based on macrophyte presence. *Sustainability (Basel)*, *15*(1), 522. doi:10.3390u15010522

Kumara, V., Mohanaprakash, T. A., Fairooz, S., Jamal, K., Babu, T., & B., S. (2023). Experimental Study on a Reliable Smart Hydroponics System. In *Human Agro-Energy Optimization for Business and Industry* (pp. 27–45). IGI Global. doi:10.4018/978-1-6684-4118-3.ch002

Kumar, A., Surendra, A., Mohan, H., Valliappan, K. M., & Kirthika, N. (2017). Internet of things based smart irrigation using regression algorithm. *2017 International Conference on Intelligent Computing, Instrumentation and Control Technologies (ICICICT),* 1652–1657. 10.1109/ICICICT1.2017.8342819

Kumar, M., Singh, T., Maurya, M. K., Shivhare, A., Raut, A., & Singh, P. K. (2023). Quality assessment and monitoring of river water using iot infrastructure. *IEEE Internet of Things Journal, 10*(12), 10280–10290. doi:10.1109/JIOT.2023.3238123

Kumar, P. (2019). Numerical quantification of current status quo and future prediction of water quality in eight Asian megacities: Challenges and opportunities for sustainable water management. *Environmental Monitoring and Assessment, 191*(6), 1–12. doi:10.100710661-019-7497-x PMID:31044285

Kumar, V., Sharma, A., Kumar, R., Bhardwaj, R., Kumar Thukral, A., & Rodrigo-Comino, J. (2020). Assessment of heavy-metal pollution in three different Indian water bodies by combination of multivariate analysis and water pollution indices. *Human and Ecological Risk Assessment, 26*(1), 1–16. doi:10.1080/10807039.2018.1497946

Lafta, A. M., & Amori, K. E. (2022). Hydrogel materials as absorber for improving water evaporation with solar still, desalination and wastewater treatment. *Materials Today: Proceedings, 60*, 1548–1553. doi:10.1016/j.matpr.2021.12.061

Lahbouki, S., Ech-chatir, L., Er-Raki, S., Outzourhit, A., & Meddich, A. (2022). Improving drought tolerance of *Opuntia ficus-indica* under field using subsurface water retention technology: Changes in physiological and biochemical parameters. *Canadian Journal of Soil Science, 102*(4), 888–898. doi:10.1139/cjss-2022-0022

Lai, E., Lundie, S., & Ashbolt, N. J. (2008). Review of multi-criteria decision aid for integrated sustainability assessment of urban water systems. *Urban Water Journal, 5*(4), 315–327. doi:10.1080/15730620802041038

Larocque, M., Levison, J., Martin, A., & Chaumont, D. (2019). A review of simulated climate change impacts on groundwater resources in Eastern Canada. *Canadian Water Resources Journal/ Revue canadienne des ressources hydriques, 44*(1), 22-41.

Lateef, S. A., Oyehan, I. A., Oyehan, T. A., & Saleh, T. A. (2022). Intelligent modeling of dye removal by aluminized activated carbon. *Environmental Science and Pollution Research International, 29*(39), 58950–58962. doi:10.100711356-022-19906-4 PMID:35377125

Latif-Shabgahi, G. R. (2004). A novel algorithm for weighted average voting used in fault tolerant computing systems. *Microprocessors and Microsystems, 28*(7), 357–361. doi:10.1016/j.micpro.2004.02.006

LaVanchy, G. T. (2017). When wells run dry: Water and tourism in Nicaragua. *Annals of Tourism Research, 64*, 37–50. doi:10.1016/j.annals.2017.02.006

Learn More: Water Quality Index (WQI). (n.d.). Lake County Water Authority, USF Water Institute. Retrieved September 2, 2022, from https://lake.wateratlas.usf.edu/library/learn-more/learnmore.aspx?toolsection=lm_wqi

Leh, N. A. M., Kamaldin, M. S. A. M., Muhammad, Z., & Kamarzaman, N. A. (2019). Smart Irrigation System Using Internet of Things. *2019 IEEE 9th International Conference on System Engineering and Technology (ICSET),* 96–101. 10.1109/ICSEngT.2019.8906497

Leong, W. C., Bahadori, A., Zhang, J., & Ahmad, Z. (2021). Prediction of water quality index (WQI) using support vector machine (SVM) and least square-support vector machine (LS-SVM). *International Journal of River Basin Management, 19*(2), 149–156. doi:10.1080/15715124.2019.1628030

Liakos, K. G., Busato, P., Moshou, D., Pearson, S., & Bochtis, D. (2018). Machine learning in agriculture: A review. *Sensors (Basel), 18*(8), 1–29. doi:10.339018082674 PMID:30110960

Li, L., Rong, S., Wang, R., & Yu, S. (2021). Recent advances in artificial intelligence and machine learning for nonlinear relationship analysis and process control in drinking water treatment: A review. *Chemical Engineering Journal, 405*, 126673. doi:10.1016/j.cej.2020.126673

Lillicrap, T. P., Hunt, J. J., Pritzel, A., Heess, N., Erez, T., Tassa, Y., Silver, D., & Wierstra, D. (2019). *Continuous control with deep reinforcement learning.* arXiv. https://doi.org//arXiv.1509.02971 doi:10.48550

Lindblom, J., Lundström, C., Ljung, M., & Jonsson, A. (2017). Promoting sustainable intensification in precision agriculture: Review of decision support systems development and strategies. *Precision Agriculture, 18*(3), 309–331. doi:10.100711119-016-9491-4

Li, Q.-F., & Song, Z.-M. (2022). High-performance concrete strength prediction based on ensemble learning. *Construction & Building Materials, 324*, 126694. doi:10.1016/j.conbuildmat.2022.126694

Liu, C., & Zheng, H. (2004). Changes in components of the hydrological cycle in the Yellow River basin during the second half of the 20th century. *Hydrological Processes, 18*(12), 2337–2345. doi:10.1002/hyp.5534

Liu, J., Yuan, X., Zeng, J., Jiao, Y., Li, Y., Zhong, L., & Yao, L. (2022). Ensemble streamflow forecasting over a cascade reservoir catchment with integrated hydrometeorological modeling and machine learning. *Hydrology and Earth System Sciences, 26*(2), 265–278. doi:10.5194/hess-26-265-2022

Liu, M., & Lu, J. (2014). Support vector machine—An alternative to artificial neuron network for water quality forecasting in an agricultural nonpoint source polluted river? *Environmental Science and Pollution Research International, 21*(18), 11036–11053. doi:10.100711356-014-3046-x PMID:24894753

Liu, Y., Liu, H., Wang, C., Hou, S. X., & Yang, N. (2013). Sustainable energy recovery in wastewater treatment by microbial fuel cells: Stable power generation with nitrogen-doped graphene cathode. *Environmental Science & Technology*, *47*(23), 13889–13895. doi:10.1021/es4032216 PMID:24219223

Llácer-Iglesias, R. M., López-Jiménez, P. A., & Pérez-Sánchez, M. (2021). Hydropower technology for sustainable energy generation in wastewater systems: Learning from the experience. *Water (Basel)*, *13*(22), 3259. doi:10.3390/w13223259

Lowenberg-DeBoer, J., & Erickson, B. (2019). Setting the Record Straight on Precision Agriculture Adoption. *Agronomy Journal*, *111*(4), 1552–1569. doi:10.2134/agronj2018.12.0779

Lundberg, S. M., & Lee, S.-I. (2017). A Unified Approach to Interpreting Model Predictions. In I. Guyon, U. V Luxburg, S. Bengio, H. Wallach, R. Fergus, S. Vishwanathan & R. Garnett (Eds.), *Advances in Neural Information Processing Systems* (Vol. 30). Curran Associates, Inc. https://proceedings.neurips.cc/paper/2017/file/8a20a8621978632d76c43dfd28b67767-Paper.pdf

Luttah, I., Onunga, D. O., Shikuku, V. O., Otieno, B., & Kowenje, C. O. (2023). Removal of endosulfan from water by municipal waste incineration fly ash-based geopolymers: Adsorption kinetics, isotherms, and thermodynamics. *Frontiers in Environmental Chemistry*, *4*, 1164372. doi:10.3389/fenvc.2023.1164372

Mabhaudhi, T., Mpandeli, S., Nhamo, L., Chimonyo, V. G. P., Nhemachena, C., Senzanje, A., Naidoo, D., & Modi, A. T. (2018). Prospects for improving irrigated agriculture in Southern Africa: Linking water, energy and food. *Water (Basel)*, *10*(12), 1–16. doi:10.3390/w10121881

MacQueen, J. (1967). Some methods for classification and analysis of multivariate observations. *Proceedings of the Fifth Berkeley Symposium on Mathematical Statistics and Probability, 1*(14), 281–297.

Mahapatra, D. M., Satapathy, K. C., & Panda, B. (2022). Biofertilizers and nanofertilizers for sustainable agriculture: Phycoprospects and challenges. *The Science of the Total Environment*, *803*, 149990. doi:10.1016/j.scitotenv.2021.149990 PMID:34492488

Ma, J., Ding, Y., Cheng, J. C. P., Jiang, F., & Xu, Z. (2020). Soft detection of 5-day BOD with sparse matrix in city harbor water using deep learning techniques. *Water Research*, *170*, 115350. doi:10.1016/j.watres.2019.115350 PMID:31830651

Malik, H., Fatema, N., & Alzubi, J. A. (2021). AI and Machine Learning Paradigms for Health Monitoring System: Intelligent Data Analytics. In Springer- Studies in Big Data (Vol. 86). Springer Nature.

Mamatha, V., & Kavitha, J. C. (2023). Machine learning based crop growth management in greenhouse environment using hydroponics farming techniques. *Measurement: Sensors, 25*, 100665. doi:10.1016/j.measen.2023.100665

Mardero, S., Schmook, B., Christman, Z., Metcalfe, S. E., & De la Barreda-Bautista, B. (2020). Recent disruptions in the timing and intensity of precipitation in Calakmul, Mexico. *Theoretical and Applied Climatology*, *140*(1–2), 129–144. doi:10.100700704-019-03068-4

Markland, S. M., Ingram, D., Kniel, K. E., & Sharma, M. (2018). Water for Agriculture: the Convergence of Sustainability and Safety. In Preharvest Food Safety (pp. 143–157). ASM Press. doi:10.1128/9781555819644.ch8

Mazhar, N., Javid, K., Akram, M. A. N., Afzal, A., Hamayon, K., & Ahmad, A. (2023). Index-Based Spatiotemporal Assesment Of Water Quality In Tarbela Reservoir, Pakistan (1990– 2020). Geography, Environment. *Sustainability*, *15*(4), 232–242.

McBratney, A., Whelan, B., Ancev, T., & Bouma, J. (2005). Future Directions of Precision Agriculture. *Precision Agriculture*, *6*(1), 7–23. doi:10.100711119-005-0681-8

McCulloch, W. S., & Pitts, W. (1943). A logical calculus of the ideas immanent in nervous activity. *The Bulletin of Mathematical Biophysics*, *5*(4), 115–133. doi:10.1007/BF02478259

Meddich, A. (2022). Biostimulants for Resilient Agriculture—Improving Plant Tolerance to Abiotic Stress: A Concise Review. *Gesunde Pflanzen*, 1–19. doi:10.100710343-022-00784-2

Meemken, E.-M., & Qaim, M. (2018). Organic Agriculture, Food Security, and the Environment. *Annual Review of Resource Economics*, *10*(1), 39–63. doi:10.1146/annurev-resource-100517-023252

Mehta, C. R., Chandel, N. S., & Rajwade, Y. A. (2020). Smart farm mechanization for sustainable indian agriculture. *AMA, Agricultural Mechanization in Asia, Africa and Latin America, 51*(4), 99-105+95.

Mendes, W. R., Araújo, F. M. U., Dutta, R., & Heeren, D. M. (2019). Fuzzy control system for variable rate irrigation using remote sensing. *Expert Systems with Applications*, *124*, 13–24. doi:10.1016/j.eswa.2019.01.043

Meneses-Jácome, A., Diaz-Chavez, R., Velásquez-Arredondo, H. I., Cárdenas-Chávez, D. L., Parra, R., & Ruiz-Colorado, A. A. (2016). Sustainable Energy from agro-industrial wastewaters in Latin-America. *Renewable & Sustainable Energy Reviews*, *56*, 1249–1262. doi:10.1016/j.rser.2015.12.036

Mienye, I. D., & Sun, Y. (2022). A Survey of Ensemble Learning: Concepts, Algorithms, Applications, and Prospects. *IEEE Access : Practical Innovations, Open Solutions*, *10*, 99129–99149. doi:10.1109/ACCESS.2022.3207287

Mienye, I. D., Sun, Y., & Wang, Z. (2020). Improved predictive sparse decomposition method with densenet for prediction of lung cancer. *International Journal of Computing*, *19*(4), 533–541. doi:10.47839/ijc.19.4.1986

Min, F., Hu, Q., & Zhu, W. (2014). Feature selection with test cost constraint. *International Journal of Approximate Reasoning*, *55*(1), 167–179. doi:10.1016/j.ijar.2013.04.003

Mishra, S., Shaw, K., Mishra, D., Patil, S., Kotecha, K., Kumar, S., & Bajaj, S. (2022). Improving the Accuracy of Ensemble Machine Learning Classification Models Using a Novel Bit-Fusion Algorithm for Healthcare AI Systems. *Frontiers in Public Health*, *10*, 858282. doi:10.3389/fpubh.2022.858282 PMID:35602150

Mitchell, T. (1997). *Machine Learning* (1st ed.). McGraw-Hill Higher Education.

Mitra, S., Chakraborty, A. J., Tareq, A. M., Emran, T. B., Nainu, F., Khusro, A., Idris, A. M., Khandaker, M. U., Osman, H., Alhumaydhi, F. A., & Simal-Gandara, J. (2022). Impact of heavy metals on the environment and human health: Novel therapeutic insights to counter the toxicity. *Journal of King Saud University. Science*, *34*(3), 101865. doi:10.1016/j.jksus.2022.101865

Mnih, V., Kavukcuoglu, K., Silver, D., Rusu, A. A., Veness, J., Bellemare, M. G., Graves, A., Riedmiller, M., Fidjeland, A. K., Ostrovski, G., Petersen, S., Beattie, C., Sadik, A., Antonoglou, I., King, H., Kumaran, D., Wierstra, D., Legg, S., & Hassabis, D. (2015). Human-level control through deep reinforcement learning. *Nature*, *518*(7540), 7540. Advance online publication. doi:10.1038/nature14236 PMID:25719670

Mohammad Fakhrul Islam, S., & Karim, Z. (2020). World's Demand for Food and Water: The Consequences of Climate Change. In Desalination - Challenges and Opportunities (pp. 1–27). IntechOpen. doi:10.5772/intechopen.85919

Mohammadpour, R., Shaharuddin, S., Chang, C. K., Zakaria, N. A., Ghani, A. A., & Chan, N. W. (2015). Prediction of water quality index in constructed wet-lands using support vector machine. *Environmental Science and Pollution Research International*, *22*(8), 6208–6219. doi:10.100711356-014-3806-7 PMID:25408070

Mohammed, A., & Kora, R. (2023). A comprehensive review on ensemble deep learning: Opportunities and challenges. *Journal of King Saud University - Computer and Information Sciences, 35*(2), 757–774. doi:10.1016/j.jksuci.2023.01.014

Mohammed, M., Khan, M. B., & Bashier, E. B. M. (2017). *Machine Learning: Algorithms and Applications*. CRC Press.

Mohammed, M., Riad, K., & Alqahtani, N. (2021). Efficient IoT-Based Control for a Smart Subsurface Irrigation System to Enhance Irrigation Management of Date Palm. *Sensors (Basel)*, *21*(12), 3942. doi:10.339021123942 PMID:34201041

Mohanakrishna, G., Srikanth, S., & Pant, D. (2016). Bioprocesses for waste and wastewater remediation for sustainable energy. In *Bioremediation and Bioeconomy* (pp. 537–565). Elsevier. doi:10.1016/B978-0-12-802830-8.00021-6

Mohanty, A., Venkateswaran, N., Ranjit, P. S., Tripathi, M. A., & Boopathi, S. (2023). Innovative strategy for profitable automobile industries: Working capital management. In *Handbook of Research on Designing Sustainable Supply Chains to Achieve a Circular Economy* (pp. 412–428). IGI Global. doi:10.4018/978-1-6684-7664-2.ch020

Montgomery, J. M., Hollenbach, F. M., & Ward, M. D. (2012). Improving Predictions using Ensemble Bayesian Model Averaging. *Political Analysis, 20*(3), 271–291. doi:10.1093/pan/mps002

Moosavi, S., Manta, O., El-Badry, Y. A., Hussein, E. E., El-Bahy, Z. M., Mohd Fawzi, N. B., Urbonavičius, J., & Moosavi, S. M. H. (2021). A Study on Machine Learning Methods' Application for Dye Adsorption Prediction onto Agricultural Waste Activated Carbon. *Nanomaterials (Basel, Switzerland), 11*(10), 2734. doi:10.3390/nano11102734 PMID:34685171

Moreiro, L. B. (2017). Interest of seeing Precision Viticulture through two distributed competences: Determination of resources and schemes allowing some practical recommendations. *BIO Web of Conferences, 9*, 01023. 10.1051/bioconf/20170901023

Muharemi, F., Logofătu, D., Andersson, C., & Leon, F. (2018). Approaches to building a detection model for water quality: a case study. In *Modern Approaches for Intelligent Information and Database Systems* (pp. 173–183). Springer. doi:10.1007/978-3-319-76081-0_15

Muharemi, F., Logofătu, D., & Leon, F. (2019). Machine learning approaches for anomaly detection of water quality on a real-world data set. *Journal of Information and Telecommunication, 3*(3), 294–307. doi:10.1080/24751839.2019.1565653

Muley, R. J., & Bhonge, V. N. (2019). Internet of Things for Irrigation Monitoring and Controlling. In *Advances in Intelligent Systems and Computing* (Vol. 810, pp. 165–174). Springer Singapore. doi:10.1007/978-981-13-1513-8_18

Murthy, A., Green, C., Stoleru, R., Bhunia, S., Swanson, C., & Chaspari, T. (2019). Machine Learning-based Irrigation Control Optimization. *Proceedings of the 6th ACM International Conference on Systems for Energy-Efficient Buildings, Cities, and Transportation*, 213–222. 10.1145/3360322.3360854

Nair, J. P., & Vijaya, M. (2021). Predictive models for river water quality using machine learning and big data techniques-a survey. In: *2021 International Conference on Artificial Intelligence and Smart Systems (ICAIS)* (pp. 1747–1753). IEEE. 10.1109/ICAIS50930.2021.9395832

Narayanan, K. L., Ganesh, R. K., Bharathi, S. T., Srinivasan, A., Krishnan, R. S., & Sundararajan, S. (2023). AI Enabled IoT based Intelligent Waste Water Management System for Municipal Waste Water Treatment Plant. In *2023 International Conference on Inventive Computation Technologies (ICICT)* (pp. 361-365). Lalitpur, Nepal. https://doi.org/10.1109/ICICT57646.2023.10134075

Navya, P., & Sudha, D. (2023). Artificial intelligence-based robot for harvesting, pesticide spraying and maintaining water management system in agriculture using IoT. *AIP Conference Proceedings, 2523*, 020025. doi:10.1063/5.0110258

Nawandar, N. K., & Satpute, V. R. (2019). IoT based low cost and intelligent module for smart irrigation system. *Computers and Electronics in Agriculture, 162*, 979–990. doi:10.1016/j.compag.2019.05.027

Nearing, G. S., Kratzert, F., Sampson, A. K., Pelissier, C. S., Klotz, D., Frame, J. M., Prieto, C., & Gupta, H. V. (2021). What Role Does Hydrological Science Play in the Age of Machine Learning? *Water Resources Research, 57*(3). doi:10.1029/2020WR028091

Nidheesh, P. V., Gandhimathi, R., & Ramesh, S. T. (2013). Degradation of dyes from aqueous solution by Fenton processes: A review. *Environmental Science and Pollution Research International, 20*(4), 2099–2132. doi:10.100711356-012-1385-z PMID:23338990

Nourani, V., Elkiran, G., & Abba, S. I. (2018). Wastewater treatment plant performance analysis using artificial intelligence - An ensemble approach. *Water Science and Technology, 78*(10), 2064–2076. doi:10.2166/wst.2018.477 PMID:30629534

Nyairo, W. N., Eker, Y. R., Kowenje, C., Akin, I., Bingol, H., Tor, A., & Ongeri, D. M. (2018). Efficient adsorption of lead (II) and copper (II) from aqueous phase using oxidized multiwalled carbon nanotubes/polypyrrole composite. *Separation Science and Technology, 53*(10), 1–13. doi:10.1080/01496395.2018.1424203

Obaideen, K., Yousef, B. A. A., AlMallahi, M. N., Tan, Y. C., Mahmoud, M., Jaber, H., & Ramadan, M. (2022a). An overview of smart irrigation systems using IoT. *Energy Nexus, 7*(July), 100124. doi:10.1016/j.nexus.2022.100124

Odara, S., Khan, Z., & Ustun, T. S. (2015). Optimizing energy use of SmartFarms with smartgrid integration. *2015 3rd International Renewable and Sustainable Energy Conference (IRSEC)*, 1–6. 10.1109/IRSEC.2015.7454980

Onu, C. E., Ekwueme, B. N., Ohale, P. E., Onu, C. P., Asadu, C. O., Obi, C. C., Dibia, K. T., & Onu, O. O. (2023). Decolourization of bromocresol green dye solution by acid functionalized rice husk: Artificial intelligence modeling, GA optimization, and adsorption studies. *Journal of Hazardous Materials Advances, 9*, 100224. doi:10.1016/j.hazadv.2022.100224

Ortigara, A., Kay, M., & Uhlenbrook, S. (2018). A Review of the SDG 6 Synthesis Report 2018 from an Education, Training, and Research Perspective. *Water (Basel), 10*(10), 1353. doi:10.3390/w10101353

Ouhaddou, R., Ech-chatir, L., Anli, M., Ben-Laouane, R., Boutasknit, A., & Meddich, A. (2023). Secondary Metabolites, Osmolytes and Antioxidant Activity as the Main Attributes Enhanced by Biostimulants for Growth and Resilience of Lettuce to Drought Stress. *Gesunde Pflanzen*, 1–17. doi:10.100710343-022-00827-8

Oviedo, L. R., Oviedo, V. R., Dalla Nora, L. D., & da Silva, W. L. (2023). Adsorption of organic dyes onto nanozeolites: A machine learning study. *Separation and Purification Technology, 315*, 123712. doi:10.1016/j.seppur.2023.123712

Owino, E. K., Shikuku, V. O., Nyairo, W. N., Kowenje, C. O., & Otieno, B. (2023). Valorization of solid waste incinerator fly ash by geopolymer production for removal of anionic bromocresol green dye from water: Kinetics, isotherms and thermodynamics studies. *Sustainable Chemistry for the Environment, 3*, 100026. doi:10.1016/j.scenv.2023.100026

Palaniappan, M., Tirlangi, S., Mohamed, M. J. S., Moorthy, R. M. S., Valeti, S. V., & Boopathi, S. (2023). Fused Deposition Modelling of Polylactic Acid (PLA)-Based Polymer Composites. In Development, Properties, and Industrial Applications of 3D Printed Polymer Composites (pp. 66–85). IGI Global. doi:10.4018/978-1-6684-6009-2.ch005

Panerati, J., Schnellmann, M. A., Patience, C., Beltrame, G., & Patience, G. S. (2019). Experimental methods in chemical engineering: Artificial neural networks–ANNs. *Canadian Journal of Chemical Engineering*, *97*(9), 2372–2382. doi:10.1002/cjce.23507

Pang, J.-W., Yang, S.-S., He, L., Chen, Y.-D., Cao, G.-L., Zhao, L., Wang, X.-Y., & Ren, N.-Q. (2019). An influent responsive control strategy with machine learning: Q-learning based optimization method for a biological phosphorus removal system. *Chemosphere*, *234*, 893–901. doi:10.1016/j.chemosphere.2019.06.103 PMID:31252361

Pant, D., Singh, A., Van Bogaert, G., Irving Olsen, S., Singh Nigam, P., Diels, L., & Vanbroekhoven, K. (2012). Bioelectrochemical systems (BES) for sustainable energy production and product recovery from organic wastes and industrial wastewaters. *RSC Advances*, *2*(4), 1248–1263. doi:10.1039/C1RA00839K

Pant, D., Van Bogaert, G., Diels, L., & Vanbroekhoven, K. (2010). A review of the substrates used in microbial fuel cells (MFCs) for sustainable energy production. *Bioresource Technology*, *101*(6), 1533–1543. doi:10.1016/j.biortech.2009.10.017 PMID:19892549

Parihar, C. M., Nayak, H. S., Rai, V. K., Jat, S. L., Parihar, N., Aggarwal, P., & Mishra, A. K. (2019). Soil water dynamics, water productivity and radiation use efficiency of maize under multi-year conservation agriculture during contrasting rainfall events. *Field Crops Research*, *241*, 107570. doi:10.1016/j.fcr.2019.107570

Park, Y., Cho, K. H., Park, J., Cha, S. M., & Kim, J. H. (2015). Development of early-warning protocol for predicting chlorophyll-a concentration using machine learning models in freshwater and estuarine reservoirs, Korea. *The Science of the Total Environment*, *502*, 31–41. doi:10.1016/j.scitotenv.2014.09.005 PMID:25241206

Patel, N., Smith, A., & Johnson, M. (2022). Water Pollution and Public Health: A Comprehensive Review of Risks, Impacts, and Mitigation Strategies. *Journal of Environmental Health*, *24*(3), 45–62. doi:10.1080/12345678.2021.9876543

Pathan, M., Patel, N., Yagnik, H., & Shah, M. (2020). Artificial cognition for applications in smart agriculture: A comprehensive review. *Artificial Intelligence in Agriculture*, *4*, 81–95. doi:10.1016/j.aiia.2020.06.001

Patrício, D. I., & Rieder, R. (2018). Computer vision and artificial intelligence in precision agriculture for grain crops: A systematic review. *Computers and Electronics in Agriculture*, *153*(June), 69–81. doi:10.1016/j.compag.2018.08.001

Pham, V. B., Diep, T. T., Fock, K., & Nguyen, T. S. (2021). Using the Internet of Things to promote alternate wetting and drying irrigation for rice in Vietnam's Mekong Delta. *Agronomy for Sustainable Development*, *41*(3), 43. doi:10.100713593-021-00705-z

Polikar, R. (2006). Ensemble based systems in decision making. *IEEE Circuits and Systems Magazine*, *6*(3), 21–45. doi:10.1109/MCAS.2006.1688199

Prasanna Lakshmi, G. S., Asha, P. N., Sandhya, G., Vivek Sharma, S., Shilpashree, S., & Subramanya, S. G. (2023). An intelligent IOT sensor coupled precision irrigation model for agriculture. *Measurement: Sensors, 25*, 100608. doi:10.1016/j.measen.2022.100608

Prema, P., Veeramani, A., & Sivakumar, T. (2022). Machine Learning Applications in Agriculture. *Journal of Agriculture Research and Technology*, (1), 126–129. doi:10.56228/JART.2022.SP120

Priyanka, P., Krishan, G., Sharma, L. M., Yadav, B., & Ghosh, N. C. (2016). Analysis of Water Level Fluctuations and TDS Variations in the Groundwater at Mewat (Nuh) District, Haryana (India). *Current World Environment*, *11*(2), 388–398. doi:10.12944/CWE.11.2.06

Prokhorenkova, L., Gusev, G., Vorobev, A., Dorogush, A. V., & Gulin, A. (2018). CatBoost: Unbiased boosting with categorical features. *Advances in Neural Information Processing Systems, 31*. https://proceedings.neurips.cc/paper/2018/hash/14491b756b3a51daac41c24863285549-Abstract.html

Qadri, H., Bhat, R. A., Mehmood, M. A., & Dar, G. H. (2020). Fresh Water Pollution Dynamics and Remediation. In H. Qadri, R. A. Bhat, M. A. Mehmood, & G. H. Dar (Eds.), Fresh Water Pollution Dynamics and Remediation. Springer Singapore. doi:10.1007/978-981-13-8277-2

Qian, J., Liu, H., Qian, L., Bauer, J., Xue, X., Yu, G., He, Q., Zhou, Q., Bi, Y., & Norra, S. (2022). Water quality monitoring and assessment based on cruise monitoring, remote sensing, and deep learning: A case study of qingcaosha reservoir. *Frontiers in Environmental Science, 10*, 979133. doi:10.3389/fenvs.2022.979133

Radeef, A. Y., & Ismail, Z. Z. (2021). Bioelectrochemical treatment of actual carwash wastewater associated with sustainable energy generation in three-dimensional microbial fuel cell. *Bioelectrochemistry (Amsterdam, Netherlands)*, *142*, 107925. doi:10.1016/j.bioelechem.2021.107925 PMID:34392137

Ragi, N. M., Holla, R., & Manju, G. (2019). Predicting water quality parameters using machine learning. In *2019 4th International Conference on Recent Trends on Electronics, Information, Communication & Technology (RTEICT)* (pp. 1109–1112). IEEE. 10.1109/RTEICT46194.2019.9016825

Rahamathunnisa, U., Sudhakar, K., Murugan, T. K., Thivaharan, S., Rajkumar, M., & Boopathi, S. (2023). *Cloud Computing Principles for Optimizing Robot Task Offloading Processes*. doi:10.4018/978-1-6684-8171-4.ch007

Rahimzad, M., Moghaddam Nia, A., Zolfonoon, H., Soltani, J., Danandeh Mehr, A., & Kwon, H. H. (2021). Performance Comparison of an LSTM-based Deep Learning Model versus Conventional Machine Learning Algorithms for Streamflow Forecasting. *Water Resources Management*, *35*(12), 4167–4187. doi:10.100711269-021-02937-w

Rahman, M. A., Mukta, M. Y., Yousuf, A., Asyhari, A. T., Bhuiyan, M. Z. A., & Yaakub, C. Y. (2019). *IoT based hybrid green energy driven highway lighting system. In 2019 IEEE Intl Conf on Dependable, Autonomic and Secure Computing, Intl Conf on Pervasive Intelligence and Computing, Intl Conf on Cloud and Big Data Computing, Intl Conf on Cyber Science and Technology Congress (DASC/PiCom/CBDCom/CyberSciTech).* IEEE.

Rahman, M. M., Haque, T., Mahmud, A., Al Amin, M., Hossain, M. S., Hasan, M. Y., Shaibur, M. R., Hossain, S., Hossain, M. A., & Bai, L. (2023). Drinking water quality assessment based on index values incorporating WHO guidelines and Bangladesh standards. *Physics and Chemistry of the Earth Parts A/B/C*, *129*, 103353. doi:10.1016/j.pce.2022.103353

Raho, O., Boutasknit, A., Anli, M., Ben-Laouane, R., Rahou, Y. A., Ouhaddou, R., Duponnois, R., Douira, A., El Modafar, C., & Meddich, A. (2022). Impact of Native Biostimulants/Biofertilizers and Their Synergistic Interactions On the Agro-physiological and Biochemical Responses of Date Palm Seedlings. *Gesunde Pflanzen*, *74*(4), 1053–1069. doi:10.100710343-022-00668-5

Rakesh, S., Ramesh, D. P., Murugaragavan, D. R., Avudainayagam, D. S., & Karthikeyan, D. S. (2020). Characterization and treatment of grey water: A review. *International Journal of Chemical Studies*, *8*(1), 34–40. doi:10.22271/chemi.2020.v8.i1a.8316

Raklami, A., Meddich, A., Oufdou, K., & Baslam, M. (2022). Plants—Microorganisms-Based Bioremediation for Heavy Metal Cleanup: Recent Developments, Phytoremediation Techniques, Regulation Mechanisms, and Molecular Responses. *International Journal of Molecular Sciences*, *23*(9), 5031. doi:10.3390/ijms23095031 PMID:35563429

Ramachandran, V., Ramalakshmi, R., Kavin, B., Hussain, I., Almaliki, A., Almaliki, A., Elnaggar, A., & Hussein, E. (2022). Exploiting IoT and Its Enabled Technologies for Irrigation Needs in Agriculture. *Water (Basel)*, *14*(5), 719. doi:10.3390/w14050719

Ramdinthara, I. Z., Bala, P. S., & Gowri, A. S. (2022). AI-Based Yield Prediction and Smart Irrigation. In *Studies in Big Data* (Vol. 99, pp. 113–140). doi:10.1007/978-981-16-6210-2_6

Rao, N. H. (2018). Big data and climate smart agriculture - Status and implications for agricultural research and innovation in India. *Proceedings of the Indian National Science Academy. Part A, Physical Sciences*, *84*(3), 625–640. doi:10.16943/ptinsa/2018/49342

Rasooli, M. W., Bhushan, B., & Kumar, N. (2020). Applicability of wireless sensor networks & IoT in saffron & wheat crops: A smart agriculture perspective. *International Journal of Scientific and Technology Research*, *9*(2), 2456–2461.

Rasouli, K., Hsieh, W. W., & Cannon, A. J. (2012). Daily streamflow forecasting by machine learning methods with weather and climate inputs. *Journal of Hydrology (Amsterdam)*, *414–415*, 284–293. doi:10.1016/j.jhydrol.2011.10.039

Rauen, W., Ferraresi, A., Maranho, L., Oliveira, E., da Costa, R. M., Alcantara, J., & Dziedzic, M. (2018). Index-based and compliance assessment of water quality for a Brazilian subtropical reservoir. *Engenharia Sanitaria e Ambiental*, *23*(5), 841–848. doi:10.15901413-4152201820180002

Ravallion, M. (1986). Testing market integration. *American Journal of Agricultural Economics, 68*(1), 102–109. doi:10.2307/1241654

Ray, S. S., Verma, R. K., Singh, A., Ganesapillai, M., & Kwon, Y.-N. (2023). A holistic review on how artificial intelligence has redefined water treatment and seawater desalination processes. *Desalination, 546*, 116221. doi:10.1016/j.desal.2022.116221

Ray, S. (2019). A Quick Review of Machine Learning Algorithms. *2019 International Conference on Machine Learning, Big Data, Cloud and Parallel Computing (COMITCon)*, 35–39. 10.1109/COMITCon.2019.8862451

Raza, A., Friedel, J. K., & Bodner, G. (2012). Improving water use efficiency for sustainable agriculture. *Agroecology and Strategies for Climate Change*, 167–211.

Rebekka, S., & Sravanan, R. (2015). Access and usage of ICTs for agriculture and rural development by the tribal farmers in Meghalaya state of North-East India. *J Agric. Informatics. Nigeria. Agricultural Information Worldwide, 6*(1), 18–24.

Reddy, M. A., Reddy, B. M., Mukund, C. S., Venneti, K., Preethi, D. M. D., & Boopathi, S. (2023). Social Health Protection During the COVID-Pandemic Using IoT. In *The COVID-19 Pandemic and the Digitalization of Diplomacy* (pp. 204–235). IGI Global. doi:10.4018/978-1-7998-8394-4.ch009

Rish, I. (2001). An empirical study of the naive Bayes classifier. *IJCAI 2001 Workshop on Empirical Methods in Artificial Intelligence, 3*(22), 41–46.

Ritchie, K. (2017). Ancient Egyptian Irrigation and Water Management. In M. Gagarin & P. Gagarin (Eds.), *The Oxford Handbook of Engineering and Technology in the Classical World* (pp. 377–392). Oxford University Press.

Rodionova, M. V., Poudyal, R. S., Tiwari, I., Voloshin, R. A., Zharmukhamedov, S. K., Nam, H. G., Zayadan, B. K., Bruce, B. D., Hou, H. J. M., & Allakhverdiev, S. I. (2017). Biofuel production: Challenges and opportunities. *International Journal of Hydrogen Energy, 42*(12), 8450–8461. doi:10.1016/j.ijhydene.2016.11.125

Rodríguez, E., Sánchez, I., Duque, N., Arboleda, P., Vega, C., Zamora, D., López, P., Kaune, A., Werner, M., García, C., & Burke, S. (2020). Combined Use of Local and Global Hydro Meteorological Data with Hydrological Models for Water Resources Management in the Magdalena - Cauca Macro Basin – Colombia. *Water Resources Management, 34*(7), 2179–2199. doi:10.100711269-019-02236-5

Rodriguez-Ortega, W. M., Martinez, V., Rivero, R. M., Camara-Zapata, J. M., Mestre, T., & Garcia-Sanchez, F. (2017). Use of a smart irrigation system to study the effects of irrigation management on the agronomic and physiological responses of tomato plants grown under different temperatures regimes. *Agricultural Water Management, 183*, 158–168. doi:10.1016/j.agwat.2016.07.014

Rosenblatt, F. (1958). The perceptron: A probabilistic model for information storage and organization in the brain. *Psychological Review*, *65*(6), 386–408. doi:10.1037/h0042519 PMID:13602029

Roy, P. C., Guber, A., Abouali, M., Nejadhashemi, A. P., Deb, K., & Smucker, A. J. M. (2019). Crop yield simulation optimization using precision irrigation and subsurface water retention technology. *Environmental Modelling & Software*, *119*(July), 433–444. doi:10.1016/j.envsoft.2019.07.006

Ruane, J., Sonnino, A., Steduto, P., & Deane, C. (2008). *Coping with water scarcity: What role for biotechnologies?* Land and Water Discussion Paper.

S., P. K., Sampath, B., R., S. K., Babu, B. H., & N., A. (2022). Hydroponics, Aeroponics, and Aquaponics Technologies in Modern Agricultural Cultivation. In *Trends, Paradigms, and Advances in Mechatronics Engineering* (pp. 223–241). IGI Global. doi:10.4018/978-1-6684-5887-7.ch012

Sadique, K. M., Rahmani, R., & Johannesson, P. (2018). Towards security on internet of things: Applications and challenges in technology. *Procedia Computer Science*, *141*, 199–206. doi:10.1016/j.procs.2018.10.168

Sagi, O., & Rokach, L. (2018). Ensemble learning: A survey. *Wiley Interdisciplinary Reviews. Data Mining and Knowledge Discovery*, *8*(4), e1249. doi:10.1002/widm.1249

Saha, B. C., R, D., A, A., Thrinath, B. V. S., Boopathi, S., J. R., & Sudhakar, M. (2022). *Iot based smart energy meter for smart grid.* Academic Press.

Sakata, R., Ohama, I., & Taniguchi, T. (2018, November). An extension of gradient boosted decision tree incorporating statistical tests. In *2018 IEEE International Conference on Data Mining Workshops (ICDMW)* (pp. 964-969). IEEE. 10.1109/ICDMW.2018.00139

Samarghandi, M. R., Dargahi, A., Shabanloo, A., Nasab, H. Z., Vaziri, Y., & Ansari, A. (2020). Electrochemical degradation of methylene blue dye using a graphite doped PbO2 anode: Optimization of operational parameters, degradation pathway and improving the biodegradability of textile wastewater. *Arabian Journal of Chemistry*, *13*(8), 6847–6864. doi:10.1016/j.arabjc.2020.06.038

Samikannu, R., Koshariya, A. K., Poornima, E., Ramesh, S., Kumar, A., & Boopathi, S. (2023). Sustainable Development in Modern Aquaponics Cultivation Systems Using IoT Technologies. In *Human Agro-Energy Optimization for Business and Industry* (pp. 105–127). IGI Global. doi:10.4018/978-1-6684-4118-3.ch006

Sampath, B. C. S., & Myilsamy, S. (2022). Application of TOPSIS Optimization Technique in the Micro-Machining Process. In Trends, Paradigms, and Advances in Mechatronics Engineering (pp. 162–187). IGI Global. doi:10.4018/978-1-6684-5887-7.ch009

Sampath, B., & Myilsamy, S. (2021). Experimental investigation of a cryogenically cooled oxygen-mist near-dry wire-cut electrical discharge machining process. *Strojniski Vestnik. Jixie Gongcheng Xuebao*, *67*(6), 322–330. doi:10.5545v-jme.2021.7161

Sampath, B., Pandian, M., Deepa, D., & Subbiah, R. (2022). Operating parameters prediction of liquefied petroleum gas refrigerator using simulated annealing algorithm. *AIP Conference Proceedings*, *2460*(1), 70003. doi:10.1063/5.0095601

Sanikhani, H., Kisi, O., Maroufpoor, E., & Yaseen, Z. M. (2019). Temperature-based modeling of reference evapotranspiration using several artificial intelligence models: Application of different modeling scenarios. *Theoretical and Applied Climatology*, *135*(1–2), 449–462. doi:10.100700704-018-2390-z

Saranya, T., Deisy, C., Sridevi, S., & Anbananthen, K. S. M. (2023). A comparative study of deep learning and Internet of Things for precision agriculture. *Engineering Applications of Artificial Intelligence*, *122*, 106034. doi:10.1016/j.engappai.2023.106034

Sarker, I. H. (2019). Context-aware rule learning from smartphone data: Survey, challenges and future directions. *Journal of Big Data*, *6*(1), 95. doi:10.118640537-019-0258-4

Sarker, I. H. (2021). Machine Learning: Algorithms, Real-World Applications and Research Directions. *SN Computer Science*, *2*(3), 160. doi:10.100742979-021-00592-x PMID:33778771

Sarker, I. H., Hoque, M. M., Uddin, M., & Alsanoosy, T. (2021). Mobile Data Science and Intelligent Apps: Concepts, AI-Based Modeling and Research Directions. *Mobile Networks and Applications*, *26*(1), 285–303. doi:10.100711036-020-01650-z

Sarker, I. H., Kayes, A. S. M., Badsha, S., Alqahtani, H., Watters, P., & Ng, A. (2020). Cybersecurity data science: An overview from machine learning perspective. *Journal of Big Data*, *7*(1), 41. doi:10.118640537-020-00318-5

Sathish, T., Sunagar, P., Singh, V., Boopathi, S., Sathyamurthy, R., Al-Enizi, A. M., Pandit, B., Gupta, M., & Sehgal, S. S. (2023). Characteristics estimation of natural fibre reinforced plastic composites using deep multi-layer perceptron (MLP) technique. *Chemosphere*, *337*(June), 139346. doi:10.1016/j.chemosphere.2023.139346 PMID:37379988

Satish, A., Nandhini, R., Poovizhi, S., Jose, P., Ranjitha, R., & Anila, S. (2017). Arduino based Smart Irrigation System using IoT. *3rd National Conference on Intelligent Information and Computing Technologies (IICT '17), December*, 1–5.

Schaible, G., & Aillery, M. (2012). Water conservation in irrigated agriculture: Trends and challenges in the face of emerging demands. *USDA-ERS Economic Information Bulletin, 99*.

Schulman, J., Wolski, F., Dhariwal, P., Radford, A., & Klimov, O. (2017). *Proximal policy optimization algorithms*. ArXiv Preprint ArXiv:1707.06347.

Selvakumar, S., Adithe, S., Isaac, J. S., Pradhan, R., Venkatesh, V., & Sampath, B. (2023). A Study of the Printed Circuit Board (PCB) E-Waste Recycling Process. In Sustainable Approaches and Strategies for E-Waste Management and Utilization (pp. 159–184). IGI Global.

Shafi, U., Mumtaz, R., García-Nieto, J., Hassan, S. A., Zaidi, S. A. R., & Iqbal, N. (2019). Precision Agriculture Techniques and Practices: From Considerations to Applications. *Sensors (Basel)*, *19*(17), 3796. doi:10.339019173796 PMID:31480709

Shah, M. A. (2015). Accelerating Public Private Partnership in Agricultural Storage Infrastructure in India. *Global Journal of Management and Business Research, 15*(A13), 23–30.

Shanmugasundaram, S., & Ramachandran, R. (2016). Rainwater Harvesting System in Meenakshi Amman Temple, Madurai: An Ancient Sustainable Water Management Practice. *Journal of Water Supply: Research & Technology - Aqua, 65*(8), 638–645. doi:10.2166/aqua.2016.108

Sharma, A., Gupta, R., & Singh, M. (2022). Sustainable Approaches for Soil Pollution Remediation: A Review of Techniques and Regulations. *Environmental Science and Pollution Research International, 29*(6), 6071–6087. doi:10.100711356-021-15987-6

Sharma, A., Jain, A., Gupta, P., & Chowdary, V. (2021). Machine Learning Applications for Precision Agriculture: A Comprehensive Review. *IEEE Access : Practical Innovations, Open Solutions, 9*, 4843–4873. doi:10.1109/ACCESS.2020.3048415

Sharshir, S. W., Algazzar, A. M., Elmaadawy, K. A., Kandeal, A. W., Elkadeem, M. R., Arunkumar, T., Zang, J., & Yang, N. (2020). New hydrogel materials for improving solar water evaporation, desalination and wastewater treatment: A review. *Desalination, 491*, 114564. doi:10.1016/j.desal.2020.114564

Shikuku, V. O., & Mishra, T. (2021). Adsorption isotherm modeling for methylene blue removal onto magnetic kaolinite clay: A comparison of two-parameter isotherms. *Applied Water Science, 11*(6), 103. doi:10.100713201-021-01440-2

Shivaprakash, K. N., Swami, N., Mysorekar, S., Arora, R., Gangadharan, A., Vohra, K., Jadeyegowda, M., & Kiesecker, J. M. (2022). Potential for Artificial Intelligence (AI) and Machine Learning (ML) Applications in Biodiversity Conservation, Managing Forests, and Related Services in India. *Sustainability (Basel), 14*(12), 7154. doi:10.3390u14127154

Shock, C. C., Barnum, J. M., & Seddigh, M. (1998). Calibration of Watermark Soil Moisture Sensors for Irrigation Management. *Proceedings of the International Irrigation Show, September*, 139–146. https://www.researchgate.net/publication/228762944

Shrestha, N. K., Du, X., & Wang, J. (2017). Assessing climate change impacts on fresh water resources of the Athabasca River Basin, Canada. *The Science of the Total Environment, 601–602*, 425–440. doi:10.1016/j.scitotenv.2017.05.013 PMID:28570976

Siddique-E-Akbor, A. H. M., Hossain, F., Sikder, S., Shum, C. K., Tseng, S., Yi, Y., Turk, F. J., & Limaye, A. (2014). Satellite Precipitation Data–Driven Hydrological Modeling for Water Resources Management in the Ganges, Brahmaputra, and Meghna Basins. *Earth Interactions, 18*(17), 1–25. doi:10.1175/EI-D-14-0017.1

Singha, B., Bar, N., & Das, S. K. (2015). The use of artificial neural network (ANN) for modeling of Pb(II) adsorption in batch process. *Journal of Molecular Liquids, 211*, 228–232. doi:10.1016/j.molliq.2015.07.002

Singh, K. S., Paul, D., Gupta, A., Dhotre, D., Klawonn, F., & Shouche, Y. (2022). Indian sewage microbiome has unique community characteristics and potential for population-level disease predictions. *The Science of the Total Environment, 160178*. Advance online publication. doi:10.1016/j.scitotenv.2022.160178 PMID:36379333

Singh, P., Pandey, P. C., Petropoulos, G. P., Pavlides, A., Srivastava, P. K., Koutsias, N., Deng, K. A. K., & Bao, Y. (2020). Hyperspectral remote sensing in precision agriculture: present status, challenges, and future trends. In *Hyperspectral Remote Sensing* (pp. 121–146). Elsevier. doi:10.1016/B978-0-08-102894-0.00009-7

Singh, P., & Singh, N. (2020). Blockchain with IoT and AI: A review of agriculture and healthcare. *International Journal of Applied Evolutionary Computation, 11*(4), 13–27. doi:10.4018/IJAEC.2020100102

Sinha, A., Shrivastava, G., & Kumar, P. (2019). Architecting user-centric internet of things for smart agriculture. *Sustainable Computing: Informatics and Systems, 23*, 88–102. doi:10.1016/j.suscom.2019.07.001

Sishodia, R. P., Ray, R. L., & Singh, S. K. (2020). Applications of Remote Sensing in Precision Agriculture: A Review. *Remote Sensing (Basel), 12*(19), 3136. doi:10.3390/rs12193136

Siyal, A. A., Shamsuddin, M. R., Khan, M. I., Rabat, N. E., Zulfiqar, M., Man, Z., Siame, J., & Azizli, K. A. (2018). A review on geopolymers as emerging materials for the adsorption of heavy metals and dyes. *Journal of Environmental Management, 224*, 327–339. doi:10.1016/j.jenvman.2018.07.046 PMID:30056352

Slimani, A. Z., A. F., Oufdou, K., & & Meddich, A. (2023). Impact of Climate Change on Water Status: Challenges and Emerging Solutions. In Water in Circular Economy (pp. 3–20). Academic Press.

Slimani, A., Raklami, A., Oufdou, K., & Meddich, A. (2022). Isolation and Characterization of PGPR and Their Potenzial for Drought Alleviation in Barley Plants. *Gesunde Pflanzen*, 1–15. doi:10.100710343-022-00709-z

Soares, C., Brazdil, P. B., & Kuba, P. (2004). A Meta-Learning Method to Select the Kernel Width in Support Vector Regression. *Machine Learning, 54*(3), 195–209. doi:10.1023/B:MACH.0000015879.28004.9b

Song, X., Zhang, J., Wang, G., He, R., & Wang, X. (2014). Development and challenges of urban hydrology in a changing environment: II: Urban stormwater modeling and management. *Shuikexue Jinzhan. Shui Kexue Jinzhan, 25*(5), 752–764.

Souza, P., Dotto, G., & Salau, N. (2018). Artificial neural network (ANN) and adaptive neuro-fuzzy interference system (ANFIS) modelling for nickel adsorption onto agro-wastes and commercial activated carbon. *Journal of Environmental Chemical Engineering, 6*(6), 7152–7160. doi:10.1016/j.jece.2018.11.013

Srivastava, A., Jain, S., Maity, R., & Desai, V. R. (2022). Demystifying artificial intelligence amidst sustainable agricultural water management. *Current Directions in Water Scarcity Research, 7*, 17–35. doi:10.1016/B978-0-323-91910-4.00002-9

Srivastava, P. K., Han, D., Ramirez, M. R., & Islam, T. (2013). Machine Learning Techniques for Downscaling SMOS Satellite Soil Moisture Using MODIS Land Surface Temperature for Hydrological Application. *Water Resources Management, 27*(8), 3127–3144. doi:10.100711269-013-0337-9

Stienen, J., Bruinsma, W. & Neuman, F. (2007). *How ICT can make a difference in agricultural livelihoods. The commonwealth ministers book-2007.* International Institute for Communication and Development.

Straits-Research. (2022). *Precision Agriculture Market Size is projected to reach USD 19.24 Billion by 2030, growing at a CAGR of 14.95%: Straits Research.* Https://Www.Globenewswire.Com/En/News-Release/2022/08/01/2489650/0/En/Precision-Agriculture-Market-Size-Is-Projected-to-Reach-USD-19-24-Billion-by-2030-Growing-at-a-CAGR-of-14-95-Straits-Research.Html

Subha, S., Inbamalar, T. M., Komala, C. R., Suresh, L. R., Boopathi, S., & Alaskar, K. (2023, February). A Remote Health Care Monitoring system using internet of medical things (IoMT). *Proceedings of 2023 3rd International Conference on Innovative Practices in Technology and Management, ICIPTM 2023.* 10.1109/ICIPTM57143.2023.10118103

Sudarmaji, A., Sahirman, S., Saparso, & Ramadhani, Y. (2019). Time based automatic system of drip and sprinkler irrigation for horticulture cultivation on coastal area. *IOP Conference Series. Earth and Environmental Science, 250*(1), 012074. doi:10.1088/1755-1315/250/1/012074

Sun, J. X., Yin, Y. L., Sun, S. K., Wang, Y. B., Yu, X., & Yan, K. (2021). Review on research status of virtual water: The perspective of accounting methods, impact assessment and limitations. *Agricultural Water Management, 243*, 106407. doi:10.1016/j.agwat.2020.106407

Sundararajan, R., & Sridharan, K. (2013). Sustainable Water Management Practices of the Chola Dynasty in Ancient Tamil Nadu. *Journal of Water Resource and Protection, 5*(11), 1117–1123. doi:10.4236/jwarp.2013.511117

Sundui, B., Ramirez Calderon, O. A., Abdeldayem, O. M., Lázaro-Gil, J., Rene, E. R., & Sambuu, U. (2021). Applications of machine learning algorithms for biological wastewater treatment: Updates and perspectives. *Clean Technologies and Environmental Policy, 23*(1), 127–143. doi:10.100710098-020-01993-x

Sun, Y., Li, Z., Li, X., & Zhang, J. (2021). Classifier Selection and Ensemble Model for Multi-class Imbalance Learning in Education Grants Prediction. *Applied Artificial Intelligence, 35*(4), 290–303. doi:10.1080/08839514.2021.1877481

Suresh Babu, C. V., Akshayah, N. S., Vinola, P. M., & Janapriyan, R. (2023). IoT-Based Smart Accident Detection and Alert System. In Handbook of Research on Deep Learning Techniques for Cloud-Based Industrial IoT (pp. 16). IGI Global. https://doi.org/ doi:10.4018/978-1-6684-8098-4.ch019

Syrmos, E., Sidiropoulos, V., Bechtsis, D., Stergiopoulos, F., Aivazidou, E., Vrakas, D., Vezinias, P., & Vlahavas, I. (2023). An Intelligent Modular Water Monitoring IoT System for Real-Time Quantitative and Qualitative Measurements. *Sustainability (Basel)*, *15*(3), 2127. doi:10.3390u15032127

Tahiri, A. I., Meddich, A., Raklami, A., Alahmad, A., Bechtaoui, N., Anli, M., Göttfert, M., Heulin, T., Achouak, W., & Oufdou, K. (2022). Assessing the potential role of compost, PGPR, and AMF in improving tomato plant growth, yield, fruit quality, and water stress tolerance. *Journal of Soil Science and Plant Nutrition*, *22*(1), 1–22. doi:10.100742729-021-00684-w

Talaviya, T., Shah, D., Patel, N., Yagnik, H., & Shah, M. (2020). Implementation of artificial intelligence in agriculture for optimisation of irrigation and application of pesticides and herbicides. *Artificial Intelligence in Agriculture*, *4*, 58–73. doi:10.1016/j.aiia.2020.04.002

Tantalaki, N., Souravlas, S., & Roumeliotis, M. (2019). Data-Driven Decision Making in Precision Agriculture: The Rise of Big Data in Agricultural Systems. *Journal of Agricultural & Food Information*, *20*(4), 344–380. doi:10.1080/10496505.2019.1638264

Tarkowska-Kukuryk, M., & Grzywna, A. (2022). Macrophyte communities as indicators of the ecological status of drainage canals and regulated rivers (eastern Poland). *Environmental Monitoring and Assessment*, *194*(3), 210. doi:10.100710661-022-09777-0 PMID:35194688

Tejoyadav, M., Nayak, R., & Pati, U. C. (2022). Multivariate water quality forecasting of river ganga using var-lstm based hybrid model. In *2022 IEEE 19th India Council International Conference (INDICON)* (pp. 1–6). IEEE. 10.1109/INDICON56171.2022.10040146

Thakur, D., Kumar, Y., Kumar, A., & Singh, P. K. (2019). Applicability of Wireless Sensor Networks in Precision Agriculture: A Review. *Wireless Personal Communications*, *107*(1), 471–512. doi:10.100711277-019-06285-2

Thomas, E., Borchard, N., Sarmiento, C., Atkinson, R., & Ladd, B. (2020). Key factors determining biochar sorption capacity for metal contaminants: A literature synthesis. *Biochar*, *2*(2), 151–163. doi:10.100742773-020-00053-3

Tien, J. M. (2017). Internet of Things, Real-Time Decision Making, and Artificial Intelligence. *Annals of Data Science*, *4*(2), 149–178. doi:10.100740745-017-0112-5

Togneri, R., Felipe dos Santos, D., Camponogara, G., Nagano, H., Custódio, G., Prati, R., Fernandes, S., & Kamienski, C. (2022). Soil moisture forecast for smart irrigation: The primetime for machine learning. *Expert Systems with Applications*, *207*(April), 1–23. doi:10.1016/j.eswa.2022.117653

Tomaszewski, L., & Kołakowski, R. (2023). Mobile Services for Smart Agriculture and Forestry, Biodiversity Monitoring, and Water Management: Challenges for 5G/6G Networks. Telecom, 4(1), 67-99. doi:10.3390/telecom4010006

Tome, S., Shikuku, V., Tamaguelon, H. D., Akiri, S., Etoh, M. A., Rüscher, C., & Etame, J. (2023). Efficient sequestration of malachite green in aqueous solution by laterite-rice husk ash-based alkali-activated materials: Parameters and mechanism. *Environmental Science and Pollution Research International, 30*(25), 67263–67277. doi:10.100711356-023-27138-3 PMID:37103713

Torğut, G., & Demirelli, K. (2018). Comparative Adsorption of Different Dyes from Aqueous Solutions onto Polymer Prepared by ROP: Kinetic, Equilibrium and Thermodynamic Studies. *Arabian Journal for Science and Engineering, 43*(7), 3503–3514. doi:10.100713369-017-2947-7

Torregrossa, D., Leopold, U., Hernández-Sancho, F., & Hansen, J. (2018). Machine learning for energy cost modelling in wastewater treatment plants. *Journal of Environmental Management, 223*, 1061–1067. doi:10.1016/j.jenvman.2018.06.092 PMID:30096746

Toubali, S., Ait-El-Mokhtar, M., Boutasknit, A., Anli, M., Ait-Rahou, Y., Benaffari, W., Ben-Ahmed, H., Mitsui, T., Baslam, M., & Meddich, A. (2022). Root Reinforcement Improved Performance, Productivity, and Grain Bioactive Quality of Field-Droughted Quinoa (*Chenopodium quinoa*). *Frontiers in Plant Science, 13*(March), 1–20. doi:10.3389/fpls.2022.860484 PMID:35371170

Touil, S., Richa, A., Fizir, M., Argente García, J. E., & Skarmeta Gómez, A. F. (2022). A review on smart irrigation management strategies and their effect on water savings and crop yield. *Irrigation and Drainage, 71*(5), 1396–1416. doi:10.1002/ird.2735

Tripathi, M., & Singal, S. K. (2019). Use of Principal Component Analysis for parameter selection for development of a novel Water Quality Index: A case study of river Ganga India. *Ecological Indicators, 96*, 430–436. doi:10.1016/j.ecolind.2018.09.025

Tschand, A. (2023). Semi-supervised machine learning analysis of crop color for autonomous irrigation. *Smart Agricultural Technology, 3*, 100116. doi:10.1016/j.atech.2022.100116

Turan, N. G., Mesci, B., & Ozgonenel, O. (2011a). Artificial neural network (ANN) approach for modeling Zn(II) adsorption from leachate using a new biosorbent. *Chemical Engineering Journal, 173*(1), 98–105. doi:10.1016/j.cej.2011.07.042

Turan, N. G., Mesci, B., & Ozgonenel, O. (2011b). The use of artificial neural networks (ANN) for modeling of adsorption of Cu(II) from industrial leachate by pumice. *Chemical Engineering Journal, 171*(3), 1091–1097. doi:10.1016/j.cej.2011.05.005

Tyralis, H., Papacharalampous, G., & Langousis, A. (2021). Super ensemble learning for daily streamflow forecasting: Large-scale demonstration and comparison with multiple machine learning algorithms. *Neural Computing & Applications, 33*(8), 3053–3068. doi:10.100700521-020-05172-3

Uddin, J., Smith, R. J., Gillies, M. H., Moller, P., & Robson, D. (2018). Smart Automated Furrow Irrigation of Cotton. *Journal of Irrigation and Drainage Engineering, 144*(5), 1–10. doi:10.1061/(ASCE)IR.1943-4774.0001282

Uddin, M. G., Nash, S., Rahman, A., & Olbert, A. I. (2023). A novel approach for estimating and predicting uncertainty in water quality index model using machine learning approaches. *Water Research, 229*, 119422. doi:10.1016/j.watres.2022.119422 PMID:36459893

Uddin, M. K. (2017). A review on the adsorption of heavy metals by clay minerals, with special focus on the past decade. *Chemical Engineering Journal, 308*, 438–462. doi:10.1016/j.cej.2016.09.029

UNESCO. (2017). *UN World Water Development Report*. Retrieved from:https://www.unwater.org/publications/un-world-water-development-report-2017

V, S. (2021). Internet of Things (IoT) based Smart Agriculture in India: An Overview. *Journal of ISMAC, 3*(1), 1–15. doi:10.36548/jismac.2021.1.001

Van Lier, J. B., Tilche, A., Ahring, B. K., Macarie, H., Moletta, R., Dohanyos, M., Hulshoff Pol, L. W., Lens, P., & Verstraete, W. (2001). New perspectives in anaerobic digestion. *Water Science and Technology, 43*(1), 1–18. doi:10.2166/wst.2001.0001 PMID:11379079

van Vliet, M. T. H., Flörke, M., & Wada, Y. (2017). Quality matters for water scarcity. *Nature Geoscience, 10*(11), 800–802. doi:10.1038/ngeo3047

van Vliet, M. T. H., Jones, E. R., Flörke, M., Franssen, W. H. P., Hanasaki, N., Wada, Y., & Yearsley, J. R. (2021). Global water scarcity including surface water quality and expansions of clean water technologies. *Environmental Research Letters, 16*(2), 024020. doi:10.1088/1748-9326/abbfc3

Vanitha, S. K. R., & Boopathi, S. (2023). Artificial Intelligence Techniques in Water Purification and Utilization. In *Human Agro-Energy Optimization for Business and Industry* (pp. 202–218). IGI Global. doi:10.4018/978-1-6684-4118-3.ch010

Veerachamy, R., Ramar, R., Balaji, S., & Sharmila, L. (2022). Autonomous Application Controls on Smart Irrigation. *Computers & Electrical Engineering, 100*(March), 107855. doi:10.1016/j.compeleceng.2022.107855

Veldkamp, T. I. E., Wada, Y., Aerts, J. C. J. H., Döll, P., Gosling, S. N., Liu, J., Masaki, Y., Oki, T., Ostberg, S., Pokhrel, Y., Satoh, Y., Kim, H., & Ward, P. J. (2017). Water scarcity hotspots travel downstream due to human interventions in the 20th and 21st century. *Nature Communications, 8*(1), 15697. doi:10.1038/ncomms15697 PMID:28643784

Venkatesh, S., Venkatesh, K., & Quaff, A. R. (2017). Dye decomposition by combined ozonation and anaerobic treatment: Cost effective technology. *Journal of Applied Research and Technology, 15*(4), 340–345. doi:10.1016/j.jart.2017.02.006

Vennila, T., Karuna, M. S., Srivastava, B. K., Venugopal, J., Surakasi, R., & B., S. (2023). New Strategies in Treatment and Enzymatic Processes. In *Human Agro-Energy Optimization for Business and Industry* (pp. 219–240). IGI Global. doi:10.4018/978-1-6684-4118-3.ch011

Verma, A., Thakur, B., Kartiyar, S., Singh, D., & Rai, M. (2013). Evaluation of ground water quality in Lucknow, Uttar Pradesh using remote sensing and geographic information systems (GIS). *Int. J. Water Res. Environ. Eng., 5*(2), 67–76. https://doi.org/https://doi.org/10.5897/IJWREE11.142

Vij, A., Vijendra, S., Jain, A., Bajaj, S., Bassi, A., & Sharma, A. (2020). IoT and Machine Learning Approaches for Automation of Farm Irrigation System. *Procedia Computer Science, 167*(2019), 1250–1257. doi:10.1016/j.procs.2020.03.440

Vollmer, D., Shaad, K., Souter, N. J., Farrell, T., Dudgeon, D., Sullivan, C. A., Fauconnier, I., MacDonald, G. M., McCartney, M. P., Power, A. G., McNally, A., Andelman, S. J., Capon, T., Devineni, N., Apirumanekul, C., Ng, C. N., Rebecca Shaw, M., Wang, R. Y., Lai, C., ... Regan, H. M. (2018). Integrating the social, hydrological and ecological dimensions of freshwater health: The Freshwater Health Index. *The Science of the Total Environment*, *627*, 304–313. doi:10.1016/j.scitotenv.2018.01.040 PMID:29426153

Vrchota, J., Pech, M., & Švepešová, I. (2022). Precision Agriculture Technologies for Crop and Livestock Production in the Czech Republic. *Agriculture*, *12*(8), 1080. doi:10.3390/agriculture12081080

Wagstaff, K., Cardie, C., Rogers, S., & Schrödl, S. (2001). Constrained k-means clustering with background knowledge. *Icml*, *1*, 577–584.

Walczak, S., & Cerpa, N. (2003). Artificial Neural Networks. In R. A. Meyers (Ed.), Encyclopedia of Physical Science and Technology (3rd ed., pp. 631–645). Academic Press. https://doi.org/doi:10.1016/B0-12-227410-5/00837-1

Wan, D., Wu, H., Li, S., & Cheng, F. (2011). Application of Data Mining in Relationship between Water Quantity and Water Quality. In H. Deng, D. Miao, J. Lei, & F. L. Wang (Eds.), *Artificial Intelligence and Computational Intelligence* (pp. 404–412). Springer Berlin Heidelberg. doi:10.1007/978-3-642-23881-9_53

Wang, J. & Kumbasar, T. (2019). Parameter optimization of interval Type-2 fuzzy neural networks based on PSO and BBBC methods. *IEEE/CAA Journal of Automatica Sinica*, *6*(1), 247–257. doi:10.1109/JAS.2019.1911348

Wang, E., Attard, S., Linton, A., McGlinchey, M., Xiang, W., Philippa, B., & Everingham, Y. (2020). Development of a closed-loop irrigation system for sugarcane farms using the Internet of Things. *Computers and Electronics in Agriculture*, *172*(March), 105376. doi:10.1016/j.compag.2020.105376

Wang, G., Jia, Q.-S., Zhou, M., Bi, J., Qiao, J., & Abusorrah, A. (2022). Artificial neural networks for water quality soft-sensing in wastewater treatment: A review. *Artificial Intelligence Review*, *55*(1), 565–587. doi:10.100710462-021-10038-8

Wang, Z., Xu, W., Jie, F., Zhao, Z., Zhou, K., & Liu, H. (2021). The selective adsorption performance and mechanism of multiwall magnetic carbon nanotubes for heavy metals in wastewater. *Scientific Reports*, *11*(1), 1–13. doi:10.103841598-021-96465-7 PMID:34413419

Ward, A. J., Hobbs, P. J., Holliman, P. J., & Jones, D. L. (2008). Optimisation of the anaerobic digestion of agricultural resources. *Bioresource Technology*, *99*(17), 7928–7940. doi:10.1016/j.biortech.2008.02.044 PMID:18406612

Ward, J. H. Jr. (1963). Hierarchical Grouping to Optimize an Objective Function. *Journal of the American Statistical Association*, *58*(301), 236–244. doi:10.1080/01621459.1963.10500845

Watkins, C. J. C. H., & Dayan, P. (1992). Q-learning. *Machine Learning*, *8*(3), 279–292. doi:10.1007/BF00992698

Wei, W., Gao, Y., Huang, J., & Gao, J. (2020). Exploring the effect of basin land degradation on lake and reservoir water quality in China. *Journal of Cleaner Production*, *268*, 122249. doi:10.1016/j.jclepro.2020.122249

Wen, Z., He, B., Kotagiri, R., Lu, S., & Shi, J. (2018, May). Efficient gradient boosted decision tree training on GPUs. In *2018 IEEE International Parallel and Distributed Processing Symposium (IPDPS)* (pp. 234-243). IEEE. 10.1109/IPDPS.2018.00033

Wilkinson, T. J., Gibson, M., & Szuchman, J. (2018). Ancient Water Systems in Mesopotamia: The Garden of Eden, Hanging Gardens of Babylon, and Tower of Babel. *Water (Basel)*, *10*(5), 552. doi:10.3390/w10050552

Wolpert, D. H. (1992). Stacked generalization. *Neural Networks*, *5*(2), 241–259. doi:10.1016/S0893-6080(05)80023-1 PMID:18276425

Wu, Q., Wu, B., & Yan, X. (2022). An intelligent traceability method of water pollution based on dynamic multi-mode optimization. *Neural Computing & Applications*. Advance online publication. doi:10.100700521-022-07002-0 PMID:35221540

Xiang, X., Li, Q., Khan, S., & Khalaf, O. I. (2021). Urban water resource management for sustainable environment planning using artificial intelligence techniques. *Environmental Impact Assessment Review*, *86*, 106515. doi:10.1016/j.eiar.2020.106515

Xiang, Y., & Jiang, L. (2009). Water quality prediction using ls-svm and particle swarm optimization. In *2009 Second International Workshop on Knowledge Discovery and Data Mining* (pp. 900–904). IEEE. 10.1109/WKDD.2009.217

Xiong, Y., Ye, M., & Wu, C. (2021). Cancer Classification with a Cost-Sensitive Naive Bayes Stacking Ensemble. *Computational and Mathematical Methods in Medicine*, *2021*, e5556992. doi:10.1155/2021/5556992 PMID:33986823

Xuan, W., Jiake, L., & Deti, X. (2010). A hybrid approach of support vector machine with particle swarm optimization for water quality prediction. In *2010 5th International Conference on Computer Science & Education* (pp. 1158–1163). IEEE. 10.1109/ICCSE.2010.5593697

Xu, B., Ge, Z., & He, Z. (2015). Sediment microbial fuel cells for wastewater treatment: Challenges and opportunities. *Environmental Science. Water Research & Technology*, *1*(3), 279–284. doi:10.1039/C5EW00020C

Xu, T., Coco, G., & Neale, M. (2020). A predictive model of recreational water quality based on adaptive synthetic sampling algorithms and machine learning. *Water Research*, *177*, 115788. doi:10.1016/j.watres.2020.115788 PMID:32330740

Xu, W., & Su, X. (2019). Challenges and impacts of climate change and human activities on groundwater-dependent ecosystems in arid areas–A case study of the Nalenggele alluvial fan in NW China. *Journal of Hydrology (Amsterdam)*, *573*, 376–385. doi:10.1016/j.jhydrol.2019.03.082

Xu, Z., Shen, J., Qu, Y., Chen, H., Zhou, X., Hong, H., Sun, H., Lin, H., Deng, W., & Wu, F. (2022). Using simple and easy water quality parameters to predict trihalomethane occurrence in tap water. *Chemosphere*, *286*, 131586. doi:10.1016/j.chemosphere.2021.131586 PMID:34303907

Yang, Q., Cao, W., Meng, W., & Si, J. (2021). Reinforcement-learning-based tracking control of waste water treatment process under realistic system conditions and control performance requirements. *IEEE Transactions on Systems, Man, and Cybernetics. Systems*, *52*(8), 5284–5294. doi:10.1109/TSMC.2021.3122802

Yang, S., Yang, D., Chen, J., Santisirisomboon, J., Lu, W., & Zhao, B. (2020). A physical process and machine learning combined hydrological model for daily streamflow simulations of large watersheds with limited observation data. *Journal of Hydrology (Amsterdam)*, *590*, 125206. doi:10.1016/j.jhydrol.2020.125206

Yang, X., Liu, Z., & Xia, J. (2023). Optimization and analysis of combined heat and water production system based on a coal-fired power plant. *Energy*, *262*, 125611. doi:10.1016/j.energy.2022.125611

Yaqoob, A. A., Ibrahim, M. N. M., Umar, K., Parveen, T., Ahmad, A., Lokhat, D., & Setapar, S. H. M. (2021). A glimpse into the microbial fuel cells for wastewater treatment with energy generation. *Desalination and Water Treatment*, *214*, 379–389. doi:10.5004/dwt.2021.26737

Yarowsky, D. (1995). Unsupervised word sense disambiguation rivaling supervised methods. *33rd Annual Meeting of the Association for Computational Linguistics*, 189–196. 10.3115/981658.981684

Yasin, Y., Ahmad, F. B. H., Ghaffari-Moghaddam, M., & Khajeh, M. (2014). Application of a hybrid artificial neural network–genetic algorithm approach to optimize the lead ions removal from aqueous solutions using intercalated tartrate-Mg–Al layered double hydroxides. *Environmental Nanotechnology, Monitoring & Management*, *1-2*, 2–7. doi:10.1016/j.enmm.2014.03.001

Yin, J., Deng, Z., Ines, A. V. M., Wu, J., & Rasu, E. (2020). Forecast of short-term daily reference evapotranspiration under limited meteorological variables using a hybrid bi-directional long short-term memory model (Bi-LSTM). *Agricultural Water Management*, *242*(February), 106386. doi:10.1016/j.agwat.2020.106386

Yuan, B., Zhang, C., Liang, Y., Yang, L., Yang, H., Bai, L., Wei, D., Wang, W., Wang, Q., & Chen, H. (2021). A Low-Cost 3D Spherical Evaporator with Unique Surface Topology and Inner Structure for Solar Water Evaporation-Assisted Dye Wastewater Treatment. *Advanced Sustainable Systems*, *5*(3), 2000245. doi:10.1002/adsu.202000245

Yupapin, P., Trabelsi, Y., Nattappan, A., & Boopathi, S. (2023). Performance Improvement of Wire-Cut Electrical Discharge Machining Process Using Cryogenically Treated Super-Conductive State of Monel-K500 Alloy. *Iranian Journal of Science and Technology. Transaction of Mechanical Engineering*, *47*(1), 267–283. doi:10.100740997-022-00513-0

Zanfei, A., Menapace, A., & Righetti, M. (2023). An artificial intelligence approach for managing water demand in water supply systems. *IOP Conference Series. Earth and Environmental Science*, *1136*(1), 012004. doi:10.1088/1755-1315/1136/1/012004

Zeng, H., Dhiman, G., Sharma, A., Sharma, A., & Tselykh, A. (2021). An IoT and Blockchain-based approach for the smart water management system in agriculture. *Expert Systems: International Journal of Knowledge Engineering and Neural Networks*, *39*(8), e12892. doi:10.1111/exsy.12892

Zhang, T., Jiang, B., Xing, Y., Ya, H., Lv, M., & Wang, X. (2022). Current status of microplastics pollution in the aquatic environment, interaction with other pollutants, and effects on aquatic organisms. *Environmental Science and Pollution Research International*, *29*(16), 16830–16859. doi:10.100711356-022-18504-8 PMID:35001283

Zhang, Z., Huang, J., Duan, S., Huang, Y., Cai, J., & Bian, J. (2022). Use of interpretable machine learning to identify the factors influencing the nonlinear linkage between land use and river water quality in the Chesapeake Bay watershed. *Ecological Indicators*, *140*, 108977. doi:10.1016/j.ecolind.2022.108977

Zhao, L., Dai, T., Qiao, Z., Sun, P., Hao, J., & Yang, Y. (2020). Application of artificial intelligence to wastewater treatment: A bibliometric analysis and systematic review of technology, economy, management, and wastewater reuse. *Process Safety and Environmental Protection*, *133*, 169–182. doi:10.1016/j.psep.2019.11.014

Zhu, X., & Ghahramani, Z. (2002). *Learning from labeled and unlabeled data with label propagation*. Citeseer. https://citeseerx.ist.psu.edu/document?repid=rep1&type=pdf&doi=8a6a114d699824b678325766be195b0e7b564705

Zhu, M., Wang, J., Yang, X., Zhang, Y., Zhang, L., Ren, H., Wu, B., & Ye, L. (2022). *A review of the application of machine learning in water quality evaluation*. Eco-Environment & Health. doi:10.1016/j.eehl.2022.06.001

Zhuo, L., & Hoekstra, A. Y. (2017). The effect of different agricultural management practices on irrigation efficiency, water use efficiency and green and blue water footprint. *Frontiers of Agricultural Science and Engineering*, *4*(2), 185. doi:10.15302/J-FASE-2017149

Zhu, X., Wang, X., & Ok, Y. S. (2019). The application of machine learning methods for prediction of metal sorption onto biochars. *Journal of Hazardous Materials*, *378*, 120727. doi:10.1016/j.jhazmat.2019.06.004 PMID:31202073

Related References

To continue our tradition of advancing academic research, we have compiled a list of recommended IGI Global readings. These references will provide additional information and guidance to further enrich your knowledge and assist you with your own research and future publications.

Abbasnejad, B., Moeinzadeh, S., Ahankoob, A., & Wong, P. S. (2021). The Role of Collaboration in the Implementation of BIM-Enabled Projects. In J. Underwood & M. Shelbourn (Eds.), *Handbook of Research on Driving Transformational Change in the Digital Built Environment* (pp. 27–62). IGI Global. https://doi.org/10.4018/978-1-7998-6600-8.ch002

Abdulrahman, K. O., Mahamood, R. M., & Akinlabi, E. T. (2022). Additive Manufacturing (AM): Processing Technique for Lightweight Alloys and Composite Material. In K. Kumar, B. Babu, & J. Davim (Ed.), *Handbook of Research on Advancements in the Processing, Characterization, and Application of Lightweight Materials* (pp. 27-48). IGI Global. https://doi.org/10.4018/978-1-7998-7864-3.ch002

Agrawal, R., Sharma, P., & Saxena, A. (2021). A Diamond Cut Leather Substrate Antenna for BAN (Body Area Network) Application. In V. Singh, V. Dubey, A. Saxena, R. Tiwari, & H. Sharma (Eds.), *Emerging Materials and Advanced Designs for Wearable Antennas* (pp. 54–59). IGI Global. https://doi.org/10.4018/978-1-7998-7611-3.ch004

Ahmad, F., Al-Ammar, E. A., & Alsaidan, I. (2022). Battery Swapping Station: A Potential Solution to Address the Limitations of EV Charging Infrastructure. In M. Alam, R. Pillai, & N. Murugesan (Eds.), *Developing Charging Infrastructure and Technologies for Electric Vehicles* (pp. 195–207). IGI Global. doi:10.4018/978-1-7998-6858-3.ch010

Aikhuele, D. (2018). A Study of Product Development Engineering and Design Reliability Concerns. *International Journal of Applied Industrial Engineering*, *5*(1), 79–89. doi:10.4018/IJAIE.2018010105

Al-Khatri, H., & Al-Atrash, F. (2021). Occupants' Habits and Natural Ventilation in a Hot Arid Climate. In R. González-Lezcano (Ed.), *Advancements in Sustainable Architecture and Energy Efficiency* (pp. 146–168). IGI Global. https://doi.org/10.4018/978-1-7998-7023-4.ch007

Al-Shebeeb, O. A., Rangaswamy, S., Gopalakrishan, B., & Devaru, D. G. (2017). Evaluation and Indexing of Process Plans Based on Electrical Demand and Energy Consumption. *International Journal of Manufacturing, Materials, and Mechanical Engineering*, *7*(3), 1–19. doi:10.4018/IJMMME.2017070101

Amuda, M. O., Lawal, T. F., & Akinlabi, E. T. (2017). Research Progress on Rheological Behavior of AA7075 Aluminum Alloy During Hot Deformation. *International Journal of Materials Forming and Machining Processes*, *4*(1), 53–96. doi:10.4018/IJMFMP.2017010104

Amuda, M. O., Lawal, T. F., & Mridha, S. (2021). Microstructure and Mechanical Properties of Silicon Carbide-Treated Ferritic Stainless Steel Welds. In L. Burstein (Ed.), *Handbook of Research on Advancements in Manufacturing, Materials, and Mechanical Engineering* (pp. 395–411). IGI Global. https://doi.org/10.4018/978-1-7998-4939-1.ch019

Anikeev, V., Gasem, K. A., & Fan, M. (2021). Application of Supercritical Technologies in Clean Energy Production: A Review. In L. Chen (Ed.), *Handbook of Research on Advancements in Supercritical Fluids Applications for Sustainable Energy Systems* (pp. 792–821). IGI Global. https://doi.org/10.4018/978-1-7998-5796-9.ch022

Arafat, M. Y., Saleem, I., & Devi, T. P. (2022). Drivers of EV Charging Infrastructure Entrepreneurship in India. In M. Alam, R. Pillai, & N. Murugesan (Eds.), *Developing Charging Infrastructure and Technologies for Electric Vehicles* (pp. 208–219). IGI Global. https://doi.org/10.4018/978-1-7998-6858-3.ch011

Araujo, A., & Manninen, H. (2022). Contribution of Project-Based Learning on Social Skills Development: An Industrial Engineer Perspective. In A. Alves & N. van Hattum-Janssen (Eds.), *Training Engineering Students for Modern Technological Advancement* (pp. 119–145). IGI Global. https://doi.org/10.4018/978-1-7998-8816-1.ch006

Armutlu, H. (2018). Intelligent Biomedical Engineering Operations by Cloud Computing Technologies. In U. Kose, G. Guraksin, & O. Deperlioglu (Eds.), *Nature-Inspired Intelligent Techniques for Solving Biomedical Engineering Problems* (pp. 297–317). Hershey, PA: IGI Global. doi:10.4018/978-1-5225-4769-3.ch015

Atik, M., Sadek, M., & Shahrour, I. (2017). Single-Run Adaptive Pushover Procedure for Shear Wall Structures. In V. Plevris, G. Kremmyda, & Y. Fahjan (Eds.), *Performance-Based Seismic Design of Concrete Structures and Infrastructures* (pp. 59–83). Hershey, PA: IGI Global. doi:10.4018/978-1-5225-2089-4.ch003

Attia, H. (2021). Smart Power Microgrid Impact on Sustainable Building. In R. González-Lezcano (Ed.), *Advancements in Sustainable Architecture and Energy Efficiency* (pp. 169–194). IGI Global. https://doi.org/10.4018/978-1-7998-7023-4.ch008

Aydin, A., Akyol, E., Gungor, M., Kaya, A., & Tasdelen, S. (2018). Geophysical Surveys in Engineering Geology Investigations With Field Examples. In N. Ceryan (Ed.), *Handbook of Research on Trends and Digital Advances in Engineering Geology* (pp. 257–280). Hershey, PA: IGI Global. doi:10.4018/978-1-5225-2709-1.ch007

Ayoobkhan, M. U. D., Y., A., J., Easwaran, B., & R., T. (2021). Smart Connected Digital Products and IoT Platform With the Digital Twin. In P. Vasant, G. Weber, & W. Punurai (Ed.), Research Advancements in Smart Technology, Optimization, and Renewable Energy (pp. 330-350). IGI Global. https://doi.org/ doi:10.4018/978-1-7998-3970-5.ch016

Baeza Moyano, D., & González Lezcano, R. A. (2021). The Importance of Light in Our Lives: Towards New Lighting in Schools. In R. González-Lezcano (Ed.), *Advancements in Sustainable Architecture and Energy Efficiency* (pp. 239–256). IGI Global. https://doi.org/10.4018/978-1-7998-7023-4.ch011

Bagdadee, A. H. (2021). A Brief Assessment of the Energy Sector of Bangladesh. *International Journal of Energy Optimization and Engineering*, *10*(1), 36–55. doi:10.4018/IJEOE.2021010103

Baklezos, A. T., & Hadjigeorgiou, N. G. (2021). Magnetic Sensors for Space Applications and Magnetic Cleanliness Considerations. In C. Nikolopoulos (Ed.), *Recent Trends on Electromagnetic Environmental Effects for Aeronautics and Space Applications* (pp. 147–185). IGI Global. https://doi.org/10.4018/978-1-7998-4879-0.ch006

Bas, T. G. (2017). Nutraceutical Industry with the Collaboration of Biotechnology and Nutrigenomics Engineering: The Significance of Intellectual Property in the Entrepreneurship and Scientific Research Ecosystems. In T. Bas & J. Zhao (Eds.), *Comparative Approaches to Biotechnology Development and Use in Developed and Emerging Nations* (pp. 1–17). Hershey, PA: IGI Global. doi:10.4018/978-1-5225-1040-6.ch001

Bazeer Ahamed, B., & Periakaruppan, S. (2021). Taxonomy of Influence Maximization Techniques in Unknown Social Networks. In P. Vasant, G. Weber, & W. Punurai (Eds.), *Research Advancements in Smart Technology, Optimization, and Renewable Energy* (pp. 351-363). IGI Global. https://doi.org/10.4018/978-1-7998-3970-5.ch017

Beale, R., & André, J. (2017). *Design Solutions and Innovations in Temporary Structures*. Hershey, PA: IGI Global. doi:10.4018/978-1-5225-2199-0

Behnam, B. (2017). Simulating Post-Earthquake Fire Loading in Conventional RC Structures. In P. Samui, S. Chakraborty, & D. Kim (Eds.), *Modeling and Simulation Techniques in Structural Engineering* (pp. 425–444). Hershey, PA: IGI Global. doi:10.4018/978-1-5225-0588-4.ch015

Ben Hamida, I., Salah, S. B., Msahli, F., & Mimouni, M. F. (2018). Distribution Network Reconfiguration Using SPEA2 for Power Loss Minimization and Reliability Improvement. *International Journal of Energy Optimization and Engineering*, *7*(1), 50–65. doi:10.4018/IJEOE.2018010103

Bentarzi, H. (2021). Fault Tree-Based Root Cause Analysis Used to Study Mal-Operation of a Protective Relay in a Smart Grid. In A. Recioui & H. Bentarzi (Eds.), *Optimizing and Measuring Smart Grid Operation and Control* (pp. 289–308). IGI Global. https://doi.org/10.4018/978-1-7998-4027-5.ch012

Beysens, D. A., Garrabos, Y., & Zappoli, B. (2021). Thermal Effects in Near-Critical Fluids: Piston Effect and Related Phenomena. In L. Chen (Ed.), *Handbook of Research on Advancements in Supercritical Fluids Applications for Sustainable Energy Systems* (pp. 1–31). IGI Global. https://doi.org/10.4018/978-1-7998-5796-9.ch001

Bhaskar, S. V., & Kudal, H. N. (2017). Effect of TiCN and AlCrN Coating on Tribological Behaviour of Plasma-nitrided AISI 4140 Steel. *International Journal of Surface Engineering and Interdisciplinary Materials Science*, *5*(2), 1–17. doi:10.4018/IJSEIMS.2017070101

Bhuyan, D. (2018). Designing of a Twin Tube Shock Absorber: A Study in Reverse Engineering. In K. Kumar & J. Davim (Eds.), *Design and Optimization of Mechanical Engineering Products* (pp. 83–104). Hershey, PA: IGI Global. doi:10.4018/978-1-5225-3401-3.ch005

Blumberg, G. (2021). Blockchains for Use in Construction and Engineering Projects. In J. Underwood & M. Shelbourn (Eds.), *Handbook of Research on Driving Transformational Change in the Digital Built Environment* (pp. 179–208). IGI Global. https://doi.org/10.4018/978-1-7998-6600-8.ch008

Bolboaca, A. M. (2021). Considerations Regarding the Use of Fuel Cells in Combined Heat and Power for Stationary Applications. In G. Badea, R. Felseghi, & I. Aschilean (Eds.), *Hydrogen Fuel Cell Technology for Stationary Applications* (pp. 239–275). IGI Global. https://doi.org/10.4018/978-1-7998-4945-2.ch010

Burstein, L. (2021). Simulation Tool for Cable Design. In L. Burstein (Ed.), *Handbook of Research on Advancements in Manufacturing, Materials, and Mechanical Engineering* (pp. 54–74). IGI Global. https://doi.org/10.4018/978-1-7998-4939-1.ch003

Calderon, F. A., Giolo, E. G., Frau, C. D., Rengel, M. G., Rodriguez, H., Tornello, M., ... Gallucci, R. (2018). Seismic Microzonation and Site Effects Detection Through Microtremors Measures: A Review. In N. Ceryan (Ed.), *Handbook of Research on Trends and Digital Advances in Engineering Geology* (pp. 326–349). Hershey, PA: IGI Global. doi:10.4018/978-1-5225-2709-1.ch009

Ceryan, N., & Can, N. K. (2018). Prediction of The Uniaxial Compressive Strength of Rocks Materials. In N. Ceryan (Ed.), *Handbook of Research on Trends and Digital Advances in Engineering Geology* (pp. 31–96). Hershey, PA: IGI Global. doi:10.4018/978-1-5225-2709-1.ch002

Ceryan, S. (2018). Weathering Indices Used in Evaluation of the Weathering State of Rock Material. In N. Ceryan (Ed.), *Handbook of Research on Trends and Digital Advances in Engineering Geology* (pp. 132–186). Hershey, PA: IGI Global. doi:10.4018/978-1-5225-2709-1.ch004

Chen, H., Padilla, R. V., & Besarati, S. (2017). Supercritical Fluids and Their Applications in Power Generation. In L. Chen & Y. Iwamoto (Eds.), *Advanced Applications of Supercritical Fluids in Energy Systems* (pp. 369–402). Hershey, PA: IGI Global. doi:10.4018/978-1-5225-2047-4.ch012

Chen, H., Padilla, R. V., & Besarati, S. (2021). Supercritical Fluids and Their Applications in Power Generation. In L. Chen (Ed.), *Handbook of Research on Advancements in Supercritical Fluids Applications for Sustainable Energy Systems* (pp. 566–599). IGI Global. https://doi.org/10.4018/978-1-7998-5796-9.ch016

Chen, L. (2017). Principles, Experiments, and Numerical Studies of Supercritical Fluid Natural Circulation System. In L. Chen & Y. Iwamoto (Eds.), *Advanced Applications of Supercritical Fluids in Energy Systems* (pp. 136–187). Hershey, PA: IGI Global. doi:10.4018/978-1-5225-2047-4.ch005

Chen, L. (2021). Principles, Experiments, and Numerical Studies of Supercritical Fluid Natural Circulation System. In L. Chen (Ed.), *Handbook of Research on Advancements in Supercritical Fluids Applications for Sustainable Energy Systems* (pp. 219–269). IGI Global. https://doi.org/10.4018/978-1-7998-5796-9.ch007

Chiba, Y., Marif, Y., Henini, N., & Tlemcani, A. (2021). Modeling of Magnetic Refrigeration Device by Using Artificial Neural Networks Approach. *International Journal of Energy Optimization and Engineering*, *10*(4), 68–76. https://doi.org/10.4018/IJEOE.2021100105

Clementi, F., Di Sciascio, G., Di Sciascio, S., & Lenci, S. (2017). Influence of the Shear-Bending Interaction on the Global Capacity of Reinforced Concrete Frames: A Brief Overview of the New Perspectives. In V. Plevris, G. Kremmyda, & Y. Fahjan (Eds.), *Performance-Based Seismic Design of Concrete Structures and Infrastructures* (pp. 84–111). Hershey, PA: IGI Global. doi:10.4018/978-1-5225-2089-4.ch004

Codinhoto, R., Fialho, B. C., Pinti, L., & Fabricio, M. M. (2021). BIM and IoT for Facilities Management: Understanding Key Maintenance Issues. In J. Underwood & M. Shelbourn (Eds.), *Handbook of Research on Driving Transformational Change in the Digital Built Environment* (pp. 209–231). IGI Global. doi:10.4018/978-1-7998-6600-8.ch009

Cortés-Polo, D., Calle-Cancho, J., Carmona-Murillo, J., & González-Sánchez, J. (2017). Future Trends in Mobile-Fixed Integration for Next Generation Networks: Classification and Analysis. *International Journal of Vehicular Telematics and Infotainment Systems*, *1*(1), 33–53. doi:10.4018/IJVTIS.2017010103

Costa, H. G., Sheremetieff, F. H., & Araújo, E. A. (2022). Influence of Game-Based Methods in Developing Engineering Competences. In A. Alves & N. van Hattum-Janssen (Eds.), *Training Engineering Students for Modern Technological Advancement* (pp. 69–88). IGI Global. https://doi.org/10.4018/978-1-7998-8816-1.ch004

Cui, X., Zeng, S., Li, Z., Zheng, Q., Yu, X., & Han, B. (2018). Advanced Composites for Civil Engineering Infrastructures. In K. Kumar & J. Davim (Eds.), *Composites and Advanced Materials for Industrial Applications* (pp. 212–248). Hershey, PA: IGI Global. doi:10.4018/978-1-5225-5216-1.ch010

Dalgıç, S., & Kuşku, İ. (2018). Geological and Geotechnical Investigations in Tunneling. In N. Ceryan (Ed.), *Handbook of Research on Trends and Digital Advances in Engineering Geology* (pp. 482–529). Hershey, PA: IGI Global. doi:10.4018/978-1-5225-2709-1.ch014

Dang, C., & Hihara, E. (2021). Study on Cooling Heat Transfer of Supercritical Carbon Dioxide Applied to Transcritical Carbon Dioxide Heat Pump. In L. Chen (Ed.), *Handbook of Research on Advancements in Supercritical Fluids Applications for Sustainable Energy Systems* (pp. 451–493). IGI Global. https://doi.org/10.4018/978-1-7998-5796-9.ch013

Daus, Y., Kharchenko, V., & Yudaev, I. (2021). Research of Solar Energy Potential of Photovoltaic Installations on Enclosing Structures of Buildings. *International Journal of Energy Optimization and Engineering*, *10*(4), 18–34. https://doi.org/10.4018/IJEOE.2021100102

Daus, Y., Kharchenko, V., & Yudaev, I. (2021). Optimizing Layout of Distributed Generation Sources of Power Supply System of Agricultural Object. *International Journal of Energy Optimization and Engineering*, *10*(3), 70–84. https://doi.org/10.4018/IJEOE.2021070104

de la Varga, D., Soto, M., Arias, C. A., van Oirschot, D., Kilian, R., Pascual, A., & Álvarez, J. A. (2017). Constructed Wetlands for Industrial Wastewater Treatment and Removal of Nutrients. In Á. Val del Río, J. Campos Gómez, & A. Mosquera Corral (Eds.), *Technologies for the Treatment and Recovery of Nutrients from Industrial Wastewater* (pp. 202–230). Hershey, PA: IGI Global. doi:10.4018/978-1-5225-1037-6.ch008

Deb, S., Ammar, E. A., AlRajhi, H., Alsaidan, I., & Shariff, S. M. (2022). V2G Pilot Projects: Review and Lessons Learnt. In M. Alam, R. Pillai, & N. Murugesan (Eds.), *Developing Charging Infrastructure and Technologies for Electric Vehicles* (pp. 252–267). IGI Global. https://doi.org/10.4018/978-1-7998-6858-3.ch014

Dekhandji, F. Z., & Rais, M. C. (2021). A Comparative Study of Power Quality Monitoring Using Various Techniques. In A. Recioui & H. Bentarzi (Eds.), *Optimizing and Measuring Smart Grid Operation and Control* (pp. 259–288). IGI Global. https://doi.org/10.4018/978-1-7998-4027-5.ch011

Related References

Deperlioglu, O. (2018). Intelligent Techniques Inspired by Nature and Used in Biomedical Engineering. In U. Kose, G. Guraksin, & O. Deperlioglu (Eds.), *Nature-Inspired Intelligent Techniques for Solving Biomedical Engineering Problems* (pp. 51–77). Hershey, PA: IGI Global. doi:10.4018/978-1-5225-4769-3.ch003

Dhurpate, P. R., & Tang, H. (2021). Quantitative Analysis of the Impact of Inter-Line Conveyor Capacity for Throughput of Manufacturing Systems. *International Journal of Manufacturing, Materials, and Mechanical Engineering*, *11*(1), 1–17. https://doi.org/10.4018/IJMMME.2021010101

Dinkar, S., & Deep, K. (2021). A Survey of Recent Variants and Applications of Antlion Optimizer. *International Journal of Energy Optimization and Engineering*, *10*(2), 48–73. doi:10.4018/IJEOE.2021040103

Dixit, A. (2018). Application of Silica-Gel-Reinforced Aluminium Composite on the Piston of Internal Combustion Engine: Comparative Study of Silica-Gel-Reinforced Aluminium Composite Piston With Aluminium Alloy Piston. In K. Kumar & J. Davim (Eds.), *Composites and Advanced Materials for Industrial Applications* (pp. 63–98). Hershey, PA: IGI Global. doi:10.4018/978-1-5225-5216-1.ch004

Drabecki, M. P., & Kułak, K. B. (2021). Global Pandemics on European Electrical Energy Markets: Lessons Learned From the COVID-19 Outbreak. *International Journal of Energy Optimization and Engineering*, *10*(3), 24–46. https://doi.org/10.4018/IJEOE.2021070102

Dutta, M. M. (2021). Nanomaterials for Food and Agriculture. In M. Bhat, I. Wani, & S. Ashraf (Eds.), *Applications of Nanomaterials in Agriculture, Food Science, and Medicine* (pp. 75–97). IGI Global. doi:10.4018/978-1-7998-5563-7.ch004

Dutta, M. M., & Goswami, M. (2021). Coating Materials: Nano-Materials. In S. Roy & G. Bose (Eds.), *Advanced Surface Coating Techniques for Modern Industrial Applications* (pp. 1–30). IGI Global. doi:10.4018/978-1-7998-4870-7.ch001

Elsayed, A. M., Dakkama, H. J., Mahmoud, S., Al-Dadah, R., & Kaialy, W. (2017). Sustainable Cooling Research Using Activated Carbon Adsorbents and Their Environmental Impact. In T. Kobayashi (Ed.), *Applied Environmental Materials Science for Sustainability* (pp. 186–221). Hershey, PA: IGI Global. doi:10.4018/978-1-5225-1971-3.ch009

Ercanoglu, M., & Sonmez, H. (2018). General Trends and New Perspectives on Landslide Mapping and Assessment Methods. In N. Ceryan (Ed.), *Handbook of Research on Trends and Digital Advances in Engineering Geology* (pp. 350–379). Hershey, PA: IGI Global. doi:10.4018/978-1-5225-2709-1.ch010

Faroz, S. A., Pujari, N. N., Rastogi, R., & Ghosh, S. (2017). Risk Analysis of Structural Engineering Systems Using Bayesian Inference. In P. Samui, S. Chakraborty, & D. Kim (Eds.), *Modeling and Simulation Techniques in Structural Engineering* (pp. 390–424). Hershey, PA: IGI Global. doi:10.4018/978-1-5225-0588-4.ch014

Fekik, A., Hamida, M. L., Denoun, H., Azar, A. T., Kamal, N. A., Vaidyanathan, S., Bousbaine, A., & Benamrouche, N. (2022). Multilevel Inverter for Hybrid Fuel Cell/PV Energy Conversion System. In A. Fekik & N. Benamrouche (Eds.), *Modeling and Control of Static Converters for Hybrid Storage Systems* (pp. 233–270). IGI Global. https://doi.org/10.4018/978-1-7998-7447-8.ch009

Fekik, A., Hamida, M. L., Houassine, H., Azar, A. T., Kamal, N. A., Denoun, H., Vaidyanathan, S., & Sambas, A. (2022). Power Quality Improvement for Grid-Connected Photovoltaic Panels Using Direct Power Control. In A. Fekik & N. Benamrouche (Eds.), *Modeling and Control of Static Converters for Hybrid Storage Systems* (pp. 107–142). IGI Global. https://doi.org/10.4018/978-1-7998-7447-8.ch005

Fernando, P. R., Hamigah, T., Disne, S., Wickramasingha, G. G., & Sutharshan, A. (2018). The Evaluation of Engineering Properties of Low Cost Concrete Blocks by Partial Doping of Sand with Sawdust: Low Cost Sawdust Concrete Block. *International Journal of Strategic Engineering*, *1*(2), 26–42. doi:10.4018/IJoSE.2018070103

Ferro, G., Minciardi, R., Parodi, L., & Robba, M. (2022). Optimal Charging Management of Microgrid-Integrated Electric Vehicles. In M. Alam, R. Pillai, & N. Murugesan (Eds.), *Developing Charging Infrastructure and Technologies for Electric Vehicles* (pp. 133–155). IGI Global. https://doi.org/10.4018/978-1-7998-6858-3.ch007

Flumerfelt, S., & Green, C. (2022). Graduate Lean Leadership Education: A Case Study of a Program. In A. Alves & N. van Hattum-Janssen (Eds.), *Training Engineering Students for Modern Technological Advancement* (pp. 202–224). IGI Global. https://doi.org/10.4018/978-1-7998-8816-1.ch010

Galli, B. J. (2021). Implications of Economic Decision Making to the Project Manager. *International Journal of Strategic Engineering*, *4*(1), 19–32. https://doi.org/10.4018/IJoSE.2021010102

Gento, A. M., Pimentel, C., & Pascual, J. A. (2022). Teaching Circular Economy and Lean Management in a Learning Factory. In A. Alves & N. van Hattum-Janssen (Eds.), *Training Engineering Students for Modern Technological Advancement* (pp. 183–201). IGI Global. https://doi.org/10.4018/978-1-7998-8816-1.ch009

Ghosh, S., Mitra, S., Ghosh, S., & Chakraborty, S. (2017). Seismic Reliability Analysis in the Framework of Metamodelling Based Monte Carlo Simulation. In P. Samui, S. Chakraborty, & D. Kim (Eds.), *Modeling and Simulation Techniques in Structural Engineering* (pp. 192–208). Hershey, PA: IGI Global. doi:10.4018/978-1-5225-0588-4.ch006

Gil, M., & Otero, B. (2017). Learning Engineering Skills through Creativity and Collaboration: A Game-Based Proposal. In R. Alexandre Peixoto de Queirós & M. Pinto (Eds.), *Gamification-Based E-Learning Strategies for Computer Programming Education* (pp. 14–29). Hershey, PA: IGI Global. doi:10.4018/978-1-5225-1034-5.ch002

Gill, J., Ayre, M., & Mills, J. (2017). Revisioning the Engineering Profession: How to Make It Happen! In M. Gray & K. Thomas (Eds.), *Strategies for Increasing Diversity in Engineering Majors and Careers* (pp. 156–175). Hershey, PA: IGI Global. doi:10.4018/978-1-5225-2212-6.ch008

Godzhaev, Z., Senkevich, S., Kuzmin, V., & Melikov, I. (2021). Use of the Neural Network Controller of Sprung Mass to Reduce Vibrations From Road Irregularities. In P. Vasant, G. Weber, & W. Punurai (Ed.), *Research Advancements in Smart Technology, Optimization, and Renewable Energy* (pp. 69-87). IGI Global. https://doi.org/10.4018/978-1-7998-3970-5.ch005

Gomes de Gusmão, C. M. (2022). Digital Competencies and Transformation in Higher Education: Upskilling With Extension Actions. In A. Alves & N. van Hattum-Janssen (Eds.), *Training Engineering Students for Modern Technological Advancement* (pp. 313–328). IGI Global. https://doi.org/10.4018/978-1-7998-8816-1.ch015A

Goyal, N., Ram, M., & Kumar, P. (2017). Welding Process under Fault Coverage Approach for Reliability and MTTF. In M. Ram & J. Davim (Eds.), *Mathematical Concepts and Applications in Mechanical Engineering and Mechatronics* (pp. 222–245). Hershey, PA: IGI Global. doi:10.4018/978-1-5225-1639-2.ch011

Gray, M., & Lundy, C. (2017). Engineering Study Abroad: High Impact Strategy for Increasing Access. In M. Gray & K. Thomas (Eds.), *Strategies for Increasing Diversity in Engineering Majors and Careers* (pp. 42–59). Hershey, PA: IGI Global. doi:10.4018/978-1-5225-2212-6.ch003

Güler, O., & Varol, T. (2021). Fabrication of Functionally Graded Metal and Ceramic Powders Synthesized by Electroless Deposition. In S. Roy & G. Bose (Eds.), *Advanced Surface Coating Techniques for Modern Industrial Applications* (pp. 150–187). IGI Global. https://doi.org/10.4018/978-1-7998-4870-7.ch007

Guraksin, G. E. (2018). Internet of Things and Nature-Inspired Intelligent Techniques for the Future of Biomedical Engineering. In U. Kose, G. Guraksin, & O. Deperlioglu (Eds.), *Nature-Inspired Intelligent Techniques for Solving Biomedical Engineering Problems* (pp. 263–282). Hershey, PA: IGI Global. doi:10.4018/978-1-5225-4769-3.ch013

Hamida, M. L., Fekik, A., Denoun, H., Ardjal, A., & Bokhtache, A. A. (2022). Flying Capacitor Inverter Integration in a Renewable Energy System. In A. Fekik & N. Benamrouche (Eds.), *Modeling and Control of Static Converters for Hybrid Storage Systems* (pp. 287–306). IGI Global. https://doi.org/10.4018/978-1-7998-7447-8.ch011

Hasegawa, N., & Takahashi, Y. (2021). Control of Soap Bubble Ejection Robot Using Facial Expressions. *International Journal of Manufacturing, Materials, and Mechanical Engineering*, *11*(2), 1–16. https://doi.org/10.4018/IJMMME.2021040101

Hejazi, T., & Akbari, L. (2017). A Multiresponse Optimization Model for Statistical Design of Processes with Discrete Variables. In M. Ram & J. Davim (Eds.), *Mathematical Concepts and Applications in Mechanical Engineering and Mechatronics* (pp. 17–37). Hershey, PA: IGI Global. doi:10.4018/978-1-5225-1639-2.ch002

Hejazi, T., & Hejazi, A. (2017). Monte Carlo Simulation for Reliability-Based Design of Automotive Complex Subsystems. In M. Ram & J. Davim (Eds.), *Mathematical Concepts and Applications in Mechanical Engineering and Mechatronics* (pp. 177–200). Hershey, PA: IGI Global. doi:10.4018/978-1-5225-1639-2.ch009

Hejazi, T., & Poursabbagh, H. (2017). Reliability Analysis of Engineering Systems: An Accelerated Life Testing for Boiler Tubes. In M. Ram & J. Davim (Eds.), *Mathematical Concepts and Applications in Mechanical Engineering and Mechatronics* (pp. 154–176). Hershey, PA: IGI Global. doi:10.4018/978-1-5225-1639-2.ch008

Henao, J., Poblano-Salas, C. A., Vargas, F., Giraldo-Betancur, A. L., Corona-Castuera, J., & Sotelo-Mazón, O. (2021). Principles and Applications of Thermal Spray Coatings. In S. Roy & G. Bose (Eds.), *Advanced Surface Coating Techniques for Modern Industrial Applications* (pp. 31–70). IGI Global. https://doi.org/10.4018/978-1-7998-4870-7.ch002

Henao, J., & Sotelo, O. (2018). Surface Engineering at High Temperature: Thermal Cycling and Corrosion Resistance. In A. Pakseresht (Ed.), *Production, Properties, and Applications of High Temperature Coatings* (pp. 131–159). Hershey, PA: IGI Global. doi:10.4018/978-1-5225-4194-3.ch006

Hrnčič, M. K., Cör, D., & Knez, Ž. (2021). Supercritical Fluids as a Tool for Green Energy and Chemicals. In L. Chen (Ed.), *Handbook of Research on Advancements in Supercritical Fluids Applications for Sustainable Energy Systems* (pp. 761–791). IGI Global. doi:10.4018/978-1-7998-5796-9.ch021

Ibrahim, O., Erdem, S., & Gurbuz, E. (2021). Studying Physical and Chemical Properties of Graphene Oxide and Reduced Graphene Oxide and Their Applications in Sustainable Building Materials. In R. González-Lezcano (Ed.), *Advancements in Sustainable Architecture and Energy Efficiency* (pp. 221–238). IGI Global. https://doi.org/10.4018/978-1-7998-7023-4.ch010

Ihianle, I. K., Islam, S., Naeem, U., & Ebenuwa, S. H. (2021). Exploiting Patterns of Object Use for Human Activity Recognition. In A. Nwajana & I. Ihianle (Eds.), *Handbook of Research on 5G Networks and Advancements in Computing, Electronics, and Electrical Engineering* (pp. 382–401). IGI Global. https://doi.org/10.4018/978-1-7998-6992-4.ch015

Ijemaru, G. K., Ngharamike, E. T., Oleka, E. U., & Nwajana, A. O. (2021). An Energy-Efficient Model for Opportunistic Data Collection in IoV-Enabled SC Waste Management. In A. Nwajana & I. Ihianle (Eds.), *Handbook of Research on 5G Networks and Advancements in Computing, Electronics, and Electrical Engineering* (pp. 1–19). IGI Global. https://doi.org/10.4018/978-1-7998-6992-4.ch001

Ilori, O. O., Adetan, D. A., & Umoru, L. E. (2017). Effect of Cutting Parameters on the Surface Residual Stress of Face-Milled Pearlitic Ductile Iron. *International Journal of Materials Forming and Machining Processes*, *4*(1), 38–52. doi:10.4018/IJMFMP.2017010103

Imam, M. H., Tasadduq, I. A., Ahmad, A., Aldosari, F., & Khan, H. (2017). Automated Generation of Course Improvement Plans Using Expert System. *International Journal of Quality Assurance in Engineering and Technology Education*, *6*(1), 1–12. doi:10.4018/IJQAETE.2017010101

Injeti, S. K., & Kumar, T. V. (2018). A WDO Framework for Optimal Deployment of DGs and DSCs in a Radial Distribution System Under Daily Load Pattern to Improve Techno-Economic Benefits. *International Journal of Energy Optimization and Engineering*, *7*(2), 1–38. doi:10.4018/IJEOE.2018040101

Ishii, N., Anami, K., & Knisely, C. W. (2018). *Dynamic Stability of Hydraulic Gates and Engineering for Flood Prevention*. Hershey, PA: IGI Global. doi:10.4018/978-1-5225-3079-4

Iwamoto, Y., & Yamaguchi, H. (2021). Application of Supercritical Carbon Dioxide for Solar Water Heater. In L. Chen (Ed.), *Handbook of Research on Advancements in Supercritical Fluids Applications for Sustainable Energy Systems* (pp. 370–387). IGI Global. https://doi.org/10.4018/978-1-7998-5796-9.ch010

Jayapalan, S. (2018). A Review of Chemical Treatments on Natural Fibers-Based Hybrid Composites for Engineering Applications. In K. Kumar & J. Davim (Eds.), *Composites and Advanced Materials for Industrial Applications* (pp. 16–37). Hershey, PA: IGI Global. doi:10.4018/978-1-5225-5216-1.ch002

Kapetanakis, T. N., Vardiambasis, I. O., Ioannidou, M. P., & Konstantaras, A. I. (2021). Modeling Antenna Radiation Using Artificial Intelligence Techniques: The Case of a Circular Loop Antenna. In C. Nikolopoulos (Ed.), *Recent Trends on Electromagnetic Environmental Effects for Aeronautics and Space Applications* (pp. 186–225). IGI Global. https://doi.org/10.4018/978-1-7998-4879-0.ch007

Karkalos, N. E., Markopoulos, A. P., & Dossis, M. F. (2017). Optimal Model Parameters of Inverse Kinematics Solution of a 3R Robotic Manipulator Using ANN Models. *International Journal of Manufacturing, Materials, and Mechanical Engineering*, *7*(3), 20–40. doi:10.4018/IJMMME.2017070102

Kelly, M., Costello, M., Nicholson, G., & O'Connor, J. (2021). The Evolving Integration of BIM Into Built Environment Programmes in a Higher Education Institute. In J. Underwood & M. Shelbourn (Eds.), *Handbook of Research on Driving Transformational Change in the Digital Built Environment* (pp. 294–326). IGI Global. https://doi.org/10.4018/978-1-7998-6600-8.ch012

Kesimal, A., Karaman, K., Cihangir, F., & Ercikdi, B. (2018). Excavatability Assessment of Rock Masses for Geotechnical Studies. In N. Ceryan (Ed.), *Handbook of Research on Trends and Digital Advances in Engineering Geology* (pp. 231–256). Hershey, PA: IGI Global. doi:10.4018/978-1-5225-2709-1.ch006

Knoflacher, H. (2017). The Role of Engineers and Their Tools in the Transport Sector after Paradigm Change: From Assumptions and Extrapolations to Science. In H. Knoflacher & E. Ocalir-Akunal (Eds.), *Engineering Tools and Solutions for Sustainable Transportation Planning* (pp. 1–29). Hershey, PA: IGI Global. doi:10.4018/978-1-5225-2116-7.ch001

Kose, U. (2018). Towards an Intelligent Biomedical Engineering With Nature-Inspired Artificial Intelligence Techniques. In U. Kose, G. Guraksin, & O. Deperlioglu (Eds.), *Nature-Inspired Intelligent Techniques for Solving Biomedical Engineering Problems* (pp. 1–26). Hershey, PA: IGI Global. doi:10.4018/978-1-5225-4769-3.ch001

Kostić, S. (2018). A Review on Enhanced Stability Analyses of Soil Slopes Using Statistical Design. In N. Ceryan (Ed.), *Handbook of Research on Trends and Digital Advances in Engineering Geology* (pp. 446–481). Hershey, PA: IGI Global. doi:10.4018/978-1-5225-2709-1.ch013

Kumar, A., Patil, P. P., & Prajapati, Y. K. (2018). *Advanced Numerical Simulations in Mechanical Engineering*. Hershey, PA: IGI Global. doi:10.4018/978-1-5225-3722-9

Kumar, G. R., Rajyalakshmi, G., & Manupati, V. K. (2017). Surface Micro Patterning of Aluminium Reinforced Composite through Laser Peening. *International Journal of Manufacturing, Materials, and Mechanical Engineering*, *7*(4), 15–27. doi:10.4018/IJMMME.2017100102

Kumar, N., Basu, D. N., & Chen, L. (2021). Effect of Flow Acceleration and Buoyancy on Thermalhydraulics of sCO2 in Mini/Micro-Channel. In L. Chen (Ed.), *Handbook of Research on Advancements in Supercritical Fluids Applications for Sustainable Energy Systems* (pp. 161–182). IGI Global. doi:10.4018/978-1-7998-5796-9.ch005

Kumari, N., & Kumar, K. (2018). Fabrication of Orthotic Calipers With Epoxy-Based Green Composite. In K. Kumar & J. Davim (Eds.), *Composites and Advanced Materials for Industrial Applications* (pp. 157–176). Hershey, PA: IGI Global. doi:10.4018/978-1-5225-5216-1.ch008

Kuppusamy, R. R. (2018). Development of Aerospace Composite Structures Through Vacuum-Enhanced Resin Transfer Moulding Technology (VERTMTy): Vacuum-Enhanced Resin Transfer Moulding. In K. Kumar & J. Davim (Eds.), *Composites and Advanced Materials for Industrial Applications* (pp. 99–111). Hershey, PA: IGI Global. doi:10.4018/978-1-5225-5216-1.ch005

Kurganov, V. A., Zeigarnik, Y. A., & Maslakova, I. V. (2021). Normal and Deteriorated Heat Transfer Under Heating Turbulent Supercritical Pressure Coolants Flows in Round Tubes. In L. Chen (Ed.), *Handbook of Research on Advancements in Supercritical Fluids Applications for Sustainable Energy Systems* (pp. 494–532). IGI Global. https://doi.org/10.4018/978-1-7998-5796-9.ch014

Li, H., & Zhang, Y. (2021). Heat Transfer and Fluid Flow Modeling for Supercritical Fluids in Advanced Energy Systems. In L. Chen (Ed.), *Handbook of Research on Advancements in Supercritical Fluids Applications for Sustainable Energy Systems* (pp. 388–422). IGI Global. https://doi.org/10.4018/978-1-7998-5796-9.ch011

Loy, J., Howell, S., & Cooper, R. (2017). Engineering Teams: Supporting Diversity in Engineering Education. In M. Gray & K. Thomas (Eds.), *Strategies for Increasing Diversity in Engineering Majors and Careers* (pp. 106–129). Hershey, PA: IGI Global. doi:10.4018/978-1-5225-2212-6.ch006

Macher, G., Armengaud, E., Kreiner, C., Brenner, E., Schmittner, C., Ma, Z., ... Krammer, M. (2018). Integration of Security in the Development Lifecycle of Dependable Automotive CPS. In N. Druml, A. Genser, A. Krieg, M. Menghin, & A. Hoeller (Eds.), *Solutions for Cyber-Physical Systems Ubiquity* (pp. 383–423). Hershey, PA: IGI Global. doi:10.4018/978-1-5225-2845-6.ch015

Madhu, M. N., Singh, J. G., Mohan, V., & Ongsakul, W. (2021). Transmission Risk Optimization in Interconnected Systems: Risk-Adjusted Available Transfer Capability. In P. Vasant, G. Weber, & W. Punurai (Ed.), *Research Advancements in Smart Technology, Optimization, and Renewable Energy* (pp. 183-199). IGI Global. https://doi.org/10.4018/978-1-7998-3970-5.ch010

Mahendramani, G., & Lakshmana Swamy, N. (2018). Effect of Weld Groove Area on Distortion of Butt Welded Joints in Submerged Arc Welding. *International Journal of Manufacturing, Materials, and Mechanical Engineering*, *8*(2), 33–44. doi:10.4018/IJMMME.2018040103

Makropoulos, G., Koumaras, H., Setaki, F., Filis, K., Lutz, T., Montowtt, P., Tomaszewski, L., Dybiec, P., & Järvet, T. (2021). 5G and Unmanned Aerial Vehicles (UAVs) Use Cases: Analysis of the Ecosystem, Architecture, and Applications. In A. Nwajana & I. Ihianle (Eds.), *Handbook of Research on 5G Networks and Advancements in Computing, Electronics, and Electrical Engineering* (pp. 36–69). IGI Global. https://doi.org/10.4018/978-1-7998-6992-4.ch003

Meric, E. M., Erdem, S., & Gurbuz, E. (2021). Application of Phase Change Materials in Construction Materials for Thermal Energy Storage Systems in Buildings. In R. González-Lezcano (Ed.), *Advancements in Sustainable Architecture and Energy Efficiency* (pp. 1–20). IGI Global. https://doi.org/10.4018/978-1-7998-7023-4.ch001

Mihret, E. T., & Yitayih, K. A. (2021). Operation of VANET Communications: The Convergence of UAV System With LTE/4G and WAVE Technologies. *International Journal of Smart Vehicles and Smart Transportation*, *4*(1), 29–51. https://doi.org/10.4018/IJSVST.2021010103

Mir, M. A., Bhat, B. A., Sheikh, B. A., Rather, G. A., Mehraj, S., & Mir, W. R. (2021). Nanomedicine in Human Health Therapeutics and Drug Delivery: Nanobiotechnology and Nanobiomedicine. In M. Bhat, I. Wani, & S. Ashraf (Eds.), *Applications of Nanomaterials in Agriculture, Food Science, and Medicine* (pp. 229–251). IGI Global. doi:10.4018/978-1-7998-5563-7.ch013

Mohammadzadeh, S., & Kim, Y. (2017). Nonlinear System Identification of Smart Buildings. In P. Samui, S. Chakraborty, & D. Kim (Eds.), *Modeling and Simulation Techniques in Structural Engineering* (pp. 328–347). Hershey, PA: IGI Global. doi:10.4018/978-1-5225-0588-4.ch011

Molina, G. J., Aktaruzzaman, F., Soloiu, V., & Rahman, M. (2017). Design and Testing of a Jet-Impingement Instrument to Study Surface-Modification Effects by Nanofluids. *International Journal of Surface Engineering and Interdisciplinary Materials Science*, *5*(2), 43–61. doi:10.4018/IJSEIMS.2017070104

Moreno-Rangel, A., & Carrillo, G. (2021). Energy-Efficient Homes: A Heaven for Respiratory Illnesses. In R. González-Lezcano (Ed.), *Advancements in Sustainable Architecture and Energy Efficiency* (pp. 49–71). IGI Global. https://doi.org/10.4018/978-1-7998-7023-4.ch003

Msomi, V., & Jantjies, B. T. (2021). Correlative Analysis Between Tensile Properties and Tool Rotational Speeds of Friction Stir Welded Similar Aluminium Alloy Joints. *International Journal of Surface Engineering and Interdisciplinary Materials Science*, *9*(2), 58–78. https://doi.org/10.4018/IJSEIMS.2021070104

Muigai, M. N., Mwema, F. M., Akinlabi, E. T., & Obiko, J. O. (2021). Surface Engineering of Materials Through Weld-Based Technologies: An Overview. In S. Roy & G. Bose (Eds.), *Advanced Surface Coating Techniques for Modern Industrial Applications* (pp. 247–260). IGI Global. doi:10.4018/978-1-7998-4870-7.ch011

Mukherjee, A., Saeed, R. A., Dutta, S., & Naskar, M. K. (2017). Fault Tracking Framework for Software-Defined Networking (SDN). In C. Singhal & S. De (Eds.), *Resource Allocation in Next-Generation Broadband Wireless Access Networks* (pp. 247–272). Hershey, PA: IGI Global. doi:10.4018/978-1-5225-2023-8.ch011

Mukhopadhyay, A., Barman, T. K., & Sahoo, P. (2018). Electroless Nickel Coatings for High Temperature Applications. In K. Kumar & J. Davim (Eds.), *Composites and Advanced Materials for Industrial Applications* (pp. 297–331). Hershey, PA: IGI Global. doi:10.4018/978-1-5225-5216-1.ch013

Mwema, F. M., & Wambua, J. M. (2022). Machining of Poly Methyl Methacrylate (PMMA) and Other Olymeric Materials: A Review. In K. Kumar, B. Babu, & J. Davim (Eds.), *Handbook of Research on Advancements in the Processing, Characterization, and Application of Lightweight Materials* (pp. 363–379). IGI Global. https://doi.org/10.4018/978-1-7998-7864-3.ch016

Mykhailyshyn, R., Savkiv, V., Boyko, I., Prada, E., & Virgala, I. (2021). Substantiation of Parameters of Friction Elements of Bernoulli Grippers With a Cylindrical Nozzle. *International Journal of Manufacturing, Materials, and Mechanical Engineering, 11*(2), 17–39. https://doi.org/10.4018/IJMMME.2021040102

Náprstek, J., & Fischer, C. (2017). Dynamic Stability and Post-Critical Processes of Slender Auto-Parametric Systems. In V. Plevris, G. Kremmyda, & Y. Fahjan (Eds.), *Performance-Based Seismic Design of Concrete Structures and Infrastructures* (pp. 128–171). Hershey, PA: IGI Global. doi:10.4018/978-1-5225-2089-4.ch006

Nautiyal, L., Shivach, P., & Ram, M. (2018). Optimal Designs by Means of Genetic Algorithms. In M. Ram & J. Davim (Eds.), *Soft Computing Techniques and Applications in Mechanical Engineering* (pp. 151–161). Hershey, PA: IGI Global. doi:10.4018/978-1-5225-3035-0.ch007

Nazir, R. (2017). Advanced Nanomaterials for Water Engineering and Treatment: Nano-Metal Oxides and Their Nanocomposites. In T. Saleh (Ed.), *Advanced Nanomaterials for Water Engineering, Treatment, and Hydraulics* (pp. 84–126). Hershey, PA: IGI Global. doi:10.4018/978-1-5225-2136-5.ch005

Nikolopoulos, C. D. (2021). Recent Advances on Measuring and Modeling ELF-Radiated Emissions for Space Applications. In C. Nikolopoulos (Ed.), *Recent Trends on Electromagnetic Environmental Effects for Aeronautics and Space Applications* (pp. 1–38). IGI Global. https://doi.org/10.4018/978-1-7998-4879-0.ch001

Nogueira, A. F., Ribeiro, J. C., Fernández de Vega, F., & Zenha-Rela, M. A. (2018). Evolutionary Approaches to Test Data Generation for Object-Oriented Software: Overview of Techniques and Tools. In M. Khosrow-Pour, D.B.A. (Ed.), Incorporating Nature-Inspired Paradigms in Computational Applications (pp. 162-194). Hershey, PA: IGI Global. https://doi.org/ doi:10.4018/978-1-5225-5020-4.ch006

Nwajana, A. O., Obi, E. R., Ijemaru, G. K., Oleka, E. U., & Anthony, D. C. (2021). Fundamentals of RF/Microwave Bandpass Filter Design. In A. Nwajana & I. Ihianle (Eds.), *Handbook of Research on 5G Networks and Advancements in Computing, Electronics, and Electrical Engineering* (pp. 149–164). IGI Global. https://doi.org/10.4018/978-1-7998-6992-4.ch005

Ogbodo, E. A. (2021). Comparative Study of Transmission Line Junction vs. Asynchronously Coupled Junction Diplexers. In A. Nwajana & I. Ihianle (Eds.), *Handbook of Research on 5G Networks and Advancements in Computing, Electronics, and Electrical Engineering* (pp. 326–336). IGI Global. https://doi.org/10.4018/978-1-7998-6992-4.ch013

Orosa, J. A., Vergara, D., Fraguela, F., & Masdías-Bonome, A. (2021). Statistical Understanding and Optimization of Building Energy Consumption and Climate Change Consequences. In R. González-Lezcano (Ed.), *Advancements in Sustainable Architecture and Energy Efficiency* (pp. 195–220). IGI Global. https://doi.org/10.4018/978-1-7998-7023-4.ch009

Osho, M. B. (2018). Industrial Enzyme Technology: Potential Applications. In S. Bharati & P. Chaurasia (Eds.), *Research Advancements in Pharmaceutical, Nutritional, and Industrial Enzymology* (pp. 375–394). Hershey, PA: IGI Global. doi:10.4018/978-1-5225-5237-6.ch017

Ouadi, A., & Zitouni, A. (2021). Phasor Measurement Improvement Using Digital Filter in a Smart Grid. In A. Recioui & H. Bentarzi (Eds.), *Optimizing and Measuring Smart Grid Operation and Control* (pp. 100–117). IGI Global. https://doi.org/10.4018/978-1-7998-4027-5.ch005

Padmaja, P., & Marutheswar, G. (2017). Certain Investigation on Secured Data Transmission in Wireless Sensor Networks. *International Journal of Mobile Computing and Multimedia Communications*, *8*(1), 48–61. doi:10.4018/IJMCMC.2017010104

Palmer, S., & Hall, W. (2017). An Evaluation of Group Work in First-Year Engineering Design Education. In R. Tucker (Ed.), *Collaboration and Student Engagement in Design Education* (pp. 145–168). Hershey, PA: IGI Global. doi:10.4018/978-1-5225-0726-0.ch007

Panchenko, V. (2021). Prospects for Energy Supply of the Arctic Zone Objects of Russia Using Frost-Resistant Solar Modules. In P. Vasant, G. Weber, & W. Punurai (Eds.), *Research Advancements in Smart Technology, Optimization, and Renewable Energy* (pp. 149-169). IGI Global. https://doi.org/10.4018/978-1-7998-3970-5.ch008

Panchenko, V. (2021). Photovoltaic Thermal Module With Paraboloid Type Solar Concentrators. *International Journal of Energy Optimization and Engineering*, *10*(2), 1–23. https://doi.org/10.4018/IJEOE.2021040101

Pandey, K., & Datta, S. (2021). Dry Machining of Inconel 825 Superalloys: Performance of Tool Inserts (Carbide, Cermet, and SiAlON). *International Journal of Manufacturing, Materials, and Mechanical Engineering*, *11*(4), 26–39. doi:10.4018/IJMMME.2021100102

Panneer, R. (2017). Effect of Composition of Fibers on Properties of Hybrid Composites. *International Journal of Manufacturing, Materials, and Mechanical Engineering*, *7*(4), 28–43. doi:10.4018/IJMMME.2017100103

Pany, C. (2021). Estimation of Correct Long-Seam Mismatch Using FEA to Compare the Measured Strain in a Non-Destructive Testing of a Pressurant Tank: A Reverse Problem. *International Journal of Smart Vehicles and Smart Transportation*, *4*(1), 16–28. doi:10.4018/IJSVST.2021010102

Paul, S., & Roy, P. (2018). Optimal Design of Power System Stabilizer Using a Novel Evolutionary Algorithm. *International Journal of Energy Optimization and Engineering*, *7*(3), 24–46. doi:10.4018/IJEOE.2018070102

Paul, S., & Roy, P. K. (2021). Oppositional Differential Search Algorithm for the Optimal Tuning of Both Single Input and Dual Input Power System Stabilizer. In P. Vasant, G. Weber, & W. Punurai (Eds.), *Research Advancements in Smart Technology, Optimization, and Renewable Energy* (pp. 256-282). IGI Global. https://doi.org/10.4018/978-1-7998-3970-5.ch013

Pavaloiu, A. (2018). Artificial Intelligence Ethics in Biomedical-Engineering-Oriented Problems. In U. Kose, G. Guraksin, & O. Deperlioglu (Eds.), *Nature-Inspired Intelligent Techniques for Solving Biomedical Engineering Problems* (pp. 219–231). Hershey, PA: IGI Global. doi:10.4018/978-1-5225-4769-3.ch010

Pioro, I., Mahdi, M., & Popov, R. (2017). Application of Supercritical Pressures in Power Engineering. In L. Chen & Y. Iwamoto (Eds.), *Advanced Applications of Supercritical Fluids in Energy Systems* (pp. 404–457). Hershey, PA: IGI Global. doi:10.4018/978-1-5225-2047-4.ch013

Plaksina, T., & Gildin, E. (2017). Rigorous Integrated Evolutionary Workflow for Optimal Exploitation of Unconventional Gas Assets. *International Journal of Energy Optimization and Engineering*, *6*(1), 101–122. doi:10.4018/IJEOE.2017010106

Popat, J., Kakadiya, H., Tak, L., Singh, N. K., Majeed, M. A., & Mahajan, V. (2021). Reliability of Smart Grid Including Cyber Impact: A Case Study. In R. Singh, A. Singh, A. Dwivedi, & P. Nagabhushan (Eds.), *Computational Methodologies for Electrical and Electronics Engineers* (pp. 163–174). IGI Global. https://doi.org/10.4018/978-1-7998-3327-7.ch013

Quiza, R., La Fé-Perdomo, I., Rivas, M., & Ramtahalsing, V. (2021). Triple Bottom Line-Focused Optimization of Oblique Turning Processes Based on Hybrid Modeling: A Study Case on AISI 1045 Steel Turning. In L. Burstein (Ed.), *Handbook of Research on Advancements in Manufacturing, Materials, and Mechanical Engineering* (pp. 215–241). IGI Global. https://doi.org/10.4018/978-1-7998-4939-1.ch010

Rahmani, M. K. (2022). Blockchain Technology: Principles and Algorithms. In S. Khan, M. Syed, R. Hammad, & A. Bushager (Eds.), *Blockchain Technology and Computational Excellence for Society 5.0* (pp. 16–27). IGI Global. https://doi.org/10.4018/978-1-7998-8382-1.ch002

Ramdani, N., & Azibi, M. (2018). Polymer Composite Materials for Microelectronics Packaging Applications: Composites for Microelectronics Packaging. In K. Kumar & J. Davim (Eds.), *Composites and Advanced Materials for Industrial Applications* (pp. 177–211). Hershey, PA: IGI Global. doi:10.4018/978-1-5225-5216-1.ch009

Ramesh, M., Garg, R., & Subrahmanyam, G. V. (2017). Investigation of Influence of Quenching and Annealing on the Plane Fracture Toughness and Brittle to Ductile Transition Temperature of the Zinc Coated Structural Steel Materials. *International Journal of Surface Engineering and Interdisciplinary Materials Science*, *5*(2), 33–42. doi:10.4018/IJSEIMS.2017070103

Robinson, J., & Beneroso, D. (2022). Project-Based Learning in Chemical Engineering: Curriculum and Assessment, Culture and Learning Spaces. In A. Alves & N. van Hattum-Janssen (Eds.), *Training Engineering Students for Modern Technological Advancement* (pp. 1–19). IGI Global. https://doi.org/10.4018/978-1-7998-8816-1.ch001

Rondon, B. (2021). Experimental Characterization of Admittance Meter With Crude Oil Emulsions. *International Journal of Electronics, Communications, and Measurement Engineering*, *10*(2), 51–59. https://doi.org/10.4018/IJECME.2021070104

Rudolf, S., Biryuk, V. V., & Volov, V. (2018). Vortex Effect, Vortex Power: Technology of Vortex Power Engineering. In V. Kharchenko & P. Vasant (Eds.), *Handbook of Research on Renewable Energy and Electric Resources for Sustainable Rural Development* (pp. 500–533). Hershey, PA: IGI Global. doi:10.4018/978-1-5225-3867-7.ch021

Sah, A., Bhadula, S. J., Dumka, A., & Rawat, S. (2018). A Software Engineering Perspective for Development of Enterprise Applications. In A. Elçi (Ed.), *Handbook of Research on Contemporary Perspectives on Web-Based Systems* (pp. 1–23). Hershey, PA: IGI Global. doi:10.4018/978-1-5225-5384-7.ch001

Sahli, Y., Zitouni, B., & Hocine, B. M. (2021). Three-Dimensional Numerical Study of Overheating of Two Intermediate Temperature P-AS-SOFC Geometrical Configurations. In G. Badea, R. Felseghi, & I. Aschilean (Eds.), *Hydrogen Fuel Cell Technology for Stationary Applications* (pp. 186–222). IGI Global. https://doi.org/10.4018/978-1-7998-4945-2.ch008

Sahoo, P., & Roy, S. (2017). Tribological Behavior of Electroless Ni-P, Ni-P-W and Ni-P-Cu Coatings: A Comparison. *International Journal of Surface Engineering and Interdisciplinary Materials Science*, *5*(1), 1–15. doi:10.4018/IJSEIMS.2017010101

Sahoo, S. (2018). Laminated Composite Hypar Shells as Roofing Units: Static and Dynamic Behavior. In K. Kumar & J. Davim (Eds.), *Composites and Advanced Materials for Industrial Applications* (pp. 249–269). Hershey, PA: IGI Global. doi:10.4018/978-1-5225-5216-1.ch011

Sahu, H., & Hungyo, M. (2018). Introduction to SDN and NFV. In A. Dumka (Ed.), *Innovations in Software-Defined Networking and Network Functions Virtualization* (pp. 1–25). Hershey, PA: IGI Global. doi:10.4018/978-1-5225-3640-6.ch001

Salem, A. M., & Shmelova, T. (2018). Intelligent Expert Decision Support Systems: Methodologies, Applications, and Challenges. In T. Shmelova, Y. Sikirda, N. Rizun, A. Salem, & Y. Kovalyov (Eds.), *Socio-Technical Decision Support in Air Navigation Systems: Emerging Research and Opportunities* (pp. 215–242). Hershey, PA: IGI Global. doi:10.4018/978-1-5225-3108-1.ch007

Samal, M. (2017). FE Analysis and Experimental Investigation of Cracked and Un-Cracked Thin-Walled Tubular Components to Evaluate Mechanical and Fracture Properties. In P. Samui, S. Chakraborty, & D. Kim (Eds.), *Modeling and Simulation Techniques in Structural Engineering* (pp. 266–293). Hershey, PA: IGI Global. doi:10.4018/978-1-5225-0588-4.ch009

Samal, M., & Balakrishnan, K. (2017). Experiments on a Ring Tension Setup and FE Analysis to Evaluate Transverse Mechanical Properties of Tubular Components. In P. Samui, S. Chakraborty, & D. Kim (Eds.), *Modeling and Simulation Techniques in Structural Engineering* (pp. 91–115). Hershey, PA: IGI Global. doi:10.4018/978-1-5225-0588-4.ch004

Samarasinghe, D. A., & Wood, E. (2021). Innovative Digital Technologies. In J. Underwood & M. Shelbourn (Eds.), *Handbook of Research on Driving Transformational Change in the Digital Built Environment* (pp. 142–163). IGI Global. https://doi.org/10.4018/978-1-7998-6600-8.ch006

Sawant, S. (2018). Deep Learning and Biomedical Engineering. In U. Kose, G. Guraksin, & O. Deperlioglu (Eds.), *Nature-Inspired Intelligent Techniques for Solving Biomedical Engineering Problems* (pp. 283–296). Hershey, PA: IGI Global. doi:10.4018/978-1-5225-4769-3.ch014

Schulenberg, T. (2021). Energy Conversion Using the Supercritical Steam Cycle. In L. Chen (Ed.), *Handbook of Research on Advancements in Supercritical Fluids Applications for Sustainable Energy Systems* (pp. 659–681). IGI Global. doi:10.4018/978-1-7998-5796-9.ch018

Sezgin, H., & Berkalp, O. B. (2018). Textile-Reinforced Composites for the Automotive Industry. In K. Kumar & J. Davim (Eds.), *Composites and Advanced Materials for Industrial Applications* (pp. 129–156). Hershey, PA: IGI Global. doi:10.4018/978-1-5225-5216-1.ch007

Shaaban, A. A., & Shehata, O. M. (2021). Combining Response Surface Method and Metaheuristic Algorithms for Optimizing SPIF Process. *International Journal of Manufacturing, Materials, and Mechanical Engineering*, *11*(4), 1–25. https://doi.org/10.4018/IJMMME.2021100101

Shafaati Shemami, M., & Sefid, M. (2022). Implementation and Demonstration of Electric Vehicle-to-Home (V2H) Application: A Case Study. In M. Alam, R. Pillai, & N. Murugesan (Eds.), *Developing Charging Infrastructure and Technologies for Electric Vehicles* (pp. 268–293). IGI Global. https://doi.org/10.4018/978-1-7998-6858-3.ch015

Shah, M. Z., Gazder, U., Bhatti, M. S., & Hussain, M. (2018). Comparative Performance Evaluation of Effects of Modifier in Asphaltic Concrete Mix. *International Journal of Strategic Engineering*, *1*(2), 13–25. doi:10.4018/IJoSE.2018070102

Sharma, N., & Kumar, K. (2018). Fabrication of Porous NiTi Alloy Using Organic Binders. In K. Kumar & J. Davim (Eds.), *Composites and Advanced Materials for Industrial Applications* (pp. 38–62). Hershey, PA: IGI Global. doi:10.4018/978-1-5225-5216-1.ch003

Shivach, P., Nautiyal, L., & Ram, M. (2018). Applying Multi-Objective Optimization Algorithms to Mechanical Engineering. In M. Ram & J. Davim (Eds.), *Soft Computing Techniques and Applications in Mechanical Engineering* (pp. 287–301). Hershey, PA: IGI Global. doi:10.4018/978-1-5225-3035-0.ch014

Shmelova, T. (2018). Stochastic Methods for Estimation and Problem Solving in Engineering: Stochastic Methods of Decision Making in Aviation. In S. Kadry (Ed.), *Stochastic Methods for Estimation and Problem Solving in Engineering* (pp. 139–160). Hershey, PA: IGI Global. doi:10.4018/978-1-5225-5045-7.ch006

Siero González, L. R., & Romo Vázquez, A. (2017). Didactic Sequences Teaching Mathematics for Engineers With Focus on Differential Equations. In M. Ramírez-Montoya (Ed.), *Handbook of Research on Driving STEM Learning With Educational Technologies* (pp. 129–151). Hershey, PA: IGI Global. doi:10.4018/978-1-5225-2026-9.ch007

Sim, M. S., You, K. Y., Esa, F., & Chan, Y. L. (2021). Nanostructured Electromagnetic Metamaterials for Sensing Applications. In M. Bhat, I. Wani, & S. Ashraf (Eds.), *Applications of Nanomaterials in Agriculture, Food Science, and Medicine* (pp. 141–164). IGI Global. https://doi.org/10.4018/978-1-7998-5563-7.ch009

Singh, R., & Dutta, S. (2018). Visible Light Active Nanocomposites for Photocatalytic Applications. In K. Kumar & J. Davim (Eds.), *Composites and Advanced Materials for Industrial Applications* (pp. 270–296). Hershey, PA: IGI Global. doi:10.4018/978-1-5225-5216-1.ch012

Skripov, P. V., Yampol'skiy, A. D., & Rutin, S. B. (2021). High-Power Heat Transfer in Supercritical Fluids: Microscale Times and Sizes. In L. Chen (Ed.), *Handbook of Research on Advancements in Supercritical Fluids Applications for Sustainable Energy Systems* (pp. 424–450). IGI Global. https://doi.org/10.4018/978-1-7998-5796-9.ch012

Sözbilir, H., Özkaymak, Ç., Uzel, B., & Sümer, Ö. (2018). Criteria for Surface Rupture Microzonation of Active Faults for Earthquake Hazards in Urban Areas. In N. Ceryan (Ed.), *Handbook of Research on Trends and Digital Advances in Engineering Geology* (pp. 187–230). Hershey, PA: IGI Global. doi:10.4018/978-1-5225-2709-1.ch005

Stanciu, I. (2018). Stochastic Methods in Microsystems Engineering. In S. Kadry (Ed.), *Stochastic Methods for Estimation and Problem Solving in Engineering* (pp. 161–176). Hershey, PA: IGI Global. doi:10.4018/978-1-5225-5045-7.ch007

Strebkov, D., Nekrasov, A., Trubnikov, V., & Nekrasov, A. (2018). Single-Wire Resonant Electric Power Systems for Renewable-Based Electric Grid. In V. Kharchenko & P. Vasant (Eds.), *Handbook of Research on Renewable Energy and Electric Resources for Sustainable Rural Development* (pp. 449–474). Hershey, PA: IGI Global. doi:10.4018/978-1-5225-3867-7.ch019

Sukhyy, K., Belyanovskaya, E., & Sukhyy, M. (2021). *Basic Principles for Substantiation of Working Pair Choice*. IGI Global. doi:10.4018/978-1-7998-4432-7.ch002

Suri, M. S., & Kaliyaperumal, D. (2022). Extension of Aspiration Level Model for Optimal Planning of Fast Charging Stations. In A. Fekik & N. Benamrouche (Eds.), *Modeling and Control of Static Converters for Hybrid Storage Systems* (pp. 91–106). IGI Global. https://doi.org/10.4018/978-1-7998-7447-8.ch004

Tallet, E., Gledson, B., Rogage, K., Thompson, A., & Wiggett, D. (2021). Digitally-Enabled Design Management. In J. Underwood & M. Shelbourn (Eds.), *Handbook of Research on Driving Transformational Change in the Digital Built Environment* (pp. 63–89). IGI Global. https://doi.org/10.4018/978-1-7998-6600-8.ch003

Terki, A., & Boubertakh, H. (2021). A New Hybrid Binary-Real Coded Cuckoo Search and Tabu Search Algorithm for Solving the Unit-Commitment Problem. *International Journal of Energy Optimization and Engineering, 10*(2), 104–119. https://doi.org/10.4018/IJEOE.2021040105

Tüdeş, Ş., Kumlu, K. B., & Ceryan, S. (2018). Integration Between Urban Planning and Natural Hazards For Resilient City. In N. Ceryan (Ed.), *Handbook of Research on Trends and Digital Advances in Engineering Geology* (pp. 591–630). Hershey, PA: IGI Global. doi:10.4018/978-1-5225-2709-1.ch017

Ulamis, K. (2018). Soil Liquefaction Assessment by Anisotropic Cyclic Triaxial Test. In N. Ceryan (Ed.), *Handbook of Research on Trends and Digital Advances in Engineering Geology* (pp. 631–664). Hershey, PA: IGI Global. doi:10.4018/978-1-5225-2709-1.ch018

Valente, M., & Milani, G. (2017). Seismic Assessment and Retrofitting of an Under-Designed RC Frame Through a Displacement-Based Approach. In V. Plevris, G. Kremmyda, & Y. Fahjan (Eds.), *Performance-Based Seismic Design of Concrete Structures and Infrastructures* (pp. 36–58). Hershey, PA: IGI Global. doi:10.4018/978-1-5225-2089-4.ch002

Vargas-Bernal, R. (2021). Advances in Electromagnetic Environmental Shielding for Aeronautics and Space Applications. In C. Nikolopoulos (Ed.), *Recent Trends on Electromagnetic Environmental Effects for Aeronautics and Space Applications* (pp. 80–96). IGI Global. https://doi.org/10.4018/978-1-7998-4879-0.ch003

Vasant, P. (2018). A General Medical Diagnosis System Formed by Artificial Neural Networks and Swarm Intelligence Techniques. In U. Kose, G. Guraksin, & O. Deperlioglu (Eds.), *Nature-Inspired Intelligent Techniques for Solving Biomedical Engineering Problems* (pp. 130–145). Hershey, PA: IGI Global. doi:10.4018/978-1-5225-4769-3.ch006

Verner, C. M., & Sarwar, D. (2021). Avoiding Project Failure and Achieving Project Success in NHS IT System Projects in the United Kingdom. *International Journal of Strategic Engineering*, *4*(1), 33–54. https://doi.org/10.4018/IJoSE.2021010103

Verrollot, J., Tolonen, A., Harkonen, J., & Haapasalo, H. J. (2018). Challenges and Enablers for Rapid Product Development. *International Journal of Applied Industrial Engineering*, *5*(1), 25–49. doi:10.4018/IJAIE.2018010102

Wan, A. C., Zulu, S. L., & Khosrow-Shahi, F. (2021). Industry Views on BIM for Site Safety in Hong Kong. In J. Underwood & M. Shelbourn (Eds.), *Handbook of Research on Driving Transformational Change in the Digital Built Environment* (pp. 120–140). IGI Global. https://doi.org/10.4018/978-1-7998-6600-8.ch005

Yardimci, A. G., & Karpuz, C. (2018). Fuzzy Rock Mass Rating: Soft-Computing-Aided Preliminary Stability Analysis of Weak Rock Slopes. In N. Ceryan (Ed.), *Handbook of Research on Trends and Digital Advances in Engineering Geology* (pp. 97–131). Hershey, PA: IGI Global. doi:10.4018/978-1-5225-2709-1.ch003

You, K. Y. (2021). Development Electronic Design Automation for RF/Microwave Antenna Using MATLAB GUI. In A. Nwajana & I. Ihianle (Eds.), *Handbook of Research on 5G Networks and Advancements in Computing, Electronics, and Electrical Engineering* (pp. 70–148). IGI Global. https://doi.org/10.4018/978-1-7998-6992-4.ch004

Yousefi, Y., Gratton, P., & Sarwar, D. (2021). Investigating the Opportunities to Improve the Thermal Performance of a Case Study Building in London. *International Journal of Strategic Engineering*, *4*(1), 1–18. https://doi.org/10.4018/IJoSE.2021010101

Zindani, D., & Kumar, K. (2018). Industrial Applications of Polymer Composite Materials. In K. Kumar & J. Davim (Eds.), *Composites and Advanced Materials for Industrial Applications* (pp. 1–15). Hershey, PA: IGI Global. doi:10.4018/978-1-5225-5216-1.ch001

Zindani, D., Maity, S. R., & Bhowmik, S. (2018). A Decision-Making Approach for Material Selection of Polymeric Composite Bumper Beam. In K. Kumar & J. Davim (Eds.), *Composites and Advanced Materials for Industrial Applications* (pp. 112–128). Hershey, PA: IGI Global. doi:10.4018/978-1-5225-5216-1.ch006

About the Contributors

Victor Shikuku is a Lecturer in the Department of Physical Sciences at Kaimosi Friends University (KAFU) in Kenya. He holds a Ph.D. in Chemistry from Maseno University (Kenya). In 2017-2018 he was a visiting scholar at CSIR-National Metallurgical Laboratory (NML) in India under the prestigious C.V Raman International Fellowship for African Researchers. In 2019, he undertook a postdoctoral research visit at Trier University in Germany, supported by the Alexander von Humboldt Foundation. He serves as the research group leader on materials research for water treatment at Kaimosi Friends University. He is also a member of the Technical Committee (TC) on Water Quality at the Kenya Bureau of Standards (KEBS). His research entails low-carbon cementitious materials (geopolymers) and their applications in water treatment and as building materials, carbonaceous materials for water purification, environmental chemistry, emerging contaminants, food safety, biodegradation of pesticides and antibiotics in soil and associated antimicrobial resistance patterns, and climate change adaptation. Dr. Shikuku has published over 40 peer-reviewed articles including, book chapters and an edited book and is a well-cited scholar.

* * *

Fatima-Zahra Akensous works on the resilience of date palm to drought, using biostimulants. She is also an Artificial Intelligence enthusiast.

Siddhartha Kumar Arjaria is B.E., MTech, and Ph.D in computer science and Engineering. He has over 14 years of experience of teaching in institutes like BITS pilani Goa Campus, IIITM Gwalior. Currently, He is working as an Assistant Professor - IT Department in Rajkiya Engineering College, Banda,U.P, India. He has published over 20 papers/Book chapters in various national/international journals/Conference Proceedings. He is a reviewer of many international journals/conferences. He has organized many FDPs/Workshops as Co-Ordinator/Convenor. His research interests include Machine learning, Data Analytics and text mining.

C.V. Suresh Babu is a pioneer in content development. A true entrepreneur, he founded Anniyappa Publications, a company that is highly active in publishing books related to Computer Science and Management. Dr. C.V. Suresh Babu has also ventured into SB Institute, a center for knowledge transfer. He holds a Ph.D. in Engineering Education from the National Institute of Technical Teachers Training & Research in Chennai, along with seven master's degrees in various disciplines such as Engineering, Computer Applications, Management, Commerce, Economics, Psychology, Law, and Education. Additionally, he has UGC-NET/SET qualifications in the fields of Computer Science, Management, Commerce, and Education. Currently, Dr. C.V. Suresh Babu is a Professor in the Department of Information Technology at the School of Computing Science, Hindustan Institute of Technology and Science (Hindustan University) in Padur, Chennai, Tamil Nadu, India. For more information, you can visit his personal blog at .

Shailendra Badal is educated in chemistry at the University of Allahabad, Prayagraj, U.P., India and is currently Assistant Professor and Head, Department of Applied science and Humanities, Rajkiya Engineering college, Atarra, Banda, U.P. India. He has over 24 year teaching experience and Dr Badal has published/presented more than 40 papers in reputed International/National Journals, Conferences and Seminars, etc.

Sampath Boopathi completed his undergraduate in Mechanical Engineering and postgraduate in the field of Computer-Aided Design. He completed his Ph.D. from Anna University and his field of research includes Manufacturing and optimization. He published 120 more research articles in Internationally Peer-reviewed journals, one Patent grant, and three published patents. He has 16 more years of academic and research experiences in the various Engineering Colleges in Tamilnadu, India.

Yadavamuthiah K. is a passionate researcher and student at Hindustan Institute of Technology and Science, specializing in technology applications for agriculture and water management. With expertise in AI, ML, IoT, and Robotics, Yadavamuthiah's primary focus is on developing innovative and sustainable solutions.

KamalaDevi M. received the Master of Technology in Computer Science and Engineering from SASTRA University, India in 2010. Currently she is an Assistant professor at SASTRA University, India. Her research interests include Data Mining and Warehousing and Distributed computing. Published few research articles in reputed journals in data mining domain.

Suryanshi Mishra received the M.Sc. degree in Mathematics from the University of Allahabad, India. She is currently a research scholar with the Department of Mathematics and Statistics at Sam Higginbottom University of Agriculture, Technology and Sciences (SHUATS), Prayagraj, India. Her research interests include Optimization and Machine Learning.

Umamaheswari P. received M.Tech in Computer Science from SASTRA Deemed University, Thanjavur, Tamil Nadu, and India in 2011 and M.B.A from Alagappa University, Karaikudi, Tamil Nadu, and India in the year 2006. She is pursuing a Doctoral degree in Computer Science and Engineering from SASTRA Deemed University. She has 17 years of teaching experience for UG and PG courses in Computer Science and presently working as an Assistant Professor in the department of computer science and engineering at SASTRA University and SRC Campus, Kumbakonam. Her research interest is in Data Mining, Machine Learning, and Deep learning. She published research papers in Book chapters, International Conferences, Scopus and Science Citation Indexed Journals.

Abhishek Singh Rathore is working as Associate Professor in Shri Vaishnav Vidyapeeth Vishwavidyalaya, Indore, India. He is alumni of IET-DAVV, Indore, India and did Ph.D from National Institute of Technology, Bhopal, India. His area of research is machine learning and healthcare.

Abirami S. is a passionate researcher and student at Hindustan Institute of Technology and Science, specializing in technology applications for water management. With the knowledge in AI, ML, and IoT. Abirami's primary focus is on developing innovative and sustainable solutions.

Satakshi received her M.Sc., M.Phil, and Ph.D. in Mathematics from IIT Roorkee, Roorkee, India. She worked as a Lecturer in the Department of Mathematics at BITS Pilani, India for two years. Currently, she is working as an Assistant Professor in the Department of Mathematics and Statistics at Sam Higginbottom University of Agriculture, Technology, and Sciences, Prayagraj. Her major research interests are in the areas of Optimization, Order Reduction, Linear Systems, and Evolutionary Techniques.

Naira Sbbar works on the resilience of olive to drought, using biostimulants.

Mushtaq Ahmad Shah is an Assistant Professor at the School of Business, Lovely Professional University, Phagwara, Punjab, India. He has a specialisation in banking, finance, and economics and has more than eight years of teaching and research experience in various academic institutions. Dr. Mushtaq holds a PhD in Infrastructure Finance in Management Studies from Guru Ghasidas Central University and a Masters of Commerce from Kashmir University. He has published 15 articles in national and international journals and attended and presented papers at various national and international conferences on banking, public-private partnerships, and behavioural finance.

Tinku Singh received the M.Tech. degree in Computer Science and Engineering from Maharshi Dayanand University, Rohtak, India. He is currently a Research Scholar with the Department of Information Technology, Indian Institute of Information Technology, Allahabad, India. His research interests include Data Science, Big Data Analytics, and Machine Learning.

Shobha Thakur obtained M.Sc. and a Ph.D. degree from Chaudhary Charan Singh University, Meerut, Uttar Pradesh, India. Currently, she is an Assistant professor in the Chemistry Department at Sam Higginbottom University of Agriculture, Technology, and Sciences (SHUATS), Prayagraj. Her major research interests are in the areas of polymer chemistry, environmental science, and the complexation of metals. She is an active member of IEEE and IEI.

Newton M. Wafula holds a Ph.D. in Computer Science, from Heinrich Heine University (Germany), a Master of Tech. in Computer Science and Engineering from University College of Engineering, Osmania University (India), and a BSc. in Electronic and Computer Engineering from Jomo Kenyatta University of Agriculture and Technology (Kenya). Dr. Masinde is a computing professional with strong communication, analytical and organizational skills gained through various training programmes as well as via work experience with a desire to master research skills in distributed computing with the aim of realizing a research and development incubator to foster young upcoming computing professionals.

Index

N

O

P

R

S

W

X

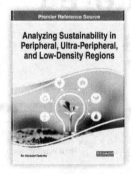

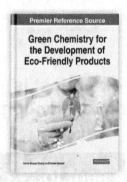

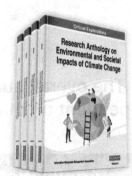

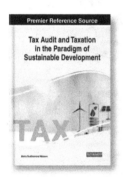

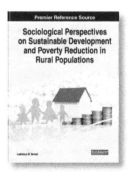

Printed in the United States
by Baker & Taylor Publisher Services

Printed in the United States
by Baker & Taylor Publisher Services